Advanced Data Mining Technologies in Bioinformatics

Table of Contents

Advanced Data Mining Technologies in Bioinformatics

Hui-Huang Hsu
Tamkang University, Taipei, Taiwan

Acquisitions Editor:	Michelle Potter
Development Editor:	Kristin Roth
Senior Managing Editor:	Amanda Appicello
Managing Editor:	Jennifer Neidig
Copy Editor:	Nicole Dean
Typesetter:	Marko Primorac
Cover Design:	Lisa Tosheff
Printed at:	Integrated Book Technology

Published in the United States of America by
Idea Group Publishing (an imprint of Idea Group Inc.)
701 E. Chocolate Avenue
Hershey PA 17033
Tel: 717-533-8845
Fax: 717-533-8661
E-mail: cust@idea-group.com
Web site: http://www.idea-group.com

and in the United Kingdom by
Idea Group Publishing (an imprint of Idea Group Inc.)
3 Henrietta Street
Covent Garden
London WC2E 8LU
Tel: 44 20 7240 0856
Fax: 44 20 7379 0609
Web site: http://www.eurospanonline.com

Library of Congress Cataloging-in-Publication Data

Advanced data mining technologies in bioinformatics / Hui-Hwang Hsu, editor.
p. cm.
Summary: "This book covers research topics of data mining on bioinformatics presenting the basics and problems of bioinformatics and applications of data mining technologies pertaining to the field"--Provided by publisher.
Includes bibliographical references and index.
ISBN 1-59140-863-6 (hardcover) -- ISBN 1-59140-864-4 (softcover) -- ISBN 1-59140-865-2 (ebook)
1. Bioinformatics. 2. Data mining. I. Hsu, Hui-Huang, 1965-
QH324.2.A38 2006
572.8'0285--dc22

2006003556

British Cataloguing in Publication Data
A Cataloguing in Publication record for this book is available from the British Library.

Preface

Bioinformatics is the science of managing, analyzing, extracting, and interpreting information from biological sequences and molecules. It has been an active research area since late 1980's. After the human genome project was completed in April 2003, this area has drawn even more attention. With more genome sequencing projects undertaken, data in the field such as DNA sequences, protein sequences, and protein structures are exponentially growing. Facing this huge amount of data, the biologist cannot simply use the traditional techniques in biology to analyze the data. In order to understand the mystery of life, instead, information technologies are needed.

There are a lot of online databases and analysis tools for bioinformatics on the World Wide Web. Information technologies have made a tremendous contribution to the field. The database technology helps collect and annotate the data. And the retrieval of the data is made easy. The networking and Web technology facilitates data and information sharing and distribution. The visualization technology allows us to investigate the RNA and protein structures more easily. As for the analysis of the data, online sequence alignment tools are quite mature and ready for use all the time. But to conquer more complicated tasks such as microarray data analysis, protein-protein interaction, gene mapping, biochemical pathways, and systems biology, sophisticated techniques are needed.

Data mining is defined as uncovering meaningful, previously unknown information from a mass of data. It is an emerging field since mid 1990's boosted by the flood of data on the Internet. It combines traditional databases, statistics, and machine learning technologies under the same goal. Computer algorithms are also important in speeding up the process while dealing with a large amount of data. State-of-the-art techniques in data mining are, for example, information retrieval, data warehousing, Bayesian learning, hidden Markov model, neural networks, fuzzy logic, genetic algorithms, and support vector machines. Generally, data mining techniques deal with three major problems, i.e., classification, clustering, and association. In analyzing biological data, these three kinds of problems can be seen quite often. The technologies in data mining have been applied to bioinformatics research in the past few years with quite a success. But more research in this field is necessary since a lot of tasks are still undergoing. Furthermore, while tremendous progress has been made over the years, many of the fundamental problems in bioinformatics are still open. Data mining will play a fundamental role in understanding the emerging problems in genomics and proteomics. This book wishes to cover advanced data mining technologies in solving such problems.

The audiences of this book are senior or graduate students majoring in computer science, computer engineering, or management information system (MIS) with interests in data mining and applications to bioinformatics. Professional instructors and researchers will also find that the book is very helpful. Readers can benefit from this book in understanding basics and problems of bioinformatics, as well as the applications of data mining technologies in tackling the problems and the essential research topics in the field.

The uniqueness of this book is that it covers important bioinformatics research topics with applications of data mining technologies on them. It includes a few advanced data mining technologies. Actually, in order to solve bioinformatics problems, there is plenty of room for improvement in data mining technologies. This book covers basic concepts of data mining and technologies from data preprocessing like hierarchical profiling, information fusion, sequence visualization, and data management, to data mining algorithms for a variety of bioinformatics problems like phylogenetics, protein threading, gene discovery, protein sequence clustering, protein-protein interaction, protein interaction networks, and gene annotations. The summaries of all chapters of the book are as follows.

Chapter I introduces the concept and the process of data mining, plus its relationship with bioinformatics. Tasks and techniques of data mining are also presented. At the end, selected bioinformatics problems related to data mining are discussed. It provides an overview on data mining in bioinformatics.

Chapter II reviews the recent developments related to hierarchical profiling where the attributes are not independent, but rather are correlated in a hierarchy. It discusses in detail several clustering and classification methods where hierarchical correlations are tackled with effective and efficient ways, by incorporation of domain specific knowledge. Relations to other statistical learning methods and more potential applications are also discussed.

Chapter III presents a method, called Combinatorial Fusion Analysis (CFA), for analyzing combination and fusion of multiple scoring systems. Both rank combination and score combination are explored as to their combinatorial complexity and computational efficiency. Information derived from the scoring characteristics of each scoring system is used to perform system selection and to decide method combination. Various applications of the framework are illustrated using examples in information retrieval and biomedical informatics.

Chapter IV introduces various visualization (i.e., graphical representation) schemes of symbolic DNA sequences, which are basically represented by character strings in conventional sequence databases. Further potential applications based on the visualized sequences are also discussed. By understanding the visualization process, the researchers will be able to analyze DNA sequences by designing signal processing algorithms for specific purposes such as sequence alignment, feature extraction, and sequence clustering.

Chapter V provides a rudimentary review of the field of proteomics as it applies to mass spectrometry, data handling and analysis. It points out the potential significance of the field suggesting that the study of nuclei acids has its limitations and that the progressive field of proteomics with spectrometry in tandem with transcription studies could potentially elucidate the link between RNA transcription and concomitant protein expression. Furthermore, the chapter describes the fundamentals of proteomics with mass spectrometry and expounds the methodology necessary to manage the vast amounts of data generated in order to facilitate statistical analysis.

Chapter VI considers the prominent problem of reconstructing the basal phylogenetic tree topology when several subclades have already been identified or are well-known by other means, such as morphological characteristics. Whereas most available tools attempt to estimate a fully resolved tree from scratch, the profile neighbor-joining (PNJ) method focuses directly on the mentioned problem and has proven a robust and efficient method for large-scale data sets, especially when used in an iterative way. The chapter also describes an implementation of this idea, the ProfDist software package, and applies the method to estimate the phylogeny of the eukaryotes.

Chapter VII provides an overview of computational problems and techniques for protein threading. Protein threading can be modeled as an optimization problem. This chapter explains the ideas employed in various algorithms developed for finding optimal or near optimal solutions. It also gives brief explanations of related problems: protein threading with constraints, comparison of RNA secondary structures, and protein structure alignment.

Chapter VIII introduces hybrid methods to tackle the major challenges of power and reproducibility of the dynamic differential gene temporal patterns. Hybrid clustering methods are developed based on resulting profiles from several clustering methods. The developed hybrid analysis is demonstrated through an application to a time course gene expression data from interferon-β-1a treated multiple sclerosis patients. The resulting integrated-condensed clusters and overrepresented gene lists demonstrate that the hybrid methods can successfully be applied.

Chapter IX discusses the issue of parameterless clustering technique for gene expression analysis. Two novel, parameterless and efficient clustering methods that fit for analysis of gene expression data are introduced. The unique feature of the methods is that they incorporate the validation techniques into the clustering process so that high quality results can be obtained. Through experimental evaluation, these methods are shown to outperform other clustering methods greatly in terms of clustering quality, efficiency, and automation on both of synthetic and real data sets.

Chapter X introduces gene selection approaches in microarray data analysis for two purposes: cancer classification and tissue heterogeneity correction. In the first part, jointly discriminatory genes which are most responsible to classification of tissue samples for diagnosis are searched for. In the second part, tissue heterogeneity correction techniques are studied. Also, non-negative matrix factorization (NMF) is employed to computationally decompose molecular signatures based on the fact that the expression values in microarray profiling are non-negative. Throughout the chapter, a real world gene expression profile data was used for experiments.

Chapter XI introduces computational methods for detecting complex disease loci with haplotype analysis. It argues that the haplotype analysis, which plays a major role in the study of population genetics, can be computationally modeled and systematically implemented as a means for detecting causative genes of complex diseases. The explanation of the system and some real examples of the haplotype analysis not only provide researchers with better understanding of current theory and practice of genetic association studies, but also present a computational perspective on the gene discovery research for the common diseases.

Chapter XII presents a Bayesian framework for improving clustering accuracy of protein sequences based on association rules. Most of the existing protein-clustering algorithms compute the similarity between proteins based on one-to-one pairwise sequence alignment instead of multiple sequences alignment. Furthermore, the traditional clustering methods are ad-hoc and the resulting clustering often converges to local optima. The experimental results manifest that the introduced framework can significantly improve the performance of traditional clustering methods.

Chapter XIII reviews high-throughput experimental methods for identification of protein-protein interactions, existing databases of protein-protein interactions, computational approaches to predicting protein-protein interactions at both residue and protein levels, various statistical and machine learning techniques to model protein-protein interactions, and applications of protein-protein interactions in predicting protein functions. Intrinsic drawbacks of the existing approaches and future research directions are also discussed.

Chapter XIV discusses the use of differential association rules to study the annotations of proteins in one or more interaction networks. Using this technique, the differences in the annotations of interacting proteins in a network can be found. The concept is extended to compare annotations of interacting proteins across different definitions of interaction networks. Both cases reveal instances of rules that explain known and unknown characteristics of the network(s). By taking advantage of such data mining techniques, a large number of interesting patterns can be effectively explored that otherwise would not be.

Chapter XV introduces the use of Text Mining in scientific literature for biological research, with a special focus on automatic gene and protein annotation. The chapter describes the main approaches adopted and analyzes systems that have been developed for automatically annotating genes or proteins. To illustrate how text-mining tools fit in biological databases curation processes, the chapter also presents a tool that assists protein annotation. At last, it presents the main open problems in using text-mining tools for automatic annotation of genes and proteins.

Chapter XVI surveys systems that can be used for annotating genomes by comparing multiple genomes and discusses important issues in designing genome comparison systems such as extensibility, scalability, reconfigurability, flexibility, usability, and data mining functionality. Further issues in developing genome comparison systems where users can perform genome comparison flexibly on the sequence analysis level are also discussed.

Acknowledgments

I would like to thank all the authors for contributing their thoughts and research results to this book. Without their contributions, this book would not be possible. They are all experts of data mining in bioinformatics. I did learn a lot from them. Further thanks go to the authors who also served as a reviewer for other chapters. Their comments and suggestions are really valuable in improving the manuscripts.

Next, I would like to acknowledge the efforts of the staffs at Idea Group Inc. Special thanks go to Renee Davies for inviting me to initiate this project, Dr. Mehdi Khosrow-Pour for reviewing and approving the project, Jan Travers for supervising the production of the book, and Kristin Roth for coordinating the production of the book. Kristin helped me a lot in answering authors' questions and providing me needed documents and solutions. She made the whole process much easier for me.

I am also grateful to two of my colleagues at Tamkang University. Prof. Timothy K. Shih is a senior professor of computer science and information engineering and a long time friend of mine. Thanks to his encouragement on the initiation of this project. I appreciate his help and friendship throughout the years. Prof. Chien-Chung Cheng of chemistry and life science has been helping me understand biological problems. Cooperation with him in the past two years has been a great pleasure.

Finally, I would like to express my appreciations to my wife Maggie. Without her support and patience, this work would not have been done smoothly.

Hui-Huang Hsu, PhD

Tamsui, Taipei, Taiwan

December 2005

Chapter I

Introduction to Data Mining in Bioinformatics

Hui-Huang Hsu, Tamkang University, Taipei, Taiwan

Abstract

Bioinformatics uses information technologies to facilitate the discovery of new knowledge in molecular biology. Among the information technologies, data mining is the core. This chapter first introduces the concept and the process of data mining, plus its relationship with bioinformatics. Tasks and techniques of data mining are then presented. At the end, selected bioinformatics problems related to data mining are discussed. Data mining aims at uncovering knowledge from a large amount of data. In molecular biology, advanced biotechnologies enable the generation of new data in a much faster pace. Data mining can assist the biologist in finding new knowledge from piles of biological data at the molecular level. This chapter provides an overview on the topic.

Introduction

Progress of information technologies has made the storage and distribution of data much easier in the past two decades. Huge amounts of data have been accumulated at a very fast pace. However, pure data are sometimes not-that-useful and meaningful because what people want is the knowledge/information hidden in the data. Knowledge/information can be seen as the patterns or characteristics of the data. It is much more valuable than data. Thus, a new technology field has emerged in the mid 1990's to deal with the discovery of knowledge/information from data. It is called *knowledge discovery in databases (KDD)* or simply *data mining (DM)* (Chen et al., 1996; Fayyad et al., 1996). Although knowledge and information can sometimes be distinguished, we will treat them as the same term in this chapter.

Data pile up like a mountain. But most of them are not that useful, just like earths and rocks in a mountain. The valuables are metals like gold, iron, or diamond. Just like the miner wants to dig out the valuables from the earth and rock, the data miner uncovers useful knowledge/information by processing a large amount of data. Formal definitions of KDD and DM have been given in different ways. Here are three examples: "Knowledge discovery in databases is the nontrivial process of identifying valid, novel, potentially useful, and ultimately understandable patterns in data" (Fayyad, 1996, p. 20). And, "Data mining is the process of extracting valid, previously unknown, comprehensible, and actionable information from large databases and using it to make crucial business decisions" (Simoudis, 1996, p. 26). Or, to be simpler, "data mining is finding hidden information in a database" (Dunhum, 2003, p. 3). Uncovering hidden information is the goal of data mining. But, the uncovered information must be:

1. **New:** Common sense or known facts are not what is searched for.
2. **Correct:** Inappropriate selection or representation of data will lead to incorrect results. The mined information needs to be carefully verified by domain experts.
3. **Meaningful:** The mined information should mean something and can be easily understood.
4. **Applicable:** The mined information should be able to be utilized in a certain problem domain.

Also, in Simoudis (1996), crucial business decision making is emphasized. That is because the cost of data mining is high and was first applied in business problems, e.g., customer relationship management, personalized advertising, and credit card fraud detection.

There is actually a slight difference between KDD and DM. Data mining is the algorithm/method used to find information from the data. Meanwhile, KDD is the whole process including data collection, data preprocessing, data mining, and information interpretation. So DM is the core of KDD. However, data preprocessing and information interpretation are also very important. Without proper preprocessing, the quality of data might be too bad to find meaningful information. Also, without correct interpretation, the mined information might be mistakenly used. Thus, the cooperation between data mining technicians and problem domain experts is very much needed when a data mining task is performed. In real life, the terms KDD and DM are used interchangeably. Data mining is more frequently referred to because it is shorter and easier to comprehend.

With the goal of uncovering information from data, data mining uses technologies from different computer and information science fields. The three major ones are *databases*, *machine learning*, and *statistics*. Database technology manages data for convenient selection of the data. Machine learning technology learns information or patterns from data in an automatic way, and statistics finds characteristics or statistical parameters of the data. Besides, *information retrieval* deals with selecting related documents by keywords. It is more related to text processing and mining. Also, *algorithms* and *parallel processing* are essential in data mining for a large amount of data. They speed up the search for useful information. But they will not be discussed in this chapter. Interested readers are referred to books of those two topics in computer science.

With the advent of high-throughput biotechnologies, biological data like DNA, RNA, and protein data are generated faster than ever. Huge amounts of data are being produced and collected. The biologist needs information technology to help manage and analyze such large and complex data sets. Database and Web technologies are used to build plenty of online data banks for data storage and sharing. Most of the data collected have been put on the World Wide Web and can be shared and accessed online. For example, *GenBank* (http://www.ncbi.nlm.nih.gov/Genbank/), *EMBL Nucleotide Sequence Database* (http://www.ebi.ac.uk/embl/), and *Protein Data Bank (PDB)* (http://www.rcsb.org/pdb/). To know the updated numbers of complete genomes, nucleotides, and protein coding sequences, the reader can check the *Genome Reviews* of EMBL-EBI (http://www.ebi.ac.uk/GenomeReviews/stats/). The reader is also referred to Protein Data Bank for the number of known protein structures. As for the analysis of the data, data mining technologies can be utilized. The mystery of life hidden in the biological data might be decoded much faster and more accurately with the data mining technologies. In the following, what kind of problems data mining can deal with, both in general and specifically in bioinformatics, are introduced. Some classical techniques of data mining are also presented.

Data Mining Process

This section gives an overview of the whole process of data mining. As emphasized in the introduction section, data mining methods is the core of the process, but other phases of the process are also very important. Each phase of data mining needs interaction with human experts because the objective is usually inexact. This might be time-consuming, but will ensure that the mined information is accurate and useful. The process consists of four phases: data collection, data preprocessing, data mining, and information interpretation. Throughout the process, visualization also plays an important role. Next, further explanations of each phase and visualization issues are presented.

Figure 1. KDD process

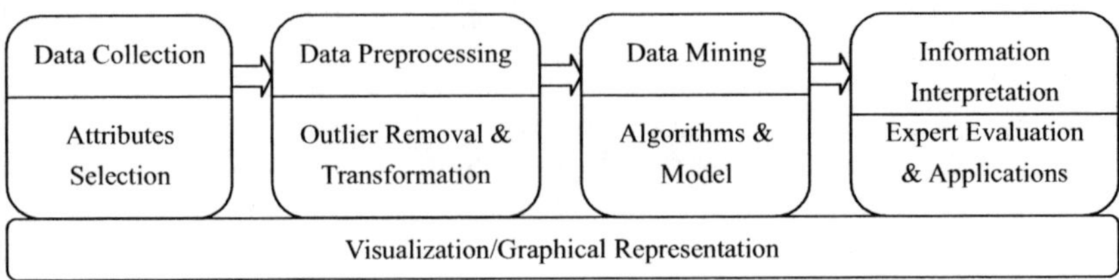

- **Data collection:** Data is the raw material for mining information. Only with proper data can useful information be mined. Domain experts are needed for the selection of data for certain problems. So after raw data of a certain problem are collected, those fields/attributes directly related to the target information are selected.
- **Data preprocessing:** After data are carefully selected, preprocessing is needed. That would include removal of erroneous data and transformation of data. Erroneous data, or so called *outliers* in statistics, should be detected and deleted. Otherwise, they would result in incorrect results. Also, transformation of data to a suitable representation is needed. For example, data can be encoded into a vector representation. Also, data can be transformed from a high dimensional space to a lower dimension to find more important features and to reduce the data mining effort at the next phase.
- **Data mining:** This phase usually involves building a model for the data. Different algorithms and techniques can be used here. However, for a certain task, suitable techniques should be chosen. The characteristics of different tasks and the state-of-the-art techniques for those tasks will be presented in the next two sections.
- **Information interpretation:** First, the mined information needs to be interpreted by human experts. The interpreted results are then evaluated by their novelty, correctness, comprehensibility, and usefulness. Only the information passing through this filtering process can be used in real applications.
- **Visualization:** Visualization technology demonstrates the data or the mined information in graphics to let the user (miner) more easily capture the patterns residing in the data or information. It can be used throughout the data mining process incorporating with data preprocessing, data mining, and information interpretation. This is very important for the presentation of data and information to the user because it enhances the comprehensibility and thus the usefulness. Readers interested in information visualization are referred to the book edited by Fayyad et al. (2002).

It should be noticed that data mining is costly. Therefore, besides going through the above mentioned steps with care, an estimate of the data mining project in advance is necessary. Although certain degree of uncertainty is always involved in a data mining project, with the cooperation of the data mining expert and the domain expert, better understanding of the problem with a clearer objective can be achieved. This increases the possibility of success of the data mining project.

Data Mining Tasks

Typical tasks of data mining are discussed in this section. The list might not be complete. We can say that any form of interesting pattern or information that can be discovered from the data can formulate a specific data mining task. But, of course, some of them might overlap with others. In the following, four major data mining tasks, namely, classification, clustering, association, and summarization, are presented with examples. Text mining handles textual data. It is not a unique task comparable with the four tasks. However, text mining is included and discussed in this section because it is essential in mining information from the huge biological/medical literature.

Classification

Classification decides the class/group for each data sample. There should be at least two classes and the classes are predefined. The input of a classification model is the attributes of a data sample and the output is the class that data sample belongs to. In machine learning, it takes *supervised learning* to build such a model. That is, a set of data with known classes (training data) is needed to estimate the parameters of the classification model. After the parameters are set, the model can be used to automatically classify any new data samples. For example, iris species can be classified based on their measurement. The input attributes are petal length, petal width, sepal length, and sepal width. The output classes are setosa, versicolor, and verginica.

Prediction is also a type of classification. To predict if it will rain tomorrow, it is like classifying the weather into two classes: Raining and Not Raining. Furthermore, to predict the value of a stock would be a classification of "many" classes within a certain range (possible values). In statistics, *regression* is a way to build a predictive model with a polynomial. Prediction is often applied to time series analysis. From the previously known patterns of change, future trends can be decided.

Another kind of classification is *deviation analysis*. It is for the detection of "significant" change. So there are two classes for such analysis: Significant change and Insignificant change. Credit card fraud detection is a case of deviation detection. When the buying behavior of a customer is suddenly changed, it might be that the credit card is stolen and needs to be detected right away to avoid possible losses.

Clustering

Clustering is to group similar data into a finite set of separate clusters/categories. This task is also referred to as *segmentation*. In machine learning, it requires *unsupervised learning*. That is, the number of clusters and the categories are not known in advance. A clustering tool simply measures the similarity of the data based on their attribute values and put similar data into the same cluster. For example, Amazon.com groups its customers into different clusters according to the books they buy. Customers in the same cluster should have the same taste in book reading. So the most bought books by the customers in one cluster can be recommended to the customers in the same cluster, who have not bought the books.

It is sometimes confusing between clustering and classification. Both of them put data/ examples into different groups. The difference is that in classification ,the groups are predefined and the task is to decide which group a new data sample should belong to. In clustering the types of groups and even the number of groups are not known and the task is to find the best way to segment all the data.

The major problem of clustering is the decision of the number of clusters. For some methods, the number of clusters needs to be specified first. For example, to use the *K-Means algorithm*, the user should input the number of clusters first. The centers of the clusters are chosen arbitrarily. Then, the data iteratively move between clusters until they converge. If the user is not satisfied with the results, another clusters number is then tried. So this kind of method is a trial-and-error process.

Another solution to the decision of the number of clusters is to use *hierarchical clustering*. There are two approaches for this: the top-down approach and the bottom-up approach. For the top-down approach, the data are separated into different clusters based on different criteria of the similarity measurement. At first, the whole data belong to a big cluster. At last, each data sample is a single cluster. There will be different levels of clustering in between the two. For the bottom-up approach, the data are group into different clusters from clusters of single data sample until all the data are of one cluster.

Association

Another task of data mining is to search for a set of data within which a subset is dependent on the rest of the set. This task is also referred to as *link analysis* or *affinity analysis*. An association rule can be written as *A* ⇨ *B* where both A and B are a data set. If such an *association rule* is identified, it means that when A is seen, B will also be seen with a high probability. For example, in a supermarket transaction database, an association rule *Diapers* ⇨ *Beer* was discovered from the transactions on Friday night. That means people who bought diapers on Friday nights also bought beer. This kind of relationship between products is very helpful to marketing decision making. And the process is called *market basket analysis*.

An association rule *A* ⇨ *B* can be identified when both the *support* and the *confidence* of the rule are larger than respective thresholds. The support of the association rule is the ratio of the number of transactions containing both A and B over the total number of transactions in the database. The confidence of the association rule is the proportion of the number of transactions containing both A and B over the total number of transactions containing A. The two thresholds are actually difficult to decide and need domain expertise. The size of the transaction database and the number of product items will affect the setting of proper thresholds. Another important issue of association rule mining is how to speed up the identification procedure of association rules. Different algorithms have been developed (Chen et al., 1996).

Summarization

A simple description of a large data set is called its *summarization* or *characterization*. It is desirable that representative information of a data set can be obtained, so that we can easily have a general view of the data set. The representative information can be a subset or general properties/characteristics of the data. For example, in a class of students, the average height 176.5 cm or a student with the height can be used to characterize the students in the class. With a diversity of data, it is somewhat difficult to find good representative information. Classical statistics can be applied to find some useful parameters to represent the data. However, summarization is not so popular or frequently used as classification, clustering, and association.

Text Mining

When the data to be mined are text instead of numerical data, the task is called *text mining*. It is originated from information retrieval (IR) of the library science. Keywords are used to find related documents from a document database. It is based on the similarity measures between the keywords and the documents. These similarity measures can also be applied to the comparison of a document with all other documents. In the Internet era, IR technique is also used for search engine to find related Web pages. It is also referred to as *Web content mining*.

For more advanced applications of text mining; classification, clustering, and association techniques are utilized. The relationship between documents or terms within the documents can be investigated via these techniques. *Natural language processing (NLP)* is also involved for semantic understanding of neighboring words (phrases or sentences). Online question-answering systems have been developed based on natural language processing and text mining. If a user asks a question in natural language, the system can search a lot of Web pages to find possible answers to it. Text mining is also important for bioinformatics. Plenty of publications in biology have been accumulated for more than a century. It is desired that new research findings can be combined or cross-referenced with "known" facts in the literature. However, in the biology literature, different terms in different documents might refer to the same thing. Also, the amount of biology literature is really huge. To overcome these problems, new technologies need to be developed and utilized. In all, text mining can facilitate the re-examination of the biology literature to link the facts that were already known.

Data Mining Techniques

Many concepts and techniques are useful for the same goal of mining hidden information from data. Among them algorithms, databases, statistics, machine learning, and information retrieval are essential. Algorithms and parallel processing techniques are mainly to accelerate the mining process. Information retrieval techniques can be applied to text mining and is introduced in the subsection of text mining. Here, we will concentrate on databases, statistics, and machine learning.

Databases and Data Warehousing

Database techniques arrange data in a structured form. This facilitates the retrieval of a subset of the data according to the user query. To organize data in a structured form definitely helps the data mining process. But in data mining tasks, the data are usually in semi-structured or unstructured forms. Also, data mining is aimed at uncovering information like "customers of a certain background usually have a good credit," not like "customers with last name Bush." The goals of databases and data mining are quite different. But the well-developed database technologies are very useful and efficient in storage and preprocessing of the collected data.

Data warehousing technologies were developed to accommodate the operational data of a company. For a large amount of data from heterogeneous databases, a data warehouse is a repository of operational databases and is used to extract needed information. It is said to contain informational data comparing to the operational data in the database. The information is a combination of historical data and can be used to support business decision making.

Statistics

Statistics has sound theoretical ground developed since the 16th century. It estimates various parameters to provide a general view of the data. It is essential in "characterizing" the data. Also, the parameters can be helpful to decide if the discovered information is significant. In the following, we introduce some basic concepts of statistics that are related to data mining (Kantardzic, 2003).

To characterize or summarize a set of data, there are some parameters that the reader should be familiar with. They are mean, median, mode, variance, and standard deviation. Also, a box plot is a graphical representation of the distribution of the data. The range of the data is divided into four parts (quartiles). Each part has the same number of data. Data outside a certain range from the median can then be viewed as outliers.

Next, we would like to discuss the *Bayesian classifier*. It is based on the well-known *Bayes theorem* — $P(Y|X) = [P(X|Y) P(Y)]/ P(X)$ where $P(Y)$ is the prior probability and $P(Y|X)$ is the posterior probability. The classification of a new data sample can be decided by estimating the probabilities of the classes (Y_i) given the new data sample (X), respectively. The class with the highest probability is the predicted class of the new data sample.

Regression is to build a mathematical model for some known temporal data and use the model to predict the upcoming values. This task would be very difficult because the "trend" of the data is usually nonlinear and very complicated. Many parameter estimates are involved if the model is nonlinear. Therefore, to simplify the problem, a linear model is usually used. A linear model is to use a straight line to estimate the trend of the data. This is called *linear regression*. For a set of data with a nonlinear nature, it can be assumed that the data trend is piecewise linear. That is, in a small period, the data trend is about a straight line.

Although statistical theories are widely used in data mining, the philosophies of data mining and statistics are different. In statistics, hypotheses are usually made first and they are proved to be valid or invalid by applying the theories on the data. On the contrary, data mining does not exactly know what it is looking for.

Machine Learning

Machine learning is a long-developed field in *artificial intelligence (AI)*. It focuses on automatic learning from a data set. A suitable model with many parameters is built first for a certain domain problem and an error measure is defined. A learning (training) procedure is then used to adjust the parameters according to the predefined error measure. The purpose is to fit the data into the model. There are different theories for the learning procedure, including gradient decent, expectation maximization (EM) algorithms, simulated annealing, and evolutionary algorithms. The learning procedure is repeated until the error measure reaches zero or is minimized. After the learning procedure is completed with the training data, the parameters are set and kept unchanged and the model can be used to predict or classify new data samples.

Different learning schemes have been developed and discussed in the machine learning literature. Important issues include the learning speed, the guarantee of convergence, and how the data can be learned incrementally. There are two categories of learning schemes: (1) *supervised learning* and (2) *unsupervised learning*. Supervised learning learns the data with an answer. Meaning, the parameters are modified according to the difference of the real output and the desired output (the expected answer). The classification problem falls into this category. On the other hand, unsupervised learning learns without any knowledge of the outcome. Clustering belongs to this category. It finds data with similar attributes and put them in the same cluster.

Various models like *neural networks (NN)*, *decision trees (DT)*, *genetic algorithms (GA)*, *fuzzy systems*, and *support vector machines (SVM)* have proved very useful in classification and clustering problems. But machine learning techniques usually handles relatively small data sets because the learning procedure is normally very time-consuming. To apply the techniques to data mining tasks, the problem with handling large data sets must be overcome.

Bioinformatics Problems Related to Data Mining

In this section, selected bioinformatics problems that are suitable for using data mining technologies are introduced. Various bioinformatics problems and their trends and possible solutions via data mining technologies are presented and discussed in the rest of the book. Here, only a few typical problems are chosen to illustrate the relationship between bioinformatics and data mining technologies. For an introduction to all kinds of bioinformatics problems, the reader is referred to Lesk (2002).

Protein Structure Prediction

With so many known proteins, the functions of most proteins are still waiting for further investigation. A protein's function depends upon its structure. If the structure of a protein is known, it would be easier for the biologist to infer the function of the protein. However, it is still costly to decide the structure of a protein via biotechnologies. On the contrary, protein sequences are relatively easy to obtain. Therefore, it is desirable that a protein's structure can be decided from its sequence through computational methods. A protein sequence is called the *primary structure* of the protein. Hydrogen bonding of the molecules results in certain substructures called the *secondary structure*. Interactions between secondary structures assemble them into the *tertiary structure*. If a protein is composed by more than one unit, it is a *quaternary structure*. Researchers had produced pretty good results in predicting secondary structures from a protein sequence. However, it is still difficult and imprecise to determine the tertiary and quaternary structures.

Two major approaches have been taken for such structure prediction. The first one is to use machine learning algorithms to learn the mapping function from the sequence segments to the secondary structures. With the learned mapping function, the secondary structures can be induced from the sequence segments of a new protein. With the knowledge of secondary structures, the tertiary structure of the protein can be further inferred. This kind of problem is exactly a prediction/classification problem in data mining. The second approach applies the physical laws of the forces between different molecules of the protein. Angle constraints of the chemical bonds are formulated and an optimization process can be used to find the optimal solution of the angles and thus the tertiary structure can be decided.

Gene Finding

A gene is a possible protein-coding region. Genes only compose a small part of the whole genome. With a complete genome, we would like to know where the genes are located. It is to identify the beginnings and endings of genes. It is also called *gene annotation*. There are also two approaches for this problem. The first one is to use machine learning techniques to learn from the patterns of subsequences of known genes and then predict/classify the genome into genes and non-genes. The second approach, called *ab initio* methods, concentrates more on the properties of the DNA sequences. Machine learning can also be applied in this approach. The two approaches can be combined to produce better prediction results.

Current computational methods on gene finding are more accurate on bacterial/prokaryotic genomes. The reason is that bacterial/prokaryotic genes are continuous. In contrast, eukaryotic genes are much more complicated because many *introns*, which will not express, reside in a gene. A splicing procedure is needed to result in a continuous gene. But, the splicing mechanism is not easy to understand. So it is still hard to annotate genes of a eukaryotic genome.

Protein-Protein Interaction

There are many different kinds of proteins within a cell of an organism. To understand the physiological functions of the cell, we need to know first how the proteins interact with each other and what the products of the interactions are. The interactions can be physical or functional. High-throughput biotechnologies were developed for such a task. Also, computational methods were investigated. With the information of protein interactions, a *protein interaction network* can be built. This kind of network is very essential in understanding the regulation of *metabolic pathways* and thus can be applied to the design of new drugs. To identify the interactions among proteins, various statistical or machine learning techniques can be used. Furthermore, to gain more information from a lot of known protein interaction networks is another important task. These are association problems in data mining.

Phylogenetics

People have been interested in finding out the relationships between different species. A tree structure can be used to illustrate such a relationship. Besides the appearances of the species, studies in anatomy and physiology have contributed to this taxonomy. However, with the knowledge of the species at the molecular level, in other words, the DNA and protein data, new findings are expected from this knowledge. And the field has advanced from the comparison of phenotypes to the comparison of genotypes. By comparing certain DNA or protein sequence data, the "distance" of two species can be determined. Species with shorter distances should be put in the same group because they are closer to each other. This is a classical problem of clustering. Hierarchical clustering can be used here to build a *phylogenetic tree* for some species with known corresponding DNA or protein sequences.

Viruses from animals cause a lot of human diseases. Such diseases have been a serious problem for human beings. Some of them are deadly and even incurable. For example, AIDS was from apes, new variant Creutzfeldt-Jacob disease (CJD) was from cows (Mad Cow Disease or Bovine Spongiform Encephalopathy, BSE), Foot-and-Mouth disease was from pigs, and Bird Flu/Avian Flu was from chickens and birds. Comparison of corresponding genes of different species might be able to help invent new methods or drugs for the control of such diseases. Thus, finding the relationship of different species at the genomic level has become an important issue.

Summary

Data can be collected much more easily nowadays and are abundant everywhere. But, what is lacking is knowledge. How to dig useful knowledge from large amount of data is really essential. Technologies of different disciplines have emerged under the same goal of mining knowledge from data in the past 10 years. They have been applied to problems of different fields with success. On the other hand, accomplishment of the human genome project in 2003 began a new era for biological research. With plenty of DNA, RNA, and protein data, it would be difficult for the biologist to uncover the knowledge residing in the data with traditional methods. Information technologies can be very helpful in the tasks. Thus, the term bioinformatics became popular. Among the technologies, data mining is the core. In this chapter, basic concepts of data mining and its importance and relationship with bioinformatics were introduced. These provide a brief overview to the topic so that the reader can go on to more advanced technologies and problems in the subsequent chapters.

References

Chen, M.-S., Han, J., & Yu, P. S. (1996). Data mining: An overview from a database perspective. *IEEE Trans. on Knowledge and Data Eng.*, *8*(6), 866-883.

Dunham, M. H. (2003). *Data mining: Introductory and advanced topics*. Upper Saddle River, NJ: Prentice Hall.

Fayyad, U. M., Piatetsky-Shapiro, G., Smyth, P., & Uthurusamy, R. (Ed.). (1996). *Advances in knowledge discovery and data mining*. Menlo Park; Cambridge, MA: AAAI/MIT Press.

Fayyad, U. M. (1996). Data mining and knowledge discovery: Making sense out of data. *IEEE Expert*, *11*(5), 20-25.

Fayyad, U., Grinstein, G. G., & Wierse, A. (Eds.). (2002). *Information visualization in data mining and knowledge discovery*. San Francisco: Morgan Kaufmann.

Kantardzic, M. (2003). *Data mining: Concepts, models, methods, and algorithms*. Piscataway, NJ: IEEE Press.

Lesk, A. M. (2002). *Introduction to bioinformatics*. New York: Oxford University Press.

Simoudis, E. (1996). Reality check for data mining. *IEEE Expert*, *11*(5), 26-33.

Chapter II

Hierarchical Profiling, Scoring and Applications in Bioinformatics

Li Liao, University of Delaware, USA

Abstract

Recently, clustering and classification methods have seen many applications in bioinformatics. Some are simply straightforward applications of existing techniques, but most have been adapted to cope with peculiar features of the biological data. Many biological data take a form of vectors, whose components correspond to attributes characterizing the biological entities being studied. Comparing these vectors, aka profiles, are a crucial step for most clustering and classification methods. We review the recent developments related to hierarchical profiling where the attributes are not independent, but rather are correlated in a hierarchy. Hierarchical profiling arises in a wide range of bioinformatics problems, including protein homology detection, protein family classification, and metabolic pathway clustering. We discuss in detail several clustering and classification methods where hierarchical correlations are tackled in effective and efficient ways, by incorporation of domain-specific knowledge. Relations to other statistical learning methods and more potential applications are also discussed.

Introduction

Profiling entities based on a set of attributes and then comparing these entities by their profiles is a common, and often effective, paradigm in machine learning. Given profiles, frequently represented as vectors of binary or real numbers, the comparison amounts to measuring "distance" between a pair of profiles. Effective learning hinges on proper and accurate measure of distances.

In general, given a set A of N attributes, $A = \{a_i | i = 1, \ldots, N\}$, profiling an entity x on A gives a mapping $p(x) \rightarrow \Re^N$, namely, p(x) is an N vector of real values. Conveniently, we also use x to denote its profile p(x), and x_i the i-*th* component of p(x). If all attributes in A can only have two discrete values 0 and 1, then $p(x) \rightarrow \{0,1\}^N$ yields a binary profile. The distance between a pair of profiles x and y is a function: $D(x, y) \rightarrow \Re$. Hamming distance is a straightforward, and also one of the most commonly used, distance measures for binary profiles; it is a simple summation of difference at each individual component:

$$D(x, y) = \Sigma_i^n d(i) \quad (1)$$

where $d(i) = | x_i - y_i |$. For example, given x = (0, 1, 1, 1, 1) and y = (1, 1, 1, 1, 1), then $D(x, y) = \Sigma_{i=1}^5 d(i) = 1+0+0+0+0 = 1$. A variant definition of d(i), which is also very commonly used, is that d(i) = 1 if $x_i = y_i$ and d(i) = -1 if otherwise. In this variant definition, $D(x, y) = \Sigma_{i=1}^5 d(i) = -1+1+1+1+1 = 3$.

The Euclidean distance, defined as $D = \sqrt{\Sigma_i^n (x_i - y_i)^2}$, has a geometric representation: a profile is mapped to a point in a vector space where each coordinate corresponds to an attribute. Besides using Euclidean metric, in vector space the distance between two profiles is also often measured as dot product of the two corresponding vectors: $x \cdot y = \Sigma_i^n x_i y_i$. Dot product is a key quantity used in Support Vector Machines (Vapnik 1997, Cristianini & Shawe-Taylor 2000, Scholkopf & Smola 2002). Many clustering methods applicable to vectors in Euclidean space can be applied here, such as K-means.

While Hamming distance and Euclidean distance are the commonly adopted measures of profile similarity, both of them imply an underlying assumption that the attributes are independent and contribute equally in describing the profile. Therefore, the distance between two profiles is simply a sum of distance (i.e., difference) between them at each attribute. These measures become inappropriate when the attributes are not equally contributing, or not independent, but rather correlated to one another. As we will see, this is often the case in the real-world biological problems.

Intuitively, nontrivial relations among attributes complicate the comparisons of profiles. An easy and pragmatic remedy is to introduce scores or weighting factors for individual attributes to adjust their apparently different contribution to the Hamming or Euclidean "distance" between profiles. That is, the value of d(i) in equation (1) now depends not only on the values of x_i and y_i, but also on the index i. Often, scoring schemes of this type are also used for situations where attributes are correlated, sometimes in a highly nonlinear way. Different scoring schemes thereby are invented in order to capture the relationships among attributes. Weighting factors in these scoring schemes are either preset *a priori* based on domain knowledge about the attributes, or fixed from the training examples, or determined by a combination of both. To put into a mathematical framework, those scoring based approaches can be viewed as approximating the correlations among attributes, which, without loss of generality, can be represented as a polynomial function. In general, a formula that can capture correlations among attributes as pairs, triples, quadruples, and so forth, may look like the following:

$$D' = \Sigma_i^n d(i) + \Sigma_{i \neq j}^n d(i)c(i,j)d(j) + \Sigma_{i \neq j \neq k}^n d(i)d(j)d(k)c(i,j,k) + \ldots \quad (2)$$

where the coefficients c(i,j), c(i,j,k), ..., are used to represent the correlations. This is much like introducing more neurons and more hidden layers in an artificial neural network approach, or introducing a nonlinear kernel functions in kernel-based methods. Because the exact relations among attributes are not known *a priori*, an open formula like equation (2) is practically useless: as the number of these coefficients grows exponentially with the profile size, solving it would be computationally intractable, and there would not be enough training examples to fit these coefficients.

However, it was found that the situation would become tractable when these correlations could be structured as a hierarchy — a quite loose requirement and readily met in many cases as we shall see later. In general, a hierarchical profile of entity x can be defined as $p(x) \rightarrow \{0,1\}^L$, where L stands for the set of leaves in a rooted tree T. A hierarchical profile is no longer a plain string of zeros and ones. Instead, it may be best represented as a tree with the leaves labeled by zeros and ones for binary profiles, or by real value numbers for real value profiles.

As the main part of this chapter, we will discuss in detail several clustering and classification methods where hierarchical profiles are coped with effectively and efficiently, noticeably due to incorporation of domain specific knowledge. Relations to other statistical learning methods, e.g., as kernel engineering, and more possible applications to other bioinformatics problems are also discussed, towards the end of this chapter.

Hierarchical Profilings in Bioinformatics

Functional Annotations

Hierarchical profiling arises naturally in many bioinformatics problems. The first example is probably from the phylogenetic profiles of proteins and using them to assign functions to proteins (Pellegrini et al., 1999). To help understand the key concepts and motivations, and also to introduce some terminologies for later discussions, we first briefly review the bioinformatics methodologies for functional annotation.

Determining protein functions, also called as functional annotation, has been and remains a central task in bioinformatics. Over the past 25 years many computational methodologies have been developed toward solving this task. The development of these computational approaches can be generally broken into four stages by both the chronological order and algorithmic sophistication (Liao & Noble, 2002). For our purpose here, we can categorize these methods into three levels according to the amount and type of information used.

The methods in level one compare a pair of proteins for sequence similarity. Among them are the landmark dynamic programming algorithm by Smith and Waterman (1980) and its heuristic variations BLAST (Altschul et al., 1990) and FASTA (Pearson, 1990). The biological reason behind these methods is protein homology; two proteins are homologous if they share a common ancestor. Therefore, homologous proteins are similar to each other in the primary sequence and keep the same function as their ancestor's, until the evolutionary divergence — mutations, deletions, or insertions from the ancestral protein (or gene, to be more precise) is significant enough to cause any change. A typical way to annotate a gene with unknown function is to search against a database of genes whose functions are already known, such as GenBank (http://www.ncbi.nlm.nih.gov/Genbank), and assign to the query gene the function of a homologous gene found in the database.

The next level's methods use multiple sequences from a family of proteins with same or similar functions, in order to collect aggregate statistic for more accurate/reliable annotation. Profiles (Gribskov et al., 1987) and hidden Markov models (Krogh et al., 1994; Durbin et al., 1998) are two popular methods to capture and represent these aggregate statistics from whole sequences for protein families. More refined and sophisticated methods using aggregate statistics are developed, such as PSI-BLAST (Altschul et al., 1997), SVM-Fisher (Jaakkola et al., 1999, 2000), Profile-Profile (Sadreyev & Grishi, 2003; Mittelman et al., 2003), SVM-Pairwise (Liao & Noble, 2002, 2003). The aggregate statistic may also be represented as patterns or motifs. Methods based on patterns and motifs include BLOCKs (Henikoff & Henikoff, 1994), MEME (Bailey & Elkan, 1995), Meta-MEME (Grundy et al., 1997), and eMotif (Nevill-Manning et al., 1998).

The third level's methods go beyond sequence similarity to utilize information such as DNA Microarray gene expression data, phylogenetic profiles, and genetic networks. Not only can these methods detect distant homologues — homologous proteins with sequence identity below 30%, but they also can identify proteins with related functions, such as those found in a metabolic pathway or a structural complex.

Given a protein, its phylogenetic profile is represented as a vector, where each component corresponds to a genome and takes a value of either one or zero, indicating respectively the presence or absence of a homologous protein in the corresponding genome. Protein phylogenetic profiles were used in Pellegrini, Marcotte, Thompson, Eisenberg, and Yeates (1999) to assign protein functions based on the hypothesis that functionally linked proteins, such as those participating in a metabolic pathway or a structural complex, tend to be preserved or eliminated altogether in a new species. In other words, these functionally linked proteins tend to co-evolve during evolution. In Pellegrini et al. (1999), 16 then-fully-sequenced genomes were used in building the phylogenetic profiles for 4290 proteins in *E. coli* genome. The phylogenetic profiles, expressed as 16-vector, were clustered as following; proteins with identical profiles are grouped and considered to be functionally linked, and two groups are called *neighbors* when their phylogenetic profiles differ by one bit. The results based on these simple clustering rules supported the functional linkage hypothesis. For instance, homologues of ribosome protein RL7 were found in 10 out of 11 eubacterial genomes and in yeast but not in archaeal genomes. They found that more than half of the *E. coli* proteins with the RL7 profile or profiles different from RL7 by one bit have functions associated with the ribosome, although none of these proteins share significant sequence similarity with the RL7 protein. That is, these proteins are unlikely to be annotated as RL7 homologues by using sequence similarity based methods.

There are some fundamental questions regarding the measure of profile similarity that can affect the results from analysis of phylogenetic profiles. Can we generalize the definition of similar profiles? In other words, can we devise a measure so we can calculate similarity for any pair of profiles? What threshold should be adopted when profile similarity is used to infer functional linkage?

A simple measure for profile similarity first brought up was Hamming distance. In Marcotte, Xenarios, van Der Bliek, and Eisenberg (2000), phylogenetic profiles are used to identify and predict subcellular locations for proteins; they found mitochondrial and non-mitochondrial proteins.

The first work that recognizes phylogenetic profiles as a kind of hierarchical profiling is Liberles, Thoren, von Heijne, and Elofsson (2002), a method that utilizes the historical evolution of two proteins to account for their similarity (or dissimilarity). Evolutionary relationships among organisms can be represented as a phylogenetic tree where leaves correspond to the current organisms and internal nodes correspond to hypothetical ancient organisms. So, rather than simply counting the presence and absence of the proteins in the current genomes, a quantity called differential parsimony is calculated that minimize the number of times when changes have to be made at tree branches to reconcile the two profiles.

Comparative Genomics

Another example of hierarchical profiling arises from comparing genomes based on their metabolic pathway profiles. A main goal of comparing genomes is to reveal the evolutionary relationships among organisms, which can be represented as phylogenetic trees.

Originally phylogenetic trees were constructed based on phenotypic — particularly morphological — features of organisms. Nowadays, molecular reconstruction of phylogenetic trees is most commonly based on comparisons of small sub-unit ribosomal RNA (16S rRNA) sequences (Woese, 1987). The small sub-unit rRNAs is orthodoxly used as gold standard for phylogeny study, mainly due to two factors: their ubiquitous presence and relative stability during evolution. However, the significance of phylogenies based on these sequences have been recently questioned with growing evidence for extensive lateral transfer of genetic material, a process which results in blurring the boundaries between species. Phylogenetic trees based on individual protein/gene sequence analysis — thus called gene tree — are not all congruent with species trees based on 16S rRNA (Eisen, 2000). Attempts of building phylogenetic trees based on different information levels, such as pathways (Dandekar et al., 1999; Forst & Schulten, 2001) or some particular molecular features such as folds (Lin & Gerstein 2000), have also led to mixed results of congruence with the 16S rRNA-based trees.

Figure 1. Binary matrix encoding the presence/absence of pathways in genomes; O_1 to O_m represent m genomes, and P_1 to P_n represent n pathways

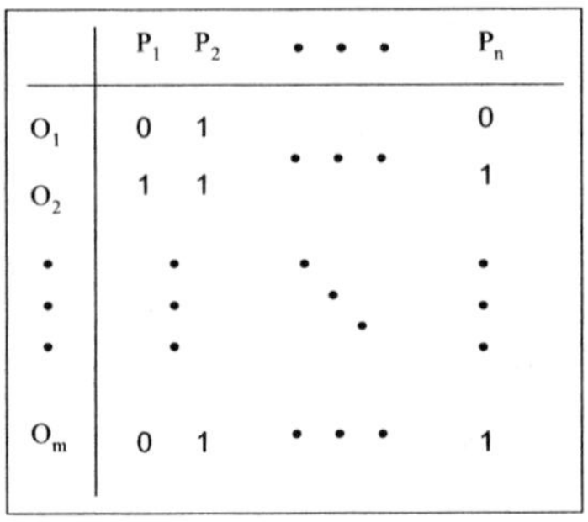

	P_1	P_2	• • •	P_n
O_1	0	1	• • •	0
O_2	1	1		1
•	•		•	•
•	•		•	•
•	•		•	•
O_m	0	1	• • •	1

From a comparative genomics perspective, it makes more sense to study evolutionary relationships based on genome-wide information, instead of a piece of the genome, be it an rRNA or a gene. As more and more genomes are fully sequenced, such genome-wide information becomes available. One particularly interesting type of information is the entire repertoire of metabolic pathways in an organism, as the cellular functions of an organism are carried out via these metabolic pathways. A metabolic pathway is a chain of biochemical reactions, together fulfilling certain cellular functions. For example, *Glycolysis* is a pathway existed in most cells, which consists of 10 sequential reactions converting glucose to pyruvate while generating the energy that the cell needs. Because most of these reactions require enzymes as catalyst, therefore in an enzyme centric scheme, pathways are represented as sequences of component enzymes. It is reasonable to set the necessary condition for a pathway to exist in an organism as that all the component enzymes of that pathway are available. Enzymes are denoted by enzyme commission (EC) numbers which specifies the substrate specificity. Most enzymes are proteins. Metabolic pathways in a completely sequenced genome are reconstructed by identifying enzyme proteins that are required for a pathway (Gaasterland & Selkov, 1995; Karp et al., 2002).

The information about presence and absence of metabolic pathways in genomes can be represented as a binary matrix, as shown in Figure 1, where an entry $(i,j) = 1/0$ represents whether pathway j is present/absent in genome i. Therefore, each row serves as a profile of the corresponding genome (termed metabolic pathway profiles), and each column serves as a profile of the corresponding pathway (termed phyletic profiles). It is reasonable to believe that valuable information about evolution and co-evolution is embedded in these profiles, and comparison of profiles would reveal, to some degree, the evolutionary relations among entities (either genomes or pathways) represented by these profiles.

Once again, the attributes used for building these profiles are not independent but correlated to each other. Because different metabolic pathways may be related to one another in terms of physiological functions, e.g., one pathway's absence/presence may be correlated with another pathway's absence/presence in a genome, these relationships among pathways, as attributes of metabolic pathway profiles, should be taken into account when comparing genomes based on their MPPs. The relationships among various pathways are categorized as a hierarchy in the WIT database (Overbeek et al., 2000), which can be found at the following URL (http://compbio.mcs.anl.gov/puma2/cgi-bin/functional_overview.cgi).

Comparing Hierarchical Profiles

In the last section, we showed that the data and information in many bioinformatics problems can be represented as hierarchical profiles. Consequently, the clustering and classification of such data and information need to account for the hierarchical correlations among attributes when measuring the profile similarity. While a generic formula like equation (2) posits to account for arbitrary correlations theoretically, its demand of exponentially growing amount of training examples and lacking of an effective learning mechanism render the formula practically useless. In hierarchical profiles, however, the structure of relations among attributes is known, and sometimes the biological interpretation of these relations is also known. As a result, the learning problems will become rather tractable.

P-Tree Approach

Realizing that the hierarchical profiles contain information not only in the vector but also in the hierarchical correlations, it is natural to first attempt at treating them as trees. How to compare trees is itself an interesting topic with wide applications, and has been the subject of numerous studies. In Liao, Kim, and Tomb, (2002), a p-tree based approach was proposed to measure the similarity of metabolic pathway profiles. The relationships among pathways are adopted from the WIT database and represented as a *Master tree*. About 3300 known metabolic pathways are collected in the WIT database and these pathways are represented as leaves in the Master tree. Then, for each genome, a *p-Tree* is derived from the Master tree by marking off leaves whose corresponding pathways are absent from the organism. In this representation, a profile is no longer a simple string of zeros and ones, where each bit is treated equally and independently. Instead, it is mapped into a p-Tree so that the hierarchical relationship among bits is restored and encoded in the tree.

The comparison of two p-Trees thus evaluates the difference between the two corresponding profiles. To take into account the hierarchy, a scoring scheme ought to weight (mis)matches at bits i and j according to their positions in the tree, such as i and j are sibling, versus i and j are located distantly in the tree. For that, the (mis)matches scores are transmitted bottom-up to the root of the Master tree in four steps: (1) overlay the two p-Trees; (2) score mismatches and matches between two p-Trees and label scores at the corresponding leaves on the master tree; (3) average scores from siblings (weight breadth) and assign the score to the parent node; (4) iterate step 3 until the root is reached.

Figure 2. A comparison of two organisms orgi and orgj, with respect to the three pathways p1, p2, and p3 (weighted scoring scheme, Case 1); Panel A featuring the master tree, where each pathway is represented by a circle and each square represents an internal node; in Panel B, the presence(absence) of a pathway is indicated by 1(0); Panel C features the p-trees, where a filled (empty) circle indicates the presence (absence) of the pathway; Panel D, the bottom-up propagation of the scores is illustrated; a cross (triangle) indicates a mismatch (match)

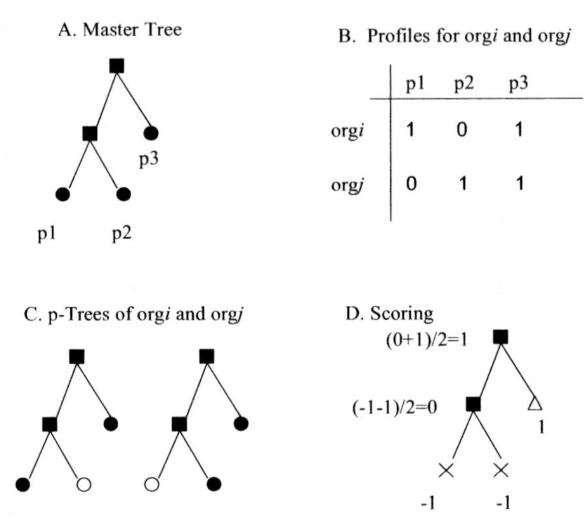

Figure 3. Case 2: A comparison of two organisms org_i and org_j, with respect to the three pathways p1, p2, and p3 (weighted scoring scheme, Case 2); the same legends as in Figure 2

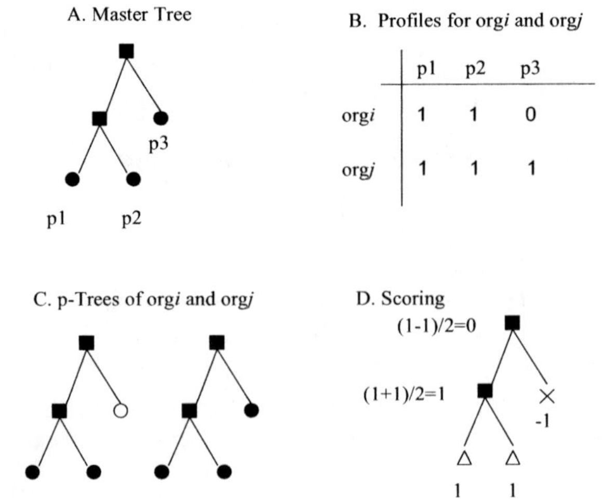

An algorithm implementing these steps is quite straightforward and has a time complexity linear with the size of the Master tree.

To demonstrate how the algorithm works, let us look at an example of two organisms, org_i and org_j, and three pathways p_1, p_2, and p_3. Two hypothetical cases are considered and are demonstrated in Figures 2 and 3 respectively. In case one, org_i contains pathways p_1 and p_3, and org_j contains p_2 and p_3. The metabolic pathway profiles for org_i and org_j are shown in panel B and their corresponding p-Trees are displayed in panel C. In the panel D, two p-Trees are superposed. Matches and mismatches are scored at the leaves and the scores are propagated up to the root. In case two, org_i contains pathways p_1 and p_2, whereas org_j contains all three pathways, and similar calculation is shown in Figure 3. In this example, given the topology of the master tree, the two cases have the same final score 0 in the p-Tree scoring scheme. This is in contrast to the variant Hamming scoring scheme, where the score for case 1 equals to $(-1-1+1) = -1$ and the score for case 2 equals to $(1+1-1) = 1$. Evidently, the p-Tree scoring scheme has taken into account the pathway relationships present in the Master tree: p_1 and p_2, being in the same branch, are likely to have a similar biochemical or physiological role, which is distinct from the role of the p_3 pathway.

Using pathway hierarchical categories in the WIT database and the scoring scheme described above, we compared 31 genomes based on their metabolic pathway profiles

Figure 4. MPP-based tree and 16S rRNA-based tree for 31 genomes; neighbor-joining program from Phylip package is used to generate these trees

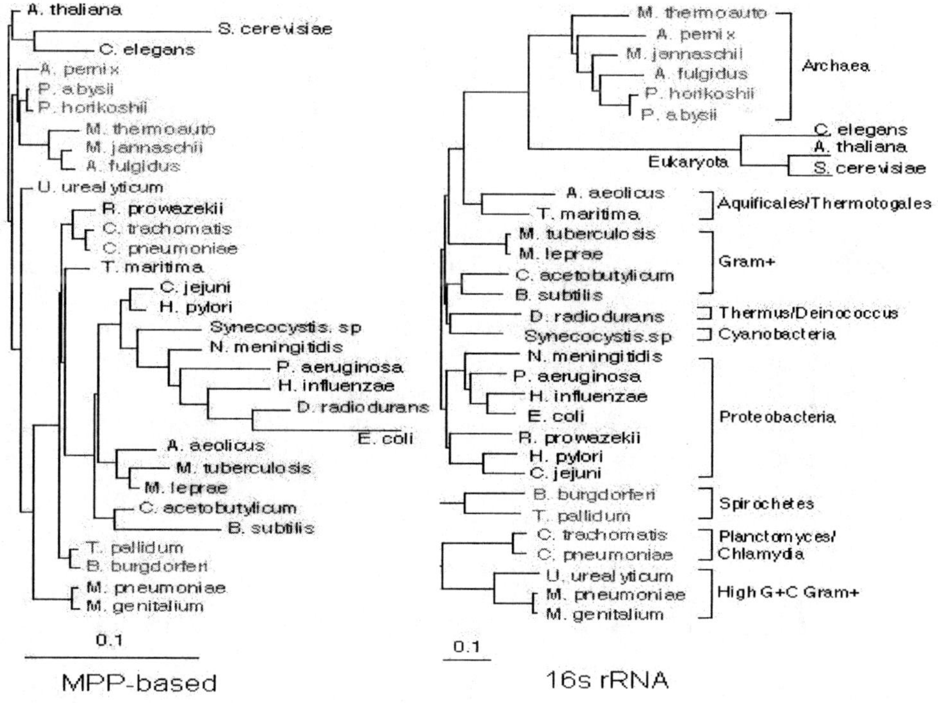

(MPP) (see Figure 4). Relations among 31 genomes are represented as a MPP-based tree and are compared with the phylogeny based on 16s rRNA. While the MPP-based tree is congruent with the 16S rRNA-based tree at several levels, some interesting discrepancies were found. For example, the extremely radiation resistant organism, *D. radiodurans* is positioned in *E. coli* metabolic lineage. The noted deviation from the classical 16s rRNA phylogeny suggests that pathways have undergone evolution transcending the boundaries of species and genera. The MPP-based trees can be used to suggest alternative production platform for metabolic engineering. A different approach to this problem on the same dataset was later proposed in Heymans and Singh (2003).

Parameterized Tree Distance

Evolution of pathways and the interaction with evolution of the host genomes can be further investigated by comparing the columns (phyletic profiles) in the binary matrix in Figure 1. The intuition is that the co-occurrence pattern of pathways in a group of organisms would coincide with the evolutionary path of these host organisms, for instance, as a result of the emergence of enzymes common to those pathways. Therefore, such patterns would give useful information, e.g., organism-specific adaptations. These co-occurrences can be detected by clustering pathways based on their phyletic profiles. Again, the phyletic profiles here are hierarchical profiles, and the distance measure between them should be calculated using the scheme that we developed above. In Zhang, Liao, Tomb, and Wang (2002), 2719 pathways selected from 31 genomes in the WIT database were clustered into 69 groups of pathways, and pathways in each group co-occur in the organisms.

Further insights were achieved by studying the evolution of enzymes that are components in co-occurred pathways. For completely sequenced genomes, sequences of these enzymes are available and can be used to build individual gene trees (in contrast to species trees). Comparisons of the gene trees of component enzymes in co-occurred pathways would serve as a basis for investigating how component enzymes evolve and whether they evolve in accordance with the pathways. For example, two pathways p1 = (e1, e2, e3) and p2 = (e4, e5) co-occur in organisms o1, o2, and o3. For component enzyme e1 of pathway p1, a gene tree T_{e1} is built with e1's homologues in organisms o1, o2 and o3, by using some standard tree reconstruction method such as neighbor joining algorithm (Felsenstein, 1989). This can be done for other component enzymes e2 to e5 as well. If, as an ideal case, gene trees T_{ei} for i = 1 to 5 are identical, we would then say that the pathway co-occurrence is congruent to speciation. When gene trees for component enzymes in co-occurred pathways are not identical, comparisons of gene trees may reveal how pathways evolve differently, possibly due to gene lateral transfer, duplication or loss. Recent work also shows that analysis of metabolic pathways may explain causes and evolution of enzymes dispensability.

To properly address tree comparisons, several algorithms were developed and tested (Zhang et al., 2002). One similarity measure is based on leaf overlap. Let T1 and T2 be two trees. Let S1 be the set of leaves of T1 and S2 be the set of leaves of T2. The leaf-overlap-based distance is defined as

$$D_x(T1, T2) = |S1 \cap S2| / |S1 \cup S2|, \tag{3}$$

where |.| denotes the set cardinality. A more elaborated metric, called parameterized distance as an extension from the editing distance between two unordered trees, was proposed to account for the structural difference between ordered and rooted trees. A parameter *c* is introduced to balance the cost incurred at different cases such as deleting, inserting, and matching subtrees. A dynamic programming algorithm is invoked to calculate the optimal distance. For example, in Figure 5, the distance between D1 and p is 1.875 and the distance between D2 and p is 1.813 when the parameter *c* is set at value of 0.5. On the other hand, the editing distance between D1 and p and between D2 and p are both 6, representing the cost of deleting the six nodes not touched by the dotted mapping lines in Figure 5. Noting that D1 differs from D2 topologically, this example shows that the parameterized distance better reflects the structural difference between trees than the editing distance. In Zhang (2002), the 523 component enzymes of these 2719 pathways were clustered, based on parameterized tree distance with c being set at 0.6, into exactly the same 69 clusters based on pathway co-occurrence. This suggests that our hypothesis about using co-occurrence to infer evolution is valid, at least approximately.

Figure 5. Illustration of parameterized distances between trees. Red dotted lines map tree P to trees D1 and D2, the parameter C = 1 gives the editing distances; the editing distance between D1 and P is 6, because six deletions of nodes other than a, b, and c in D1 will make D1 identical to tree P, although D1 and D2 are topologically different, their editing distances to P are both equal to 6

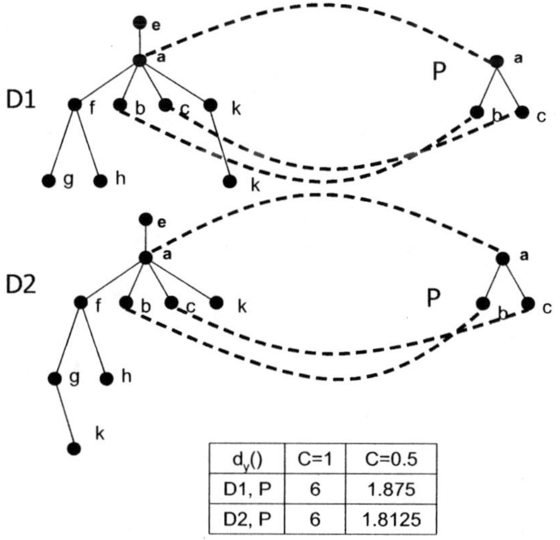

$d_y()$	C=1	C=0.5
D1, P	6	1.875
D2, P	6	1.8125

Tree Kernel Approach

As shown before, use of phylogenetic profiles for proteins has led to methods more sensitive to detect remote homologues. In the following we discuss classifiers using support vector machines to explore similarity among phylogenetic profiles as hierarchical profiles.

As a powerful statistical learning method, support vector machines (SVMs) have recently been applied with remarkable success in bioinformatics problems, including remote protein homology detection, microarrray gene expression analysis, and protein secondary structure prediction. SVMs have been applied to problems in other domains, such as face detection and text categorization. SVMs possess many nice characteristics a good learning method shall have: it is expressive; it requires fewer training examples; it has an efficient learning algorithm; it has an elegant geometric interpretation; and above all, it generalizes well. The power of SVMs comes partly from the data representation, where an entity, e.g., a protein, is represented by a set of attributes instead of a single score. However, how those attributes contribute to distinguishing a true positive (filled dot in Figure 6) from a true negative (empty circle) may be quite complex. In other words, the boundary line between the two classes, if depicted in a vector space, can be highly nonlinear (dashed line in the left panel of Figure 6), and nonetheless, it is the goal of a classifier to find this boundary line. The SVMs method will find a nonlinear mapping that transform the data from the original space, called input space, into a higher dimensional space, called feature space, where the data can be linearly separable (right panel of Figure 6). The learning power of SVMs comes mostly from the use of kernel functions, which define how the dot product between two points in the feature space can be calculated as a function of their corresponding vectors in the input space. As the dot product between vectors is the only quantity needed to find a class boundary in the

Figure 6. Schematic illustration of nonlinear mapping of data from input space to feature space for a SVM

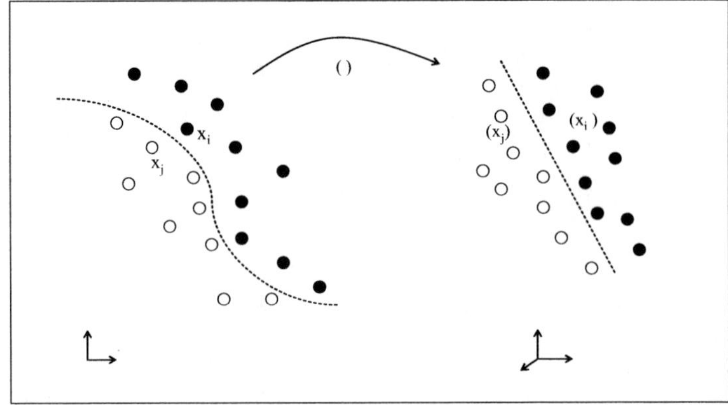

feature space, kernel functions therefore contain sufficient information, and more importantly they avoid explicit mapping to high dimensional feature space; high dimensionality often poses difficult problems for learning such as overfitting, thus termed the curse of dimensionality. The other mechanism adopted by SVMs is to pick the boundary line that has the maximum margin to both classes. A maximum margin boundary line has low Vapnik-Chervonenkis dimension, which ensures good generalization. Because of the central role played by kernel functions, how to engineer kernel functions to incorporate domain specific information for better performance has been an active research activity. It is worth noting that, as compared to other similar learning methods such as artificial neural networks, the SVMs require fewer training examples, which is a great advantage in many bioinformatics applications.

Vert (2002) proposed a tree kernel to compare not just the profiles themselves but also their global patterns of inheritance reflected in the phylogenetic tree. In other words, the tree kernel takes into account phylogenetic histories for two genes — when in evolution they transmitted together or not – rather than by just comparing phylogenetic profiles organism per organism, at the leaves of the phylogenetic tree. A kernel function is thus defined as

$$K(x, y) = \Sigma_{i=1 \text{ to } D}\ \Phi_i(x)\, \Phi_i(y) \tag{4}$$

where $\Phi_i(x)$ is an inheritance pattern *i* for profile x. An inheritance pattern of x gives an explanation of x, that is, the presence (1) or absence (0) of gene x at each current genome is the result of a series of evolutionary events happened at the ancient genomes. If we assume a gene is possessed by the ancestral genome, at each branch of the phylogenetic tree which corresponds to a speciation, the gene may either be retained or get lost. An inheritance pattern corresponds to a series of assignments of retain or loss at all branches such that the results at leaves match the profile of x. Because of the stochastic property of these evolutionary events, we cannot be certain whether a gene is retained or lost. Rather, the best we know may be the probability of either case. Let $F_i(x) = \Phi(x \mid i)$, which is the probability that profile x can be interpreted by the inheritance pattern i, then $K(x, y) = \Sigma_{i=1 \text{ to } D} P(x|i)\, P(y|i)$ is the joint probability that both profiles x and y are resulted from all possible pattern i. Intuitively, not all possible inheritance patterns occur at the same frequency. Let P(i) be the probability that pattern *i* actually occurs during the evolution, then the so called tree kernel is refined as

$$K(x, y) = \Sigma_{i=1 \text{ to } D}\ P(i)\, P(x|i)\, P(y|i) \tag{5}$$

The formalism of using joint probability as kernels first appeared in other applications, such as convolution kernels (Watkins, 1999; Haussler, 1999). Because the number of patterns D grows exponentially with the size of the phylogenetic tree, an efficient algorithm is needed to compute the kernel. Such an algorithm was developed in Vert (2002), which uses post-order traversals of the tree and has a time complexity linear with respect to the tree size.

To test the validity of the tree kernel method, phylogenetic profiles were generated for 2465 yeast genes, whose accurate functional classification are already known, by BLAST search against 24 fully-sequenced genomes. For each gene, if a BLAST hit with E-value less than 1.0 is found in a genome, the corresponding bit for that genome is then assigned as 1, otherwise is assigned as 0. The resulting profile is a 24-bit string of zeros and ones. Two assumptions were made in calculating the tree kernel for a pair of genes x and y. Although their exact probabilities may never be known, it is reasonable to assume that losing a gene or obtaining a new gene is relatively rare as compared to keeping the status quo. In Vert, the probability that an existing gene is retained at a tree branch (i.e., speciation) is set at 0.9, and the probability that a new gene is created at a branch is set at 0.1. It was further assumed that such a distribution remains the same at all branches for all genes. Even with these crude assumptions, in the cross validation experiments on those 2465 yeast genes, the tree kernel's classification accuracy already significantly exceed that of a kernel using just the dot product $x \cdot y = \Sigma_{i=1 \text{ to } 24} x_i y_i$.

Extended Phylogenetic Profiles

The tree-kernel approach's improvement at classification accuracy is mainly due to engineering the kernel functions. In Narra and Liao (2004, 2005) further improvement is attained by both data representations and kernel engineering. As the reader should have been convinced by now, these phylogenetic profiles contain more information than just the string of zeros and ones; the phylogenetic tree provides relationships among the bits in these profiles. In Narra and Liao, a two-step procedure is adopted to extend phylogenetic profiles with extra bits encoding the tree structure: (1) a score is assigned at each internal tree node; (2) the score labeled tree is then flatten into an extended vector. For an internal tree node in a phylogenetic tree, as it is interpreted as ancestor of the nodes underneath it, one way to assign a score for it is to take the average of the scores from its children nodes. This scoring scheme works top-down recursively until the leaves are reached: the score at a leaf is just the value of the corresponding component in the hierarchical profile. The same scoring scheme was also used in p-tree approach. Unlike p-Tree approach that keeps just the score the root node and thus inevitably causes information loss, the scores at all internal nodes are retained and then mapped into a vector via a post-order tree traversal. This vector is then concatenated to the original profile vector forming an extended vector, which is called tree-encoded profile. The scheme works for both binary and real-valued profiles. In order to retain information, real value profiles for yeast genes are used; the binary profiles for tree-kernel are derived real value profiles by imposing a cutoff at E-values

Given a pair of tree-encoded profiles x and y, the polynomial kernel is used classification:

$$K(x, y) = [1 + s\, D(x, y)]^d \tag{6}$$

where s and d are two adjustable parameters. Unlike ordinary polynomial kernels, D(x, y) is not the dot product of vectors x and y, but rather, a generalized Hamming distance for real value vectors:

$$D(x, y) = \Sigma_{i=1 \text{ to } n} (S(|x_i - y_i|)) \quad (7)$$

where the *ad hoc* function S has value 7 for a match, 5 for a mismatch by a difference less then 0.1, 3 for a mismatch by a difference less than 0.3, and 1 for a mismatch by a difference less than 0.5. The values in the *ad hoc* function S are assigned based on the E-value distribution of the protein dataset. The methods are tested on the same data set by using the same cross-validation protocols as in Vert. The classification accuracy of using the extended phylogenetic profiles with E-values and polynomial kernel generally outperforms the tree-kernel approach at most of the 133 functional classes of 2465 yeast genes in Vert.

More Applications and Future Trends

We have seen in the last two sections some problems in bioinformatics and computational biology where relationships can be categorized as hierarchy, and how such hierarchical structure can be utilized to facilitate the learning. Because of the central role played by evolution theory in biology, and the fact that phylogeny is natively expressed as a hierarchy, it is no surprise that hierarchical profiling arises in many biological problems.

In Siepel and Haussler (2004), methods are developed that combine phylogenetic and hidden Markov models for biosequence analysis. Hidden Markov models, first studied as a tool for speech recognition, were introduced to bioinformatics field around 1994 (Krogh et al., 1994). Since then, hidden Markov models have been applied to problems in bioinformatics and computational biology including gene identification, protein homology detection, secondary structure prediction, and many more.

In sequence modeling, hidden Markov models essentially simulate processes along the length of the sequence, mostly ignoring the evolutionary process at each position. On the other hand, in phylogenetic analysis, the focus is on the variation across sequences at each position, mostly ignoring correlations from position to position along their length. From a hierarchical profiling viewpoint, each column in a multiple sequence alignment is a profile, not binary but 20ary, and provides a sampling for the current sequences. It is a very attractive idea to combine these two apparently orthogonal models. In Siepel and Haussler, a simple and efficient method was developed to build higher-order states in the HMM, which allows for context-sensitive models of substitution, leading to significant improvements in the fit of a combined phylogenetic and hidden Markov model. Their work promises to be a very useful tool for some important biosequence analysis applications, such as gene finding and secondary structure prediction.

Hierarchical relations also exist in applications where phylogeny is not the subject. For example, in Holme, Huss, and Jeong (2003), biochemical networks are decomposed into sub-networks. Because of the inherent non-local features possessed by these networks, a hierarchical analysis was proposed to take into account the global structure while doing decomposition. In another work (Gagner et al., 2004), the traditional way of representing metabolic networks as a collection of connected pathways is questioned: such representation suffers the lack of rigorous definition, yielding pathways of disparate content and size. Instead, they proposed a hierarchical representation that emphasizes the gross organization of metabolic networks in largely independent pathways and sub-systems at several levels of independence.

While hierarchical relations are widely existed, there are many applications where the topology of the data relation can not be described as a hierarchy but rather as a network. Large scale gene expression data are often analyzed by clustering genes based on gene expression data alone, though a priori knowledge in the form of biological networks is available. In Hanish, Zien, Zimmer, and Lengauer (2002), a co-clustering method is developed that makes use of this additional information and has demonstrated considerably improvements for exploratory analysis.

Summary

As we have seen, hierarchical profiling can be applied to many problems in bioinformatics and computational biology, from remote protein homology detection, to genome comparisons, to metabolic pathway clustering, wherever the relations among attribute data possess structure as a hierarchy. These hierarchical relations may be inherent or as an approximation to more complex relationships. To properly deal with such relationships is essential for clustering and classification of biological data. It is advisable to heed the relations that bear artificially hierarchical structure; hierarchical profiling on these relations may yield misleading results.

In many cases, the hierarchy, and the biological insight wherein embodied, can be integrated into the framework of data mining. It consequently facilitates the learning and renders meaningful interpretation of the learning results. We have reviewed the recent developments in this respect. Some methods treat hierarchical profile scoring as a tree comparison problem, some as an encoding problem, and some as a graphical model with a Bayesian interpretation. The latter approach is of particular interest, since most biological data are stochastic by nature.

A trend is seen in bioinformatics that combines different methods and models so the hybrid method can achieve a better performance. In the tree-kernel method, the hierarchical profiling and scoring are incorporated as kernel engineering task of the support vector machines. In sequence analysis, the phylogenetic techniques and hidden Markov models are combined to account for the relationships exhibited in sequences that either method alone can not handle properly. As more and more biological data with complex relationships being generated, it is reasonable to believe the hierarchical profiling will see more applications, either serve by itself as a useful tool for analyzing these data, or serve as a prototype for developing more sophisticated and powerful data mining tools.

References

Altschul, S. F., Gish, W., Miller, W., Myers, E., & Lipman, D. J. (1990). Basic local alignment search tool. *Journal of Molecular Biology, 215*, 403-410.

Altschul, S. F., Madden, T. L., Schäffer, A. A., Zhang, J., Zhang, Z., Miller, W., et al. (1997). Gapped BLAST and PSI-BLAST: A new generation of protein database search programs. *Nucleic Acids Res., 25*, 3389-3402.

Bailey, T. L., & Elkan, C. P. (1995). Unsupervised learning of multiple motifs in biopolymers using EM. *Machine Learning, 21*(1-2), 51-80.

Burges, C. J. C. (1998). A tutorial on support vector machines for pattern recognition. *Data Mining and Knowledge Discovery, 2*, 121-167.

Cristianini, N., & Shawe-Taylor, J. (2000). *An introduction to support vector machines and other kernel-based learning methods*. Cambridge, UK: Cambridge University Press.

Dandekar, T., Schuster, S., Snel, B., Huynen, M., & Bork, P. (1999). Pathway alignment: application to the comparative analysis of glycolytic enzymes. *Biochemical Journal, 343*, 115-124.

Durbin, R., Eddy, S., Krogh, A., & Mitchison, G. (1998). *Biological sequence analysis: Probabilistic models of proteins and nucleic acids*. Cambridge, UK: Cambridge University Press.

Eisen, J. A. (2000). Horizontal gene transfer among microbial genomes: New insights from complete genome analysis. *Current Opinion in Genetics & Development, 10*, 606-611.

Felsenstein, J. (1989). PHYLIP — Phylogeny Inference Package (Version 3.2). *Cladistics, 5*, 164-166.

Forst, C. V., & Schulten, K. (2001). Phylogenetic analysis of metabolic pathways. *Journal of Molecular Evolution, 52*, 471-489.

Gaasterland, T., & Selkov, E. (1995). Reconstruction of metabolic networks using incomplete information. In *Proceedings of the Third International Conference on Intelligent Systems for Molecular Biology* (pp. 127-135). Menlo Park, CA: AAAI Press.

Gagneur, J., Jackson, D., & Casar, G. (2003). Hierarchical analysis of dependence in metabolic networks. *Bioinformatics, 19*, 1027-1034.

Gribskov, M., McLachlan, A. D., & Eisenberg, D. (1987). Profile analysis: Detection of distantly related proteins. *Proc. Natl. Acad. Sci. USA, 84* (pp. 4355-4358).

Grundy, W. N., Bailey, T. L., Elkan, C. P., & Baker, M. E. (1997). Meta-MEME: Motif-based hidden Markov Models of biological sequences. *Computer Applications in the Biosciences, 13*(4), 397-406.

Hanisch, D., Zien, A., Zimmer, R., & Lengauer, T. (2002). Co-clustering of biological networks and gene expression data. *Bioinformatics, 18*, S145-S154.

Haussler, D. (1999). *Convolution kernels on discrete structures* (Technical Report UCSC-CRL-99-10). Santa Cruz: University of California.

Henikoff, S., & Henikoff, J.G. (1994). Protein family classification based on search a database of blocks. *Genomics*, *19*, 97-107.

Heymans, M., & Singh, A. J. (2003). Deriving phylogenetic trees from the similarity analysis of metabolic pathways. *Bioinformatics, 19*, i138-i146.

Holme, P., Huss, M., & Jeong, H. (2003). Subnetwork hierarchies of biochemical pathways. *Bioinformatics*, *19*, 532-538.

Jaakkola, T., Diekhans, M., & Haussler, D. (1999). Using the Fisher Kernel Method to detect remote protein homologies. In *Proceedings of the Seventh International Conference on Intelligent Systems for Molecular Biology* (pp. 149-158). Menlo Park, CA: AAAI Press.

Jaakkola, T., Diekhans, M., & Haussler, D. (2000). A discriminative framework for detecting remote protein homologies. *Journal of Computational Biology*, *7*, 95-114.

Karp, P.D., Riley, M., Saier, M., Paulsen, I. T., Paley, S. M., & Pellegrini-Toole, A. (2000). The EcoCyc and MetaCyc databases. *Nucleic Acids Research*, *28*, 56-59.

Krogh, A., Brown, M., Mian, I. S., Sjolander, K., & Haussler, D. (1994). Hidden Markov models in computational biology: Applications to protein modeling. *Journal of Molecular Biology*, *235*, 1501-1531.

Liao, L., Kim, S., & Tomb, J-F. (2002). Genome comparisons based on profiles of metabolic pathways", In *The Proceedings of The Sixth International Conference on Knowledge-Based Intelligent Information & Engineering Systems* (pp. 469-476). Crema, Italy: IOS Press.

Liao, L., & Noble, W. S. (2003). Combining pairwise sequence similarity and support vector machines for detecting remote protein evolutionary and structural relationships. *Journal of Computational Biology*, *10*, 857-868.

Liberles, D. A., Thoren, A., von Heijne, G., & Elofsson, A. (2002). The use of phylogenetic profiles for gene predictions. *Current Genomics*, *3*, 131-137.

Lin, J., & Gerstein, M. (2002). Whole-genome trees based on the occurrence of folds and orthologs: Implications for comparing genomes on different levels. *Genome Research, 10*, 808-818.

Marcotte, E. M., Xenarios, I., van Der Bliek, A. M., & Eisenberg, D. (2000). Localizing proteins in the cell from their phylogenetic profiles. *Proc. Natl. Acad. Sci. USA*, 97, (pp. 12115-12120).

Mittelman, D., Sadreyev, R., & Grishin, N. (2003). Probabilistic scoring measures for profile-profile comparison yield more accurate shore seed alignments. *Bioinformatics*, *19*, 1531-1539.

Narra, K., & Liao, L. (2004). Using extended phylogenetic profiles and support vector machines for protein family classification. In *The Proceedings of the Fifth International Conference on Software Engineering, Artificial Intelligence, Networking, and Parallel/Distributed Computing* (pp. 152-157). Beijing, China: ACIS Publication.

Narra, K., & Liao, L. (2005). Use of extended phylogenetic profiles with E-values and support vector machines for protein family classification. *International Journal of Computer and Information Science*, *6*(1).

Nevill-Manning, C. G., Wu, T. D., & Brutlag, D. L.(1998). Highly specific protein sequence motifs for genome analysis. *Proc. Natl. Acad. Sci. USA*, *95*(11), 5865-5871.

Noble, W. (2004). Support vector machine applications in computational biology. In B. Scholkopf, K. Tsuda, & J-P. Vert. (Eds.), *Kernel methods in computational biology* (pp. 71-92). Cambridge, MA: The MIT Press.

Overbeek, R., Larsen, N., Pusch, G. D., D'Souza, M., Selkov Jr., E., Kyrpides, N., Fonstein, M., Maltsev, N., & Selkov, E. (2000). WIT: Integrated system for high throughput genome sequence analysis and metabolic reconstruction. *Nucleic Acids Res.*, *28*, 123-125.

Pearson, W. (1990). Rapid and sensitive sequence comparison with FASTP and FASTA. *Meth. Enzymol.*, *183*, 63-98.

Pellegrini, M., Marcotte, E. M., Thompson, M. J., Eisenberg, D., & Yeates, T. O. (1999). Assigning protein functions by comparative genome analysis: Protein phylogenetic profiles. *Proc. Natl. Acad. Sci. USA*, *96*, (pp. 4285-4288).

Rabiner, L. R. (1989). A tutorial on hidden Markov models and selected applications in speech recognition. *Proc. IEEE*, *77*, 257-286.

Sadreyev, R., & Grishin, N. (2003). Compass: A tool for comparison of multiple protein alignments with assessment of statistical significance. *Journal of Molecular Biology*, *326*, 317-336.

Scholkopf, B., & Smola, A. J. (2001). *Learning with kernels: Support vector machines, learning).* Cambridge, MA: The MIT Press.

Siepel, A., & Haussler, D. (2004). Combining phylogenetic and hidden Markov Models in biosequence analysis. *J. Comput. Biol.*, *11*(2-3), 413-428.

Smith, T. F., & Waterman, M. S.(1981). Identification of common molecular subsequences. *Journal of Molecular Biology*, *147*, 195-197.

Vapnik, V. (1998). *Statistical Learning Theory: Adaptive and learning systems for signal processing, communications, and control.* New York: Wiley.

Vert, J-P. (2002). A tree kernel to analyze phylogenetic profiles. *Bioinformatics*, *18*, S276-S284.

Watkins, C. (1999). Dynamic alignment kernels. In A. J. Smola, P. Bartlett, B. SchÄolkopf, & C. Schuurmans (Ed.), *Advances in large margin classifiers*. Cambridge, MA: The MIT Press.

Woese, C. (1987). Bacterial evolution. *Microbial Rev.*, *51*, 221-271.

Zhang, K., Wang, J. T. L., & Shasha, D. (1996). On the editing distance between undirected acyclic graphs. *International Journal of Foundations of Computer Science*, *7*, 43-58.

Zhang, S., Liao, L., Tomb, J-F., Wang, J. T. L. (2002). Clustering and classifying enzymes in metabolic pathways: Some preliminary results. In *ACM SIGKDD Workshop on Data Mining in Bioinformatics,* Edmonton, Canada (pp. 19-24).

Chapter III

Combinatorial Fusion Analysis: Methods and Practices of Combining Multiple Scoring Systems

D. Frank Hsu, Fordham University, USA

Yun-Sheng Chung, National Tsing Hua University, Taiwan

Bruce S. Kristal, Burke Medical Research Institute and Weill Medical College of Cornell University, USA

Abstract

Combination methods have been investigated as a possible means to improve performance in multi-variable (multi-criterion or multi-objective) classification, prediction, learning, and optimization problems. In addition, information collected from multi-sensor or multi-source environment also often needs to be combined to produce more accurate information, to derive better estimation, or to make more knowledgeable decisions. In this chapter, we present a method, called Combinatorial Fusion Analysis (CFA), *for analyzing combination and fusion of* multiple scoring. *CFA characterizes each Scoring system as having included a Score function, a Rank function, and a Rank/score function. Both rank combination and score combination are explored as to their combinatorial complexity and computational efficiency.*

Information derived from the scoring characteristics of each scoring system is used to perform system selection and to decide method combination. In particular, the rank/ score graph defined by Hsu, Shapiro and Taksa (Hsu et al., 2002; Hsu & Taksa, 2005) is used to measure the diversity between scoring systems. We illustrate various applications of the framework using examples in information retrieval and biomedical informatics.

Introduction

Many problems in a variety of applications domains such as information retrieval, social / welfare / preference assignments, internet/intranet search, pattern recognition, multi-sensor surveillance, drug design and discovery, and biomedical informatics can be formulated as multi-variable (multi-criterion or multi-objective) classification, prediction, learning, or optimization problems. To help obtain the maximum possible (or practical) accuracy in calculated solution(s) for these problems, many groups have considered the design and integrated use of multiple, (hopefully) complementary scoring schemes (algorithms or methods) under various names such as multiple classifier systems (Ho, 2002; Ho, Hull, & Srihari, 1992, 1994; Melnik, Vardi, & Zhang 2004; Xu, Krzyzak, & Suen, 1992; Kittler & Alkoot, 2003), social choice functions (Arrow, 1963; Young, 1975; Young & Levenglick, 1978), multiple evidences, Web page scoring systems or meta searches (Aslam, Pavlu, & Savell, 2003; Fagin, Kumar, & Sivakumar, 2003; Diligenti, Gori, & Maggini, 2004), multiple statistical analysis (Chuang, Liu, Brown, et al., 2004; Chuang, Liu, Chen, Kao, & Hsu, 2004; Kuriakose et al., 2004), cooperative multi-sensor surveillance systems (Collins, Lipton, Fujiyoshi, & Kanade, 2001; Hu, Tan, Wang, & Maybank, 2004), multi-criterion ranking (Patil & Taillie, 2004), hybrid systems (Duerr, Haettich, Tropf, & Winkler, 1980; Perrone & Cooper, 1992), and multiple scoring functions and molecular similarity measurements (Ginn, Willett, & Bradshaw, 2000; Shoichet, 2004; Yang, Chen, Shen, Kristal, & Hsu, 2005). For convenience and without loss of generality, we use the term **multiple scoring systems (MSS)** to denote all these aforementioned schemes, algorithms, or methods. We further note the need for the word "hopefully" above — there are limited practical means of predicting which combinations will be fruitful — the problem we address in the remainder of this report.

The main purpose in constructing multiple scoring systems is to combine those MSS's in order to improve the efficiency and effectiveness or increase the sensitivity and specificity of the results. This purpose has been met; it has been demonstrated that combining MSS's can improve the optimization results. Combination of multiple scoring systems has been studied under different names such as classification ensemble (Ho, 2002; Ho et al., 1992, 1994; Kittler & Alkoot, 2003; Tumer & Ghosh, 1999; Xu et al., 1992), evidence combination (Belkin, Kantor, Fox, & Shaw, 1995; Chuang, Liu, Brown, et al., 2004; Chuang, Liu, Chen, et al., 2004), data / information fusion (Dasarathy, 2000; Hsu & Palumbo, 2004; Hsu et al., 2002; Hsu & Taksa, 2005; Ibraev, Ng, & Kantor, 2001; Kantor, 1998; Kuriakose et al., 2004; Lee, 1997; Ng & Kantor, 1998, 2000), rank aggregation (Dwork, Kumar, Naor, & Sivakumar, 2001; Fagin et al., 2003), consensus scoring (Ginn et al., 2000; Shoichet, 2004; Yang et al., 2005), and cooperative surveillance (Collins et al., 2001; Hu et al., 2004). In addition, combination of MSS's has been also used in conjunction with other machine learning or evolutional computation approaches such as neural network and evolutional optimization (Garcia-Pedrajas, Hervas-Martinez, & Ortiz-Boyer, 2005; Jin & Branke, 2005). We use the term **combinatorial fusion** to denote all the aforementioned methods of combination.

Combination of multiple approaches (multiple query formulation, multiple retrieval schemes or systems) to solving a problem has been shown to be effective in data fusion in **information retrieval (IR)** and in internet meta search (Aslam et al., 2003; Belkin et al., 1995; Dwork et al., 2001; Fagin et al., 2003; Hsu et al., 2002; Hsu & Taksa, 2005; Lee, 1997; Ng & Kantor, 1998, 2000; Vogt & Cottrell, 1999). In performing classifier combination in the **pattern recognition (PR)** domain, rules are used to combine the output of multiple classifiers. The objective is to find methods (or rules) for building a hybrid classifier that would outperform each of the individual classifiers (Ho, 2002; Ho et al., 1992, 1994; Melnik et al., 2004). In **protein structure prediction (PSP)**, results from different features are combined to improve the accurate predictions of secondary classes or 3-D folding patterns (C.-Y. Lin et al., 2005; K.-L. Lin et al., 2005a, 2005b). Biology may well represent a major area of future needs with respect to combinatorial fusion and related concepts. The last decade has seen an explosion in two data-driven concepts, so-called -omics level studies and *in silico* approaches to modeling.

Omics approaches are approaches that attempt to take snapshots of an organism at a specific level, for example, simultaneously measuring all the metabolites in a tissue and reconstructing pathways. The -omics levels studies range from the four major areas (e.g., genomics — the omics field of DNA analysis; transcriptomics — the omics field of RNA analysis; proteomics—the omics field of protein analysis; metabolomics — the omics field of metabolite analysis) to very specific subareas, such as glycomics (omics approaches to glycated proteins). Omics approaches represent a shift for two reasons: (1) the amount of data inherent either prevents or forces modifications in traditional data analysis approaches (e.g., *t*-tests being replaced by *t*-tests with false discovery rate calculations); and (2) these omics level approaches lead to data-driven and/or hypothesis-generating analyses, not (at least generally) to the testing of specific hypotheses. Most importantly, by offering greatly improved ability to consider systems as a whole and identify unexpected pieces of information, these approaches have opened new areas of investigation and are, perhaps too optimistically, expected to offer fundamentally new insights into biological mechanisms and fundamentally new approaches to issues such as diagnostics and medical classification.

In silico simulations are also becoming a major focus of some biological studies. There are at least two broad areas of simulation studies that are already playing major roles in biological investigations:

- *in silico* ligand-receptor (or drug-target) binding studies, and;
- *in silico* simulations of physiological systems.

In each of these cases, the great advantage lies not in the qualitative change empowered by technological advances, but in the quantitative savings of time and money. For example, obtaining and testing a small chemical library for binding or inhibition can readily cost between $1 and $100 and up (per compound), considering assay costs and obtaining (or synthesizing) the compounds. In contrast, once basic algorithms and binding site models are in place, *in silico* screening is limited only by computational costs, which are dropping exponentially. Similarly, *in silico* models of physiological systems can be somewhat costly to develop, but they are capable of identifying potential targets for intervention and to determine that other targets cannot work, saving tens or hundreds of millions of dollars in failed trials.

Fully utilizing the potential power of these omics and *in silico* based approaches, however, is heavily dependent on the quantity and quality of the computer resources available. The increases in hardware capacity, readily available software tools, database technology, and imaging and scanning techniques, have given the biomedical research community large scale and diversified data sets and the ability to begin to utilize these sets. The problem of how to manipulate, analyze, and interpret information from these biomedical data is a challenging and daunting task. Gradually biologists are discovering tools such as clustering, projection analyses, neural nets, genetic algorithms, genetic programs, and other machine learning and data mining techniques built in other fields, and these tools have found their way to the biomedical informatics domain.

A major issue here lies in the choice of appropriate tools; there is, for example, no clear "informatics pipeline" for omics level studies. One can readily pick up many of software tools, but their applicability for a given problem can be difficult to determine *apriori*. For example, clustering is often powerful for microarray experiments, but we have shown it is very limiting within some metabolomics experiments (Shi, et al., 2002a, 2002b). Similarly, principal components analyses and its supervised cousin, **Soft Independent Modeling of Class Analogy (SIMCA)** seem to work very well within defined cohorts, but they breakdown in complex multi-cohort studies (Shi et al., 2002a, 2002b, 2004), a problem apparently solvable using discriminant-based projection analyses (Paolucci, Vigneau-Callahan, Shi, Matson, & Kristal, 2004). Thus, the choice of an analysis method must often be determined empirically, in a slow, laborious step-wise manner. For the purpose of this article, we will break these biological studies into two broad areas, one of description (i.e., how do we best understand the data in front of us) and one of prediction (i.e., how can we use this information to make predictions about, for example, which ligand will bind or which person will become ill — questions which in many ways are mathematically equivalent).

The report focuses on mathematical issues related to the latter of these two broad issues, i.e., "can we use CFA to improve prediction accuracy?" The goal is a complex zero-sum game — we ideally want to further save time by reducing both false positives and false negatives, while simultaneously increasing accuracy on continuous measurements (e.g. binding strengths). In practice, it is almost certain that some trade-offs will have to be made. To be useful we must, at a minimum, identify an approach which enables some *a priori* decisions to be made about whether such fusion approaches are likely to succeed. Otherwise we have done nothing but to add a layer of complexity between where we are and where we need to be, without removing the time-consuming, laborious, and inherently limited stages of empirical validation. The system we choose to focus on is in **virtual screening (VS)**, the use of *in silico* approaches to identify potentially optimal binding ligands. VS is an area in which consensus scoring has been used in drug design and discovery and molecular similarity measurement for years, and in which data fusion approaches have recently been a major focus of efforts. (see Shoichet, 2004; Yang et al., 2005; and their references).

In this chapter, we present a method called combinatorial fusion analysis (CFA) which uses the **Cayley network Cay(S_n, T_n)** on the symmetric group S_n with generating set T_n (Biggs & White, 1979; Grammatikakis, Hsu, & Kraetzl, 2001; Heydemann, 1997; Marden, 1995). We study the fusion and combination process in the set $\mathbf{R}^n$, called score space, where $\mathbf{R}$ is the set of real numbers, and in the set S_n, called rank space. In the next section, we define rank and score functions and describe the concept of a rank/score graph defined by Hsu, Shapiro and Taksa (Hsu et al., 2002; Hsu & Taksa, 2005). The combination of two scoring systems (each with rank and score functions) is discussed and analyzed in the context of a Cayley network. The next section also entails the property of combined scoring system, performance evaluation and diversity issues, and various kinds of fusion algorithms on rank functions. The section "Data Mining Using Combinatorial Fusion" deals with mining and searching databases and the Web, and includes examples from application domains in biomedical informatics, in particular in virtual screening of chemical libraries and protein structure prediction. Then, in section "Conclusion and Discussion", we summarize our results with some discussion and future work.

Combinatorial Fusion

Multiple Scoring Systems

Successfully turning raw data into useful information and then into valuable knowledge requires the application of scientific methods to the study of the storage, retrieval, extraction, transmission, diffusion, fusion/combination, manipulation, and interpretation at each stage of the process. Most scientific problems are multi-faceted and can be quantified in a variety of ways. Among many methodologies and approaches to solve complex scientific problems and deal with large datasets, we only mention three: (a) classification, (b) clustering, and (c) similarity measurement. Hybrid methods combining (a), (b), and (c) have been used.

Large data sets collected from multi-sensor devices or multi-sources or generated by experiments, surveys, recognition and judging systems are stored in a **data grid** $G(n, m, q)$ with n objects in $D = \{d_1, d_2, \ldots, d_n\}$, m features/attributes/indicators/cues in $G = \{a_1, a_2, \ldots, a_m\}$ and, possibly, q temporal traces in $T = \{t_1, t_2, \ldots, t_q\}$. We call this three-dimensional grid the **data space** (see Figure 1).

Since both m and q can be very big and the size of the datasets may limit the utility of single informatics approaches, it is difficult to use/design a single method/system because of the following reasons:

1. **Different methods/systems are appropriate for different features / attributes / indicators / cues and different temporal traces.** There have been a variety of

Figure 1. Data space $G(n, m, q)$

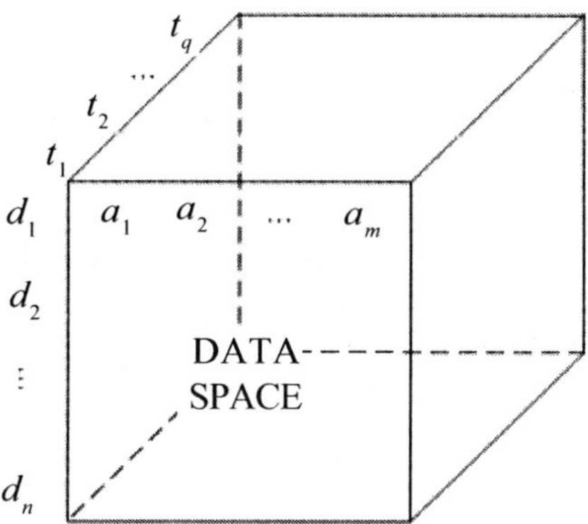

different methods/systems used/proposed in the past decades such as statistical analysis and inference (e.g., *t*-test, non-parametric *t*-test, linear regression, analysis of variance, Bayesian systems), machine learning methods and systems (e.g., neural networks, self-organizing map, support vector machines), and evolutional computations (e.g., genetic algorithms, evolutionary programming).

2. **Different features / attributes / indicators / cues may use different kinds of measurements.** Many different measurements have been used such as variates, intervals, ratio-scales, and binary relations between objects. So in the data grid, each column on the *D-G* plane a_j can be an assignment of a score or a rank to each object d_i. For example, $M(i,j)$ is the score or rank assigned to object d_i by feature/attribute/indicator/cue a_j.
3. **Different methods/systems may be good for the same problem with different data sets generated from different information sources/experiments.** When different data sets generated from different information sources or experiments, different methods/systems should be used according to the style of the source and the nature of the experiment.
4. **Different methods/systems may be good for the same problem with the same data sets generated or collected from different devices/sources.** Even when the data sets are the same in the same experiments, different methods/systems should be adopted according to a variety of multi-sensor/multi-sources.

Due to the complexity of the problem involved, items (3) and (4) indicate that each single system/method, when applied to the problem, can be improved in performance to some extent, but it is difficult to become perfect. Item (1) indicates that performance of a single system / method may be optimal for some features/attributes/indicators/cues, but may downgrade its performance for other features/attributes/indicators/cues.

Recently, it has been demonstrated that combination of multiple systems/methods improves the performance of accuracy, precision, and true positive rate in several domains such as pattern recognition (Brown et al, 2005; Duerr et al., 1980; Freund, Iyer, Schapire, & Singer, 2003; Garcia-Pedrajas et al., 2005; Ho, 2002; Ho et al., 1992, 1994; Jin & Branke, 2005; Kittler & Alkoot, 2003; Perrone & Cooper, 1992; Triesch & von der Malsburg, 2001; Tumer & Ghosh, 1999; Xu et al., 1992), microarray gene expression analysis (Chuang, Liu, Brown, et al., 2004; Chuang, Liu, Chen, et al., 2004; Kuriakose et al., 2004), information retrieval (Aslam et al., 2003; Belkin et al., 1995; Diligenti et al., 2004; Dwork et al., 2001; Fagin et al., 2003; Hsu et al., 2002; Hsu & Taksa, 2005; Kantor, 1998; Lee, 1997; Ng & Kantor, 1998, 2000), virtual screening and drug discovery (Shoichet, 2004; Yang et al., 2005; and references in both), and protein structure prediction (C.-Y. Lin et al., 2005; K.-L. Lin et al., 2005a, 2005b).

There have been special meetings (such as the Workshop on Multiple Classifier Systems), conferences (such as International Conference on Information Fusion), societies (such as International Society of Information Fusion), and journals (e.g.: Information Fusion [Dasarathy, 2000]) dedicated to the scientific study of fusion/combination. The main title of the featured article "The Good of the Many Outweighs the Good of the One," by Corne, Deb, Fleming, & Knowles (2003) in IEEE Neural Networks Society typifies the scientific and philosophical merits of and motivations for fusion/combination.

Each system/method offers the ability to study different classes of outcomes, e.g., class assignment in a classification problem or similarity score assignment in the similarity measurement problem. In this chapter, we view the outcome of each system/method as a scoring system A which assigns (a) an object as a class among all objects in D, (b) a score to each object in D, and (c) a rank number to each object in D. These three outcomes were described as the abstract, score, and rank level respectively by Xu et al. (1992). We now construct the **system grid** $H(n, p, q)$ with n objects in $D = \{d_1, d_2, \ldots, d_n\}$, p systems

Figure 2. System space $H(n, p, q)$

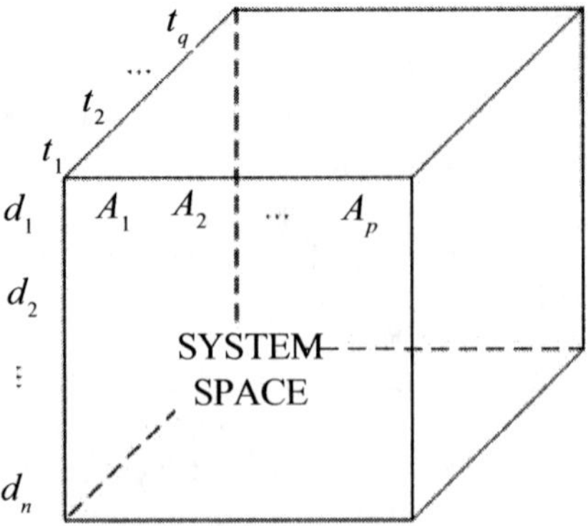

in $H = \{A_1, A_2, \ldots, A_p\}$, and possibly, q temporal traces in $T = \{t_1, t_2, \ldots, t_q\}$. We call this three dimensional grid the **system space** for the multiple scoring systems (see Figure 2).

In the next section, we will define score function, rank function and the rank/score function for each scoring system A. In the section following the next, rank and score combination are defined and studied. Section "Method of Combinatorial Fusion" deals with performance evaluation criteria and diversity between and among different scoring systems.

Score Function, Rank Function and the Rank/Score Graph

Let $D = \{d_1, d_2, \ldots, d_n\}$ be a set of n objects and $N = \{1, 2, 3, \ldots, n\}$ be the set of all positive integers less than or equal to n. Let **R** be the set of real numbers. We now state the following three functions that were previously defined and studied by Hsu, Shapiro and Taksa (2002) and Hsu and Taksa (2005).

Definition 1

(a) **Score Function:** A score function s is a function from D to **R** in which the function s assigns a score (a real number) to each object in D. In a more formal way, we write $s: D \rightarrow \mathbf{R}$ such that for every d_i in D, there exists a real number $s(d_i)$ in **R** corresponding to d_i.

(b) **Rank Function:** A rank function r is a function from D to N such that the function $r: D \rightarrow N$ maps each element d_i in D to a natural number $r(d_i)$ in N. The number $r(d_i)$ stands for the rank number $r(d_i)$ assigned to the object d_i.

(c) **Rank/Score Function:** Given r and s as rank and score function on the set D of objects respectively, the rank/score function f is defined to be $f: N \rightarrow \mathbf{R}$ such that $f(i) = (s \circ r^{-1})(i) = s(r^{1}(i))$. In other words, the score function $s = f \circ r$ is the composite function of the rank/score function and the rank function.

(d) **Rank/Score Graph:** The graph representation of a rank/score function.

We note that in several application domains, one has to normalize the score function values before any combination can be performed. Hence it is quite natural to define the two functions s and f in the way that each of them has $[0, 1] = \{x \mid x \text{ in } \mathbf{R}, 0 \leq x \leq 1\}$ instead of **R** as their function range. Other intervals of real numbers can be also used depending on the situation and environment. We also note that since the rank function r' defined by Hsu, Shapiro and Taksa (see Hsu et al., 2002; Hsu & Taksa, 2005) is the inverse of r defined above, the rank/score function f would be such that $f = s \circ r'$ instead of $f = s \circ r'^{-1}$.

At this point, we would like to comment on some perspectives regarding rank vs. score function. Although these two functions both deal with the set of objects under study (in the case of PR, IR, VS and PSP, these would be classes, documents, chemical compounds, and classes or folding patterns, respectively), their emphases are different. The Score function deals more with the detailed data level while Rank function is more relevant for or related to the decision level. In theory, the Score function depends more on the variate data in the parametric domain while the Rank function depicts more on the ordinal data in the non-parametric fashion. The comparison can go on for a long time as score vs. rank, data level vs. decision level, variate data vs. ordinal data, and parametric vs. non-parametric. Historically and from the discipline perspective, scores are used in sciences, engineering, finance, and business, while ranks are used in social choices, ordinal data analysis and decision science. However, in biomedical informatics, since the data collected is large (and of multiple dimension) and the information we are seeking from biological and physiological systems is complex (and multi-variable), the information we find (or strive to find) from the relation between score and rank function would become valuable in biological, physiological, and pharmaceutical study.

The concept of a rank/score graph which depicts the graph representation of a rank/score function has at least three characteristics and advantages:

Remark 1

(a) **Efficiency:** When a score function s_A is assigned resulting from scoring system A by either lab work or field study (conducted in vivo or in vitro), treating s_A as an array and sorting the array of scores into descending order would give rise to the rank function r_A. The rank/score function can be obtained accordingly. If there are n objects, this transformation takes $O(n \log n)$ steps.

(b) **Neutrality:** Since the rank/score function f is defined from N to **R** or from N to [0, 1], it does not depend on the set of objects $D = \{d_1, d_2, \ldots, d_n\}$. Hence the rank/score function f_A of a scoring system A exhibits the behavior of the scoring system (scorer or ranker) A and is independent of who (or which object) has what rank or what

Figure 3. Score function, Rank function and Rank/Score function of A and B

D	d_1	d_2	d_3	d_4	d_5	d_6	d_7	d_8	d_9	d_{10}
$s_A(d_i)$	4	10	4.2	3	6.4	6.2	2	7	0	1
$r_A(d_i)$	6	1	5	7	3	4	8	2	10	9

(a) Score and Rank function for A

D	d_1	d_2	d_3	d_4	d_5	d_6	d_7	d_8	d_9	d_{10}
$s_B(d_i)$	4	7	3	1	10	8	5	6	9	2
$r_B(d_i)$	7	4	8	10	1	3	6	5	2	9

(b) Score and Rank function for B

N	1	2	3	4	5	6	7	8	9	10
$f_A(i)$	10	7	6.4	6.2	4.2	4	3	2	1	0
$f_B(i)$	10	9	8	7	6	5	4	3	2	1

(c) Rank/Score function for A and B

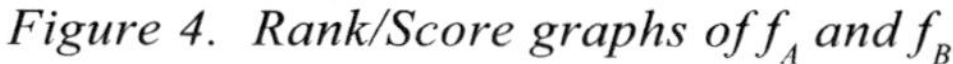

Figure 4. Rank/Score graphs of f_A *and* f_B

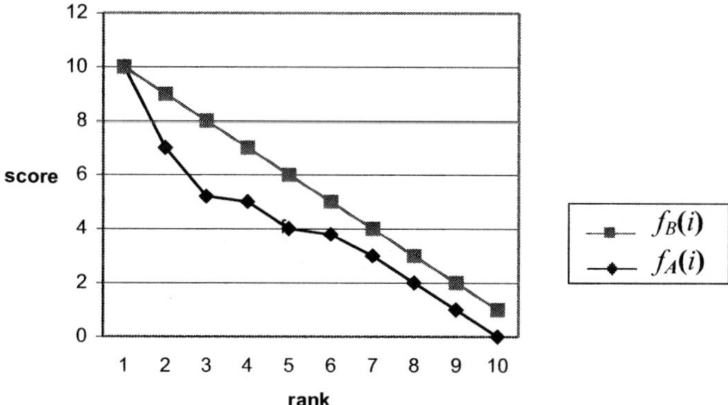

score. The rank/score function f also fills the gap of relationship between the three sets D, N and **R**.

(c) **Visualization:** The graph of the rank/score function f_A can be easily and clearly visualized. From the graph of f_A, it is readily concluded that the function f_A is a non-increasing monotonic function on N. The thrust of this easy to visualize property is that comparison (or difference) on two functions f_A and f_B can be recognized by drawing the graph of f_A and f_B on the same coordinate system.

Two examples of score functions, rank functions (derived from score function), rank/score functions with respect to scoring systems A and B are illustrated in Figure 3 and the rank / score graphs of f_A and f_B are included in Figure 4, where $D = \{d_i \mid i = 1 \text{ to } 10\}$ and $s(d_i)$ is in [0, 10].

Rank and Score Combination

As mentioned earlier, the combinations (or fusions) of multiple classifiers, multiple evidences, or multiple scoring functions in the PR, IR, VS, and PSP domain has gained tremendous momentum in the past decade. In this section, we deal with combinations of two functions with respect to both score and rank combinations. The following definitions were used by Hsu, Shapiro, and Taksa (2002) and Hsu and Taksa (2005).

Definition 2

(a) **Score Combinations:** Given two score functions s_A and s_B, the score function of the score combined function s_F is defined as $s_F(d) = \frac{1}{2}[s_A(d) + s_B(d)]$ for every object d in D.

Figure 5. Score and Rank function of E and F

D	d_1	d_2	d_3	d_4	d_5	d_6	d_7	d_8	d_9	d_{10}
$s_E(d_i)$	6.5	2.5	6.5	8.5	2	3.5	7	3.5	6	9
$r_E(d_i)$	6	2	4	9	1	3	8	7	5	10

(a) Score and rank function of E

D	d_1	d_2	d_3	d_4	d_5	d_6	d_7	d_8	d_9	d_{10}
$s_F(d_i)$	4.0	8.5	3.6	2.0	8.2	7.1	3.5	6.5	4.5	1.5
$r_F(d_i)$	6	1	7	9	2	3	8	4	5	10

(b) Score and rank functions of F

(b) **Rank Combinations:** Given two rank functions r_A and r_B, the score function of the rank combined function s_E is defined as $s_E(d) = \frac{1}{2}[r_A(d)+r_B(d)]$ for every object d in D.

Since each of the scoring systems A and B has their score and rank function s_A, r_A and s_B, r_B respectively, each of the combined scoring systems E and F (by rank and by score combination) has s_E, r_E and s_F, r_F, respectively. These functions can be obtained as follows. The score function of the rank combination s_E is obtained from r_A and r_B using rank combination. Sorting s_E into ascending order gives rise to r_E. The score function of the score combination s_F is obtained from s_A and s_B using score combination. Sorting s_F into descending order gives rise to r_F. Hence for scoring systems E and F, we have the score function and rank function s_E, r_E and s_F, r_F respectively.

We note the difference between converting from s_E to r_E and that from s_F to r_F. Since the scoring systems E and F are rank and score combination of A and B respectively, the transformation from s_E to r_E is by sorting into ascending order while that from s_F to r_F is by sorting into descending order. The fusion architecture given by Hsu, Shapiro and Taksa (Figure 5 in Hsu & Taksa, 2005) depicts the framework for the fusion method we use. Figure 5(a) and (b) show the rank and score function of the rank and score combination E and F (both related to the scoring systems A and B in Figure 3) respectively. The reader might have noticed that in the combination (rank or score), a simple average combination was used. Combination using weighted proportion on A and B is studied in Hsu and Palumbo (2004).

Recall that the main purpose of doing fusion/combination is whether or not the fused scoring system can outperform each individual scoring system in isolation. The concept of "performance" has to be defined for a given study (or, arguably, a given class of studies), so that it is meaningful to say "outperform". Recall also that our fusion framework is for application domains such as PR, IR, VS and PSP. In PR, each classifier produces a ranking of a set of possible classes. When given an image, each of the classifiers including the combined gives rise to a rank function and the class that was ranked at the top is predicted by the classifier as the identity of the image. So **the performance** (or **accuracy**) **of the classifier** A, written $P(A)$, is the percentage of times that this classification gives a correct prediction. From the perspective of biology and medicine, $P(A)$ may be defined as the reduction of false negatives (e.g., in cancer diagnostic) or the reduction of false positives (e.g., in screens of related compounds where follow-up is expensive).

In **IR**, a query Q is given. Then each of the ranking algorithms calculates similarity between the query and the documents in the database of n documents $D = \{d_1, d_2 ..., d_n\}$. A score function s_A for algorithm A is assigned and a rank function r_A is obtained. **The performance of the algorithm for the query**, written $P_q(A)$, is defined as the precision of A at q with respect to the query Q. More specifically, the following is defined and widely used in the information retrieval community.

Definition 3

(a) Let $Rel(Q)$ be the set of all documents that are judged to be relevant with respect to the query Q. Let $|Rel(Q)| = q$ for some q, $0 \le q \le n$. On the other hand, let $A_{(k)} = \{d \mid d \text{ in } D \text{ and } r_A(d) \le k\}$.

(b) **Precision of** A **at** q. The performance of the scoring system A with respect to the query Q is defined to be $P_q(A) = |Rel(Q) \cap A_{(q)}| / q$, where $q = |Rel(Q)|$.

In **VS**, molecular compound libraries are searched for the discovery of novel lead compounds for drug development and/or therapeutical treatments. Let L be the total number of active ligands and T the total number of compounds in the database. Let L_h be the number of active ligands among the T_h highest ranking compounds (i.e., the hit list). Then the **goodness-of-hit (GH) score** for a scoring system A is defined as (see Yang et al., 2005):

$$GH(A) = \left(\frac{L_h(3L + T_h)}{4T_h L}\right)\left(1 - \frac{T_h - L_h}{T - L}\right).$$

The GH score ranges from 0.0 to 1.0, where 1.0 represents a perfect hit list. The GH score as defined in Yang et al. (2005) contains a coefficient to penalize excessive hit list size. We will come back to this topic in section "Virtual Screening and Drug Discovery".

In the **protein structure prediction problem (PSP)**, one wishes to extract structural information from the sequence databases as an alternative to determine the 3-D structure of a protein using the X-ray diffraction or NMR (resource and labor intensive, expensive, and, in practice, often difficult or impossible, particularly when one deals with variably modified proteins). Given a protein sequence, the objective is to predict its secondary structure (class) or its 3-D structures (folding patterns). The standard performance evaluation for the prediction of the n_i protein sequences $T_i = \{t_1, t_2, ..., t_{n_i}\}$ for the ith class or ith folding pattern is the percentage accuracy rate $Q_i = Q(T_i) = (p_i / n_i) \times 100$, where n_i is the number of testing proteins in the ith class or ith folding pattern and p_i is the number of proteins within the n_i protein sequences correctly predicted. The overall prediction accuracy rate Q is defined as $Q = \sum_{i=1}^{k} q_i Q_i$, where $q_i = n_i / N$, N is the total number of proteins tested (i.e., $N = \sum_{i=1}^{k} n_i$) and k is the number of classes or folding patterns.

Now we come back to the fundamental issue of when the combined scoring system E or F outperforms its individual scoring system A and B. The following two central questions regarding combination and fusion were asked by Hsu, Shapiro, and Taksa (2002) and Hsu and Taksa (2005):

Remark 2

(a) For what scoring systems A and B and with what combination or fusion algorithm, $P(E)$ (or $P(F)$) $\geq \max\{P(A), P(B)\}$, and

(b) For what A and B, $P(E) \geq P(F)$?

Four important issues are central to CFA: (1) What is the best fusion algorithm / method to use? (2) Does the performance of E or F depend very much (or how much) on the relationship between A and B? (3) Given a limited number of primary scoring systems, can we optimize the specifics of the fusion? and (4) Can we answer any or all of the previous three issues without resorting to empirical validation? The general issue of combination algorithm / method will be discussed in the next section. In this section, we simply use the average combination regardless of rank or score combination.

Arguably, issues (1) and (2) may be considered primary issues, and issues (3) and (4) secondary or derivative. We propose that issue (2) is as, if not more, important as issue (1). It has been observed and reported extensively and intensively that the combined scoring system E or F performs better than each individual scoring system A and B when A and B are "different", "diverse", or "orthogonal". In particular, Vogt and Cottrell (1999) studied the problem of predicting the performance of a linearly combined system and stated that the linear combination should only be used when the individual systems involved have high performance, a large overlap of relevant documents, and a small overlap of non-relevant documents. Ng and Kantor (2000) identified two predictive variables for the effectiveness of the combination: (a) the output dissimilarity of A and B, and (b) the ratio of the performance of A and B. Then Hsu, Shapiro and Taksa (2002) and Hsu and Taksa (2005) suggested using the difference between the two rank/score functions f_A and f_B as the diversity measurement to predict the effectiveness of the combination. This diversity measurement has been used in microarray gene expression analysis (Chuang, Liu, Brown, et al., 2004; Chuang, Liu, Chen, et al., 2004) and in virtual screening (Yang et al., 2005). We will discuss in more details the use of graphs for rank/score functions f_A and f_B in the diversity measurement between A and B in section "Virtual Screening and Drug Discovery".

Method of Combinatorial Fusion

As defined in section "Score Function, Rank Function and the Rank/Score Graph", a rank function r_A of a scoring system A is an one-one function from the set $D = \{d_1, d_2, \ldots, d_n\}$ of n objects to the set N of positive integers less than or equal to n. When considering r_A as a permutation of N, we have $r_A(D) = [r_A(d_1), r_A(d_2), \ldots, r_A(d_n)]$. Without loss of clarity, we write r instead of r_A.

Since the set of all permutations of the set N is a group with the composition $\alpha \circ \beta$ as the binary operation, called symmetric group S_n, $r(D)$ as a permutation of N is an element in the group S_n. Each element a in S_n has two different ways of representation as $[\alpha(1), \alpha(2), \ldots, \alpha(n)] = [\alpha_1, \alpha_2, \ldots, \alpha_n] = [\alpha_1\ \alpha_2\ \alpha_3 \ldots \alpha_n]$ called **standard representation** and as the product of disjoint cycles each consisting of elements from N called **cycle representation**. The general concept of a Cayley graph (or network) can be found in the book and article (Grammatikakis et al., 2001; Heydemann, 1997).

Definition 4

(a) **Example of Permutations:** The permutation $a(i) = 2i$ for $i = 1, 2, 3$ and $2i - 7$ for $i = 4, 5, 6$ in S_6 can be written as $[2\ 4\ 6\ 1\ 3\ 5]$ and $(124)(365)$. The permutation $b(i) = i$ for $i = 1, 4, 5, 6$ and $b(2) = 3$, $b(3) = 2$ can be written as $b = [1\ 3\ 2\ 4\ 5\ 6] = (23)$. Note that in the cycle representation, we ignore the cycles that are singletons.

(b) In the group S_n, the set of $n - 1$ adjacent transpositions such as $b = (23)$ in S_6 is denoted as T_n. In other words, T_n consists of all cycles of length 2 which are adjacent transpositions and $T_n = \{(1\ 2), (2\ 3), \ldots, (n-1\ n)\}$ is a subset of S_n. With this in mind, we can define a Cayley network based on S_n and T_n:

Cayley network Cay (S_n, T_n)**:** The Cayley network Cay(S_n, T_n) is a graph $G(V, E)$ with the node set $V = S_n$ and arc set $E = \{(\alpha, \alpha \circ t) \mid \alpha \text{ in } S_n \text{ and } t \text{ in } T_n\}$.

The concept of a Cayley network extends the group structure in S_n to the graph structure in Cay(S_n, T_n). By doing so, a distance measure between any two permutations (and hence any two rank functions) is well defined in the context of applications that will prove to be very useful in biomedical informatics. In fact, it has been mentioned by Hsu, Shapiro and Taksa (2002) and Hsu and Taksa (2005) that the graph distance in Cay(S_n, T_n) is the same as Kendall's tau distance in the rank correlation analysis (RCA) (see e.g., Kendall &Gibbons, 1990; Marden, 1995). This striking coincidence supports the importance and usefulness of using Cayley networks as a framework for fusion and combination. Moreover, we point out that the combinatorial fusion we proposed and the rank correlation studied by many researchers in the past bear similarity but have differences. They are very similar because they all study ranks although one treats ranks as a function and the other treats them as ordinal data or the order of the values of a random variable. On the other hand, they are quite different. The CFA (combinatorial fusion analysis) views the set S_n as a rank space aiming to produce a dynamic process and reasonable algorithms to reach a better combined rank function (or in general, combined scoring system). The RCA (rank correlation analysis) views the set S_n as a population space aiming to calculate the static correlation and significant P-value to reach a hypothesis testing result.

Suppose we are given the following p rank functions A_j obtained from the data set $D = \{d_1, d_2, \ldots, d_n\}, A_j, j = 1, 2, \ldots, p$:

$$A_j = (a_{1j}, a_{2j}, a_{3j}, \ldots, a_{nj})^t,$$

where V^t is the transpose of the vector V. Let M_r be the matrix, called **rank matrix**, with dimension $n \times p$ such that $M_r(i,j) = M(i,j) =$ the rank assigned to the object d_i by scoring system A_j. Now, we describe the rank version of the combinatorial fusion problem.

Definition 5

Combinatorial Fusion Problem (rank version): Given p nodes $A_j, j = 1, 2, \ldots, p$ in the Cayley network Cay(S_n, T_n) with respect to n objects $D = \{d_1, d_2, \ldots, d_n\}$, find a node A^* in S_n which "performs" as good as or better than the best of A_j's in the sense of performance as defined as accuracy, precision or goodness-of-hit in IR, VS and PSP described previously.

There are several ways to find the candidates for the node A^* when given the p nodes $A_j, j = 1, 2, \ldots, p$ in S_n. We briefly describe, in Definition 6, the following six types of methods / algorithms to fuse the given p nodes and generate the candidate node. All of these approaches aim to construct a score function which, when sorted, would lead to a rank function.

Definition 6

(a) **Voting:** Scoring function $s^*(d)$, d in D. The score of the object d_i, $s^*(d_i)$, is obtained by a voting scheme among the p values $M(i,j)$, $j = 1, 2, \ldots, p$. These include max, min, and median.

(b) **Linear Combination:** These are the cases that $s^*(d_i)$ is a weighted linear combination of the $M(i,j)$'s, i.e., $s^*(d_i) = \sum_{j=1}^{p} w_j \cdot M(i,j)$ for some weighted function so that $\sum_{j=1}^{p} w_j = 1$. When $w_j = 1/p$, $s^*(d_i)$ is the average of the ranks $M(i,j)$'s, $j = 1, 2, \ldots, p$.

(c) **Probability Method:** Two examples are the Bayes rule that uses the information from the given p nodes $A_j, j = 1, 2, \ldots, p$ to predict the node A^*, and the Markov Chain method that calculates a stochastic transition matrix.

(d) **Rank Statistics:** Suppose that the p rank functions are obtained by the p scoring systems or observers who are ranking the n objects. We may ask: what is the true ranking of each of the n objects? Since the real (or true) ranking is difficult to come by, we may ask another question: What is the best estimate of the true ranking when we are given the p observations? Rank correlation among the p rank functions can be calculated as $W = 12S / [p^2(n^3 - n)]$, where $S = \sum_{i=1}^{n} R_i^2 - np^2(n+1)^2/4$ and $R_i = \sum_{j=1}^{p} M(i,j)$. The significance of an observed value of W is then tested in the $(n!)^p$ possible sets of rank functions.

(e) **Combinatorial Algorithm:** For each of the n objects and its set of p elements $\{M(i, j) \mid j = 1, 2, \ldots, p\} = C$ as the candidate set for $s^*(d_i)$, one combinatorial algorithm considers the power set 2^C and explores all the possible combinatorial combinations. Another algorithm treats the n objects as n vectors $d_i = (a_{i1}, a_{i2}, \ldots, a_{ip})$ where $a_{ij} = M(i,j)$, $i = 1, 2, \ldots, n$ and $1 \leq a_{ij} \leq n$. It then places these n vectors in the context of a **partially ordered set (Poset)** L consisting of all the n^p vectors $(a_{i1}, a_{i2}, \ldots, a_{ip})$, $i = 1, 2, \ldots, n$ and a_{ij} in N. The scores $s^*(d_i)$, $i = 1, 2, \ldots, n$ is then calculated based on the relative position of the vector d_i in the Poset L.

(f) **Evolutional Approaches:** Genetic algorithms and other machine learning techniques such as neural networks and support vector machines can be used on the p rank functions to process a (large) number of iterations in order to produce a rank function that is closest to the node A^*.

The voting schemes in Definition 6 (a) have been used in social choice functions (Arrow, 1963; Young & Levenglick, 1978). Linear combination and average linear combination in 6(b), due to their simplicity, have been widely used in many application domains (Kendall & Gibbons, 1990; Kuriakose et al., 2004; Hsu & Palumbo, 2004; Hsu et al., 2002; Hsu & Taksa, 2005; Vogt & Cottrell, 1999). In fact, the concept of Borda count, used by Jean-Charles de Borda of the L'Academie Royale des Sciences in 1770, is equivalent to the average linear combination. Dwork et al. (2001) used Markov chain method to aggregate the rank functions for the Web. As described in Definition 6(d), the significance of S depends on the distribution of S in the $(n!)^p$ possible set of rank functions. Due to the manner that S is defined, it may be shown that the average linear combination gives a "best" estimate in the sense of Spearman's rho distance (see Kendall & Gibbons, 1990, Chapter 6). Combinatorial algorithms stated in Definition 6(e) have been used in Mixed Group Ranks and in Rank and Combine method by researchers (Chuang, Liu, Brown, et al., 2004; Chuang, Liu, Chen, et al., 2004; Melnik et al., 2004). Although genetic algorithms such as GemDOCK and GOLD were used to study the docking of ligands into a protein, the authors in Yang et al. (2005) use linear combination and the rank/score graph as a diversity measurement. We will discuss the application in more details in next section.

Definition 6 lists six different groups of methods/algorithms/approaches for performing combination. Here we return to the second issue raised by Remark 2. That is: What are the predictive variables / parameters / criteria for effective combination? In accordance with Remark 2, we focus on two functions A and B (i.e. $p = 2$) at this moment although the methods / algorithms / approaches in Definition 6, are able to deal with the multiple functions ($p \geq 2$). We summarize, in Definition 7, the two variables for the prediction of effective combination among two scoring systems A and B (Chuang, Liu, Brown, et al., 2004; Chuang, Liu, Chen, et al., 2004; Hsu et al., 2002; Hsu & Taksa, 2005; Ng & Kantor, 2000; Vogt & Cottrell, 1999; Yang et al., 2005).

Definition 7

(a) **The performance ratio,** P_l / P_h, measures the relative performance of A and B where P_l and P_h are the lower performance and higher performance among $\{P(A), P(B)\}$ respectively.

(b) **The bi-diversity between A and B,** $d_2(A, B)$, measures the "difference / dissimilarity / diversity" between the two scoring systems A and B.

We note that in order to properly use diversity $d_2(A, B)$ as a predictive parameter for effective combination of functions A and B, $d_2(A, B)$ might be defined to reflect different combination algorithms and different domain applications. However, for the diversity measurement to be effective, it has to be universal at least among a variety of data sets in applications domain. Diversity measurement between two scoring systems A and B, $d_2(A, B)$ have been defined and used (see Chuang, Liu, Brown, et al., 2004; Chuang, Liu, Chen, et al., 2004; Ng & Kantor, 2000; Yang et al., 2005). We summarize in Definition 8.

Definition 8

Diversity Measure: The bi-diversity (or 2-diversity) measure $d_2(A, B)$ between two scoring systems A and B can be defined as one of the following:

(a) $d_2(A, B) = d(s_A, s_B)$, the distance between score functions s_A and s_B. One example of $d(s_A, s_B)$ is the covariance of s_A and s_B, $\mathrm{Cov}(s_A, s_B)$, when s_A and s_B are viewed as two random variables,

(b) $d_2(A, B) = d(r_A, r_B)$, the distance between rank functions r_A and r_B. One example of $d(r_A, r_B)$ is the Kendall's tau distance as we defined in S_n, and

(c) $d_2(A, B) = d(f_A, f_B)$, the distance between rank/score functions f_A and f_B.

We note that diversity measure for multiple classifier systems in pattern recognition and classification has been studied extensively (Kuncheva, 2005).

Definition 9

In the data space $G(n, m, q)$, $m = 2$, defined in Figure 1, given a temporal step t_i in $T = \{t_1, t_2, \ldots, t_q\}$ and the two scoring systems A and B, we define:

(a) $d_{t_i}(A, B) = \sum_j | f_A(j) - f_B(j) |$, where j is in $N = \{1, 2, \ldots, n\}$, as the function value of the **diversity score function** $d_x(A, B)$ for t_i;

(b) if we let i vary and fix the system pair A and B, then $s_{(A,B)}(x)$ is the diversity score function, defined as $s_{(A,B)}(t_i) = d_{t_i}(A, B)$, from $T = \{t_1, t_2, \ldots, t_q\}$ to **R**;

(c) sorting $s_{(A,B)}(x)$ into ascending order leads to the **diversity rank function** $r_{(A,B)}(x)$ from T to $\{1, 2, \ldots, q\}$; and then

(d) the **diversity rank/score function** $f_{(A,B)}(j)$ can be obtained as $f_{(A,B)}(j) = (s_{(A,B)} \circ r^{-1}_{(A,B)})(j) = s_{(A,B)}(r^{-1}_{(A,B)}(j))$, where j is in $\{1, 2, ..., q\}$;

(e) the **diversity rank/score graph** (or diversity graph) is the graph representation of the diversity rank/score function $f_{(A,B)}(j)$ from $\{1, 2, ..., q\}$ to **R**.

We note the difference between the rank/score function and the diversity rank/score function. In the definition of rank/score function f_A: $N \rightarrow [0, 1]$ (see Definition 1(c)), the set N is different from the set D which is in turn the set of objects (classes, documents, ligands, and classes or folding patterns). The set N is used as the index set for the rank function values. The rank/score function f_A so defined describes the scoring (or ranking) behavior of the scoring system A and is independent of the objects under consideration. The diversity rank/score function (see Definition 9(d)) $f_{(A,B)}(j)$ is defined from $Q = \{1, 2, ..., q\}$ to **R** (or [0, 1]). The set Q is different from the set $T = \{t_1, t_2, ..., t_q\}$ which is the set of temporal steps under study. The set Q is used as the index set for the diversity rank function values. The diversity rank/score function $f_{(A,B)}$ so defined describes the diversity trend of the pair of scoring systems A and B and is independent of the specific temporal step t_i for some i under study.

Data Mining Using Combinatorial Fusion

In this section, we present three examples of data mining using combinatorial fusion as defined in the previous section. These three examples are from applications in information retrieval (IR) (Hsu et al, 2002; Hsu & Taksa, 2005; Ng & Kantor, 1998, 2000), consensus scoring in virtual screening (VS) (Yang et al., 2005), and protein structure prediction (PSP) (C.-Y. Lin et al., 2005; K.-L. Lin et al., 2005a, 2005b). But before we concentrate on special cases, we will further discuss the relation between rank and score functions as defined in the previous section.

Rank/Score Transfer

Let $M_r(i,j)$ be the **rank matrix** defined before, where M_{ij} in $M_r(i,j)$ is the rank assigned to the object d_i by scoring system A_j. Let $M_s(i,j)$ be the $n \times p$ **score matrix** defined similarly with respect to the p score functions so that M_{ij} (without ambiguity) in $M_s(i,j)$ is the score value assigned to the object d_i by scoring system A_j. The algorithms and approaches described in Definition 6 can be applied to the rank matrix M_r. However, some of these algorithms have been also applied to the score matrix M_s. Up to this point in this chapter, we have emphasized rank combination algorithms and considered ranks of objects as the basic data of a given situation / experiment / work, regardless of the manner in which they were obtained. However, in many situations, the ranking takes place according to the score values of a variable or variate. It is, therefore, of considerable interest to study the relationship between $M_r(i,j)$ and $M_s(i,j)$. As we mentioned before, $M_r(i,j)$ can be derived from $M_s(i,j)$ by sorting each column, A_j, into descending order and assigning higher value with higher rank (i.e., smaller number). One of the interesting questions is that: Is there any difference between the information represented by $M_s(i,j)$ and that by $M_r(i,j)$? In 1954, A. Stuart showed the following (see Kendall & Gibbons, 1990, Chapters 9, 10):

Remark 3 (Correlation between Scores and Ranks)

With $n = 25$, the correlation between scores and ranks for a scoring system (ranker/scorer) A is as high as 0.94 under the assumption of normal distribution and 0.96 for the uniform distribution among the score values. These values increase when the sample size (i.e., the number n) increases and reach the limits of 0.98 and 1, for normal and uniform distribution, respectively.

In light of this close relationship between ranks $M_r(i,j)$ and scores $M_s(i,j)$, we might expect that operating on $M_r(i,j)$ and on $M_s(i,j)$ would draw the same conclusion. This appears to be so in a number of special cases. But in general, it has to be approached with certain amount of caution and care. It is clear, for example, that a few points with comparatively high residuals (i.e. poor correlations) would not have major effects on the overall correlation and correlation structure of the dataset, but these outliers may well be the key target of the investigation. We list the special features of transforming from $M_s(i,j)$ to $M_r(i,j)$.

Remark 4 (Rank/Score Transfer)

When transforming from a score function s_A to a rank function r_A on n objects, we have:

(a) the dimension of sample space is reduced from $\mathbf{R}^n$ for the score space to N^n (and then $N!$ because of permutation) for the rank space, and

(b) the score function values have been standardized to the scale while the mean is fixed for every rank function.

Remark 4(a) states that dimension reduction is obtained by a rank/score transform process that gives rise to the concept of a rank/score function, and the rank/score graph has at least three advantages: efficiency, neutrality, and visualization (see Remark 1). Remark 4(b) states that in $M_r(i, j)$, the mean of each column (a rank function on the n objects) is fixed to be $(n + 1)/2$. The same phenomenon is also true for some score data under certain cases of specific functions and study objectives. However, we note that when non-parametric and ordinal rank data is used, emphasis is not on the mean of the data. Rather it is on the discrete order and position each of the rank data is placed.

We recall that fully ranked data on n objects are considered as rank functions on the symmetric group of n elements S_n. Since S_n does not have a natural linear ordering, graphical methods such as histograms and bar graphs may not be appropriate for displaying ranked data in S_n. However, a natural partial ordering on S_n is induced in the Cayley network Cay(S_n, T_n). Moreover, since a polytope is the convex hull of a finite set of points in $\mathbf{R}^{n-1}$, the $n!$ nodes in Cay(S_n, T_n) constitute a permutation polytope when regarded as vectors in $\mathbf{R}^n$ (see Marden, 1995; McCullagh, 1992; Thompson, 1992). In fact,

Figure 6. Nodes in $\mathrm{Cay}(S_3, T_3)$ *and* $\mathrm{Cay}(S_4, T_4)$, *(a) six nodes in* $\mathrm{Cay}(S_3, T_3)$, *and (b) 24 nodes in* $\mathrm{Cay}(S_4, T_4)$

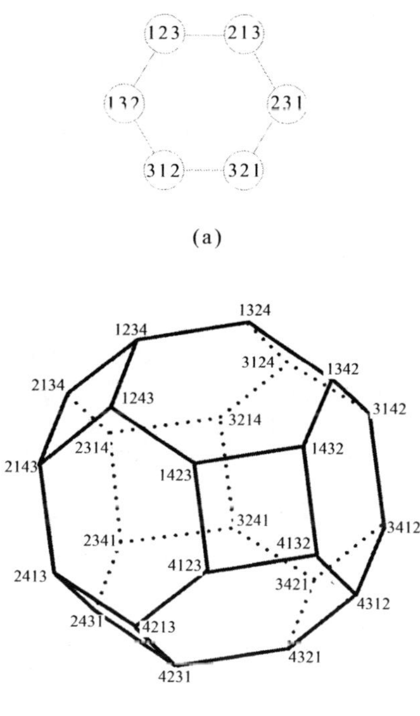

the n! nodes of $\mathrm{Cay}(S_n, T_n)$ lie on the surface of a sphere in $\mathbf{R}^{n-1}$. The six nodes and twenty four nodes of $\mathrm{Cay}(S_3, T_3)$ and $\mathrm{Cay}(S_4, T_4)$ are exhibited in Figure 6(a) and 6(b), respectively.

Information Retrieval

We now turn to the application of these data mining techniques to the information retrieval domain. We use as an example the study by Ng and Kantor (1998, 2000) (We call this the NK-study). Their exploratory analysis considered data from TREC competition with 26 systems and 50 queries for each system on a large but fixed database consisting of about 1000 documents. The results from these 26 systems are then fused in a paired manner. As such, there are $[(26 \times 25) / 2] \times 50 = 16,250$ cases of data fusion in the training data set. In 3,623 of these cases, the performance measures, as P_{100}, of the combined system is better than the best of the two original systems. We refer to these as **positive cases**. There are 9,171 **negative cases** where the performance of the combined system is worse than the best of the two original systems. In order to understand these two outcomes, two predictive variables are used. The first is the ratio of P_{100}, $r = P_l / P_h$ (see Definition 7(a)). The second variable is the normalized dissimilarity $z = d(r_A, r_B)$ (see

Definition 8(b)). We summarize the results of the NK-study as follows. See Figure 7 for an illustration.

Remark 5 (NK-study)

The results of the NK-study shows that (a) the positive cases tend to lie above the diagonal line $r + z = 1$, and (b) the negative cases are more likely to scatter around the line $r + z = 1$.

Remark 5 gives the general trend as to where the positive cases and negative cases should fall. There are very few negative cases with small r and small z and comparatively very few cases with high r and high z. Since the negative cases all spread around the line $r + z = 1$, z approaches 0 as r approaches 1 and vice versa. This means that for the negative cases, when the performances P_{100} of the two IR systems are about the same, their rank functions are similar to each other. For the positive cases, it was found that there are very few cases with small r and z and comparatively few cases with large r and z. But as Remark 5(a) indicated, the positive cases are more likely to lie above the diagonal $r + z = 1$. This indicates that systems with dissimilar (i.e., diverse) rank functions but comparable performance are more likely to lead to effective fusion.

Virtual Screening and Drug Discovery

We now turn to the application of the data mining technique to biomedical informatics. In particular, we discuss in more details the study by Yang et al. (2005) (we call this paper the YCSKH-study). The study explores criteria for a recently developed virtual screening technique called "Consensus Scoring" (CS). It also provides a CS procedure for improving the enrichment factor in CS using combinatorial fusion analysis (CFA) techniques and explores diversity measures on scoring characteristics between individual scoring functions.

In structure-based virtual screening, a docking algorithm and a scoring function are involved (see Shoichet, 2004, and its references). The primary purpose for a docking program is to find out the most favorable combination of orientation and conformation (**Pose**). It also requires a comparison of the best pose (or top few poses) of a given ligand with those of the other ligands in chemical data base such that a final ranking or ordering can be obtained. In essence, VS uses computer-based methods to discover new ligands on the basis of biological structure. Although it was once popular in the 1970s and 1980s, it has since struggled to meet its initial promise. Drug discovery remains dominated by empirical screening in the past three decades. Recent successes in predicting new ligands and their receptor-bound structure have re-invigorated interest in VS, which is now widely used in drug discovery.

Although VS of molecular compound libraries has emerged as a powerful and inexpensive method for the discovery of novel lead compounds, its major weakness — the inability to consistently identify true positive (leads) — is likely due to a lack of understanding of the chemistry involved in ligand binding and the subsequently

Figure 7. The two predictive variables proposed in NK-study — r and z, and the regions that positive and negative cases are most likely to scatter around

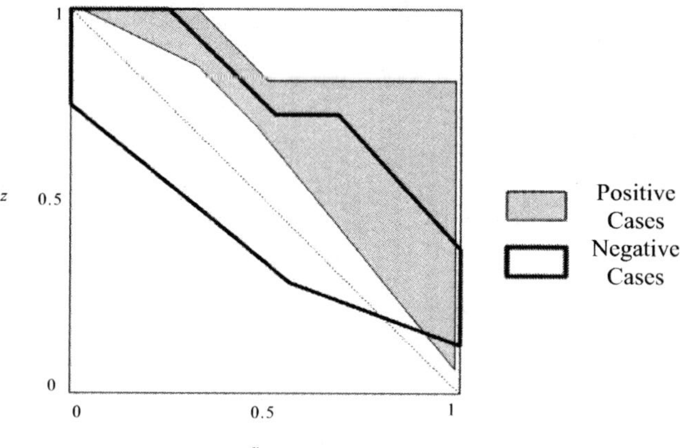

imprecise scoring algorithms. It has been demonstrated that consensus scoring (CS), which combines multiple scoring functions, improves enrichment of true positions. Results of VS using CS have largely focused on empirical study. The YCSKH-study is one attempt to provide theoretical analysis using combinatorial fusion (Yang et al., 2005).

The YCSKH-study developed a novel CS system that was tested for five scoring systems ($A, B, C, D,$ and E) with two evolutionary docking algorithms (GemDOCK and GOLD) on four targets: thymidine kinase (TK), human dihydrofolate reductase (DHFR), and estrogen receptors (ER) of antagonists and agonists (ERA). Their scoring systems consist of both rank-based and score-based CS systems (RCS and SCS). They used the GH (goodness-of-hit) score to evaluate the performance of each individual and combined systems. That is, $P(A)$ = GH score of the system A. Two predicative variables are used: (a) the performance ratio $PR(A, B) = P_l/P_h$ and (b) the diversity measure $d_2(f_A, f_B)$ as defined in Definitions 7 and 8 (see Yang et al., 2005).

$$PR(A,B) = P_l / P_h = \min\{P(A), P(B)\} / \max\{P(A), P(B)\},$$

and

$$d_2(f_A, f_B) = \left\{ \sum_{j=1}^{n} (f_A(j) - f_B(j))^2 / n \right\}^{1/2},$$

where $P(A)$ and $P(B)$ are the performances of the two scoring systems A and B to be combined.

Figure 8. Positive and negative cases w.r.t. P_l / P_h *and* $d_2(f_A, f_B)$

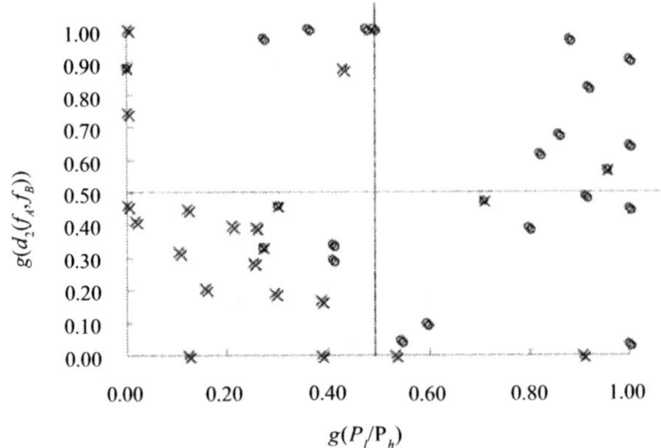

Let $g(x)$ denote the normalization function for the two predictive variables $PR(A, B)$ and $d_2(f_A, f_B)$ so that $g(x)$ is in [0, 1]. Their results regarding bi-diversity and combination of two scoring systems are summarized in the following remark where we use x and y as the coordinates for $g(P_l / P_h)$ and $g(d_2(f_A, f_B))$ respectively (see Figure 8).

Remark 6 (YCSKH-study)

The YCSKH-study shows that numbers of positive and negative cases split into roughly half and (a) most of the positive cases are located above the line $x + y = 1$ while none of the few cases below the line have both $x \leq 0.30$ and $y \leq 0.30$., and (b) most of the negative cases tend to be located below the line $x + y = 1$ while only one of the few cases above the line have both $x \geq 0.50$ and $y \geq 0.50$. The exceptional case has $g(d_2(f_A, f_B)) \approx 0.60$ and $g(PR(A, B)) \approx 0.95$ but both P_l and P_h are very small.

Remark 6 reconfirms that combining two scoring systems (rank function or score function) improves performance only if (a) each of the individual scoring systems has relatively high performance and (b) the scoring characteristics of each of the individual scoring systems are quite different. This suggests that the two systems to be combined have to be fairly diverse so that they can complement each other, and the performance of each system has to be good, although we cannot yet quantitate/constrain the quality of "good" outside of our specific study.

Protein Structure Prediction

Following their previous work establishing a hierarchical learning architecture (HLA), two indirect coding features, and a gate function to differentiate proteins according to their classes and folding patterns, C.-Y. Lin et al. (2005) and K.-L. Lin et al. (2005a, 2005b) have used combinatorial fusion to improve their prediction accuracy on the secondary class structure and/or 3-D folding patterns. Using 8 and 11 features respectively (i.e., scoring systems) and neural networks as a multi-class classifier to build HLA, they adopted the radical basis function network (RBFN) model to predict folding pattern for 384 proteins. In other words, in the system space $M(n,p,q)$ defined in Figure 2, the number of scoring systems $p = 8$ or 11 and the number of temporal steps $q = 384$.

The work by C. Y. Lin et al. (2005) using 8 features has an overall prediction rate of 69.6% for the 27 folding categories, which improves previous results by Ding and Dubchak (2001) of 56% and by Huang et al. (2003) of 65.5%. The work by K.-L. Lin et al. (2005a, 2005b), using CFA to facilitate feature selection using 11 features, achieves a prediction accuracy of 70.9% for 27 folding patterns.

Both works utilize the concept of diversity rank/score graph (Definition 9(e)) to select features to combine. In C.-Y. Lin et al. (2005), the scoring systems (features in this case) E, F, G, H are selected from the eight features {A, B, C, D, E, F, G, H}. Then the diversity rank/score functions for the six pairs of the four scoring systems E, F, G, H, are calculated (see Figure 9), where the diversity rank/score graph for the pair (E, H) is found to have the highest overall value across the $q = 384$ protein sequences tested. Hence E and H are considered to have the highest diversity among the six pairs of scoring systems. The best result was obtained by combining these two scoring systems E and H.

The work by K.-L. Lin et al. (2005a, 2005b) uses 11 features and has selected H, I, K out of 11 features {A, B, C, D, E, F, G, H, I, J, K} because of their higher individual performance. The diversity rank/score function is then calculated for any pair among H, I, K (see Figure 10). From the graphs of these three functions, they conclude that the pair H, I has the highest diversity across the $q = 384$ protein sequences tested. This pair of scoring systems H and I are then used to perform combination to achieve the desired result.

Conclusion and Discussion

In this chapter, we have described a method, called Combinatorial Fusion Analysis (CFA), for combining multiple scoring systems (MSS) each of which is obtained from a set of homogeneous or heterogeneous features/attributes/indicators/cues. The method (CFA) is based on information obtained from multi-variable (multi-criterion/multi-objective) classification, prediction, learning or optimization problems, or collected from multi-sensor / multi-source environments or experiments. We distinguished between the data space (see Figure 1) which consists of using features/attributes/indicators/cues to describe objects (e.g., pattern classes / documents / molecules / folding patterns), and the system space (see Figure 2), which consists of using different scoring systems (statistical methods, learning systems, combinatorial techniques, or computational algorithms) to assign a score and a rank to each of the objects under study. In the extreme case when each scoring system is a feature / attribute / indicator / cue, the system space coincides with the data space.

We use the concepts of score functions, rank functions and rank/score functions as defined by Hsu, Shapiro and Taksa (2002) and Hsu and Taksa (2005) to represent a scoring

Figure 9. Diversity graphs for the six pairs of scoring systems E, F, G and H

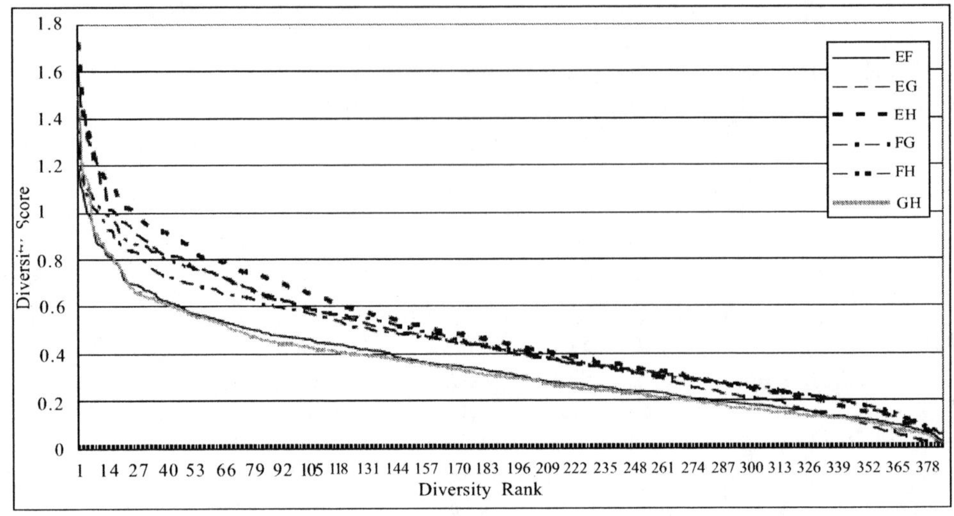

Figure 10. Diversity graphs for the three pairs of scoring systems H, I and J.

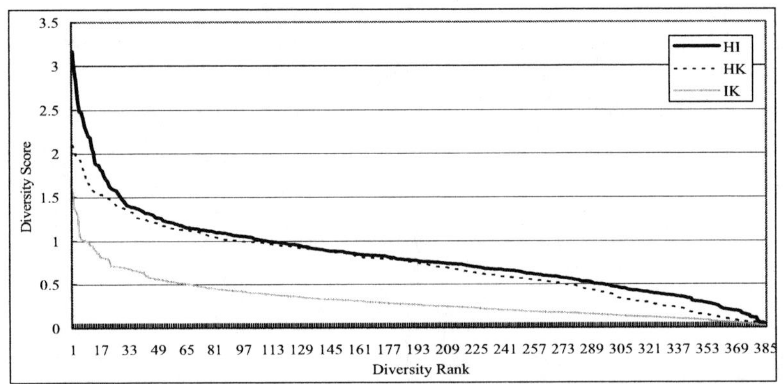

system. Rank / score transfer and correlation between scores and ranks are described and examined. We also described various performance measurements with respect to different application domains. Various combination/fusion methods/algorithms of combining multiple scoring systems have been explored. Theoretical analysis that gives insights into system selection and methods of combination/fusion has been provided. In particular, we observe that combining multiple scoring systems improves the performance only if (a) each of the individual scoring systems has relatively high performance, and (b) the individual scoring systems are distinctive (different, diverse or orthogonal). We have initiated study on these two issues (a) and (b) for the special case of two scoring systems *A* and *B*. Two predictive parameters are used. The first parameter is the performance ratio, P_l/P_h, which measures the relative performance of A and B where P_l and P_h are the lower performance and higher performance respectively. The second parameter deals with the bi-diversity between A and B, $d_2(A, B)$, which measures the degree of difference/dissimilarity/diversity between the two scoring systems A and B.

The bi-diversity (or 2-diversity) measure $d_2(A, B)$ between two scoring systems A and B is defined as one of the following three possibilities: (1) $d_2(A, B) = d(s_A, s_B)$, the distance between the score functions s_A and s_B of scoring system A and B; (2) $d_2(A, B) = d(r_A, r_B)$, the distance between the rank functions r_A and r_B; and (3) $d_2(A, B) = d(f_A, f_B)$, the distance between rank/score functions f_A and f_B. Diversity measures have been studied extensively in pattern recognition and classification (Kuncheva, 2005; Brown et al., 2005). Diversity measure defined in the form of rank function was used in information retrieval (Ng & Kantor, 1998, 2000). The work of Hsu and Taksa (2005) and Yang et al. (2005) used rank/score functions to measure diversity between two scoring systems in their study of comparing rank vs. score combination and consensus scoring criteria for improving enrichment in virtual screening and drug discovery. For the protein structure prediction problem, the rank/score functions of A and B, f_A and f_B, are used for each protein sequence p_i, where $d_{p_i}(A,B) = \sum_j |f_A(j) - f_B(j)|, j$ in $N = \{1, 2, \ldots, n\}$, is the diversity score function for p_i (C.-Y. Lin et al., 2005; K.-L. Lin et al., 2005a, 2005b).

Considering all protein sequences p_i in $P = \{p_1, p_2, \ldots, p_q\}$, the diversity score function, $s_{(A,B)}(x)$ written as $s_{(A,B)}(p_i) = d_{p_i}(A,B)$, is a function from P to $\mathbf{R}$. Consequently, sorting $s_{(A,B)}(x)$ into descending order leads to a diversity rank function $r_{(A,B)}(x)$ from P to $Q = \{1, 2, \ldots, q\}$. Hence the diversity rank/score function (or diversity function) $f_{(A,B)}(j)$ defined as $f_{(A,B)}(j) = (s_{(A,B)} \circ r^{-1}_{(A,B)})(j) = s_{(A,B)}(r^{-1}_{(A,B)}(j))$, where j is in Q, is a function from Q to $\mathbf{R}$. The diversity function and its graph play important roles in system selection and method combination in the PSP problem. We note that the diversity function so defined for PSP problem can be applied to other classification or prediction problems as well.

We illustrate the method of combinatorial fusion using examples from three different domains IR, VS, and PSP. In all three applications, multiple scoring systems are used. The issue of bi-diversity was discussed. The diversity rank/score function was calculated in the protein structure prediction problem.

In summary, we have discussed the method of combinatorial fusion analysis developed and used recently in pattern recognition (PR), information retrieval (IR), virtual screening (VS), and protein structure prediction (PSP). Our current work has generated several issues and topics worthy of further investigation. Among them, we list four:

1. We have so far emphasized more on bi-diversity (i.e., 2-diversity). How about tri-diversity (i.e., 3-diversity)? How about higher level diversity measurement? Can this be (or is this best) treated as a single optimization or a sequential series of bi-diversity problems?

2. Our diversity score function $d_{t_i}(A,B)$ for the feature pair or scoring systems A and B with respect to temporal step t_i is defined using the variation of the rank/score functions between A and B (i.e., $d(f_A, f_B)$). In general, the variation of the score functions $d(s_A, s_B)$ or the rank functions $d(r_A, r_B)$ could be used to define the diversity score function $d_{t_i}(A,B)$.

3. There have been several results concerning the combination of multiple scoring systems in the past few years. Three important issues we would like to study are: (a) selecting the most appropriate scoring systems to combine, (b) finding the best way to combine (Definition 6), and (c) establishing the predictive variables for effective combination. But due to page limitation and chapter constraint, we have included only three examples in IR, VS and PSP with respect to all three issues (a), (b,) and (c) described above.
4. We will explore more examples combining multiple heterogeneous scoring systems in the future. For example, we are working on multiple scoring systems which are hybrid of classifier systems, prediction systems, and learning systems (such as neural networks, support vector machines, or evolutionary algorithms).

Acknowledgments

D.F. Hsu thanks Fordham University, National Tsing Hua University, and Ministry of Education of Taiwan, for a Faculty Fellowship and for support and hospitality during his visit to NTHU in Spring 2004. B.S. Kristal acknowledges support from NIH and a NY State SCORE grant.

References

Arrow, K.J. (1963). *Social choices and individual values*. New York: John Wiley.

Aslam, J.A, Pavlu, V., & Savell, R. (2003). A unified model for metasearch, pooling, and system evaluation. In O. Frieder (Ed.), *Proceedings of the Twelfth International Conference on Information and Knowledge Management* (pp. 484-491). New York: ACM Press.

Belkin, N.J., Kantor, P.B., Fox, E.A., & Shaw, J.A. (1995). Combining evidence of multiple query representation for information retrieval. *Information Processing & Management, 31*(3), 431-448.

Biggs, N.L., & White, T. (1979). *Permutation groups and combinatorial structures* (LMS Lecture Note Series, Vol. 33). Cambridge: Cambridge University Press.

Brown, G., Wyatt, J., Harris, R., & Yao, X. (2005). Diversity creation methods: a survey and categorization. *Information Fusion, 6*, 5-20.

Chuang, H.Y., Liu, H.F., Brown, S., McMunn-Coffran, C., Kao, C.Y., & Hsu, D.F. (2004). Identifying significant genes from microarray data. In *Proceedings of IEEE BIBE'04* (pp. 358-365). IEEE Computer Society.

Chuang, H.Y., Liu, H.F., Chen, F.A., Kao, C.Y., & Hsu, D.F. (2004). Combination method in microarray analysis. In D.F. Hsu et al. (Ed.), *Proceedings of the 7th International*

Symposium on Parallel Architectures, Algorithms and Networks (I-SPAN'04) (pp. 625-630). IEEE Computer Society.

Collins, R.T., Lipton, A.J., Fujiyoshi, H., & Kanade, T. (2001). Algorithms for cooperative multisensor surveillance. *Proceedings of the IEEE, 89* (10), 1456-1477.

Corne, D.W., Deb, K., Fleming, P.J., & Knowles, J.D. (2003). The good of the many outweighs the good of the one: evolutional multi-objective optimization [Featured article]. *IEEE Neural Networks Society*, 9-13.

Dasarathy, B.V. (2000). Elucidative fusion systems—an exposition. *Information Fusion, 1*, 5-15.

Diligenti, M., Gori, M., & Maggini, M. (2004). A unified probabilistic framework for web page scoring systems, *IEEE Trans. on Knowledge and Data Engineering, 16*(1), 4-16.

Ding, C.H.Q., & Dubchak, I. (2001). Multi-class protein fold recognition using support vector machines and neural networks. *Bioinformatics, 17* (4), 349-358.

Duerr, B., Haettich, W., Tropf, H., & Winkler, G. (1980). A combination of statistical and syntactical pattern recognition applied to classification of unconstrained handwritten numerals. *Pattern Recognition, 12*(3), 189-199.

Dwork, C., Kumar, R., Naor, M., & Sivakumar, D. (2001). Rank aggregation methods for the web. In *Proceedings of the Tenth International World Wide Web Conference, WWW 10* (pp. 613-622). New York: ACM Press.

Fagin, R., Kumar, R., & Sivakumar, D. (2003). Comparing top *k*-lists. *SIAM Journal on Discrete Mathematics, 17*, 134-160.

Freund, Y., Iyer, R., Schapire, R.E., & Singer, Y. (2003). An efficient boosting algorithm for combining preferences. *Journal of Machine Learning Research, 4*, 933-969.

Garcia-Pedrajas, N., Hervas-Martinez, C., & Ortiz-Boyer, D. (2005). Cooperative coevolution of artificial neural network ensembles for pattern classification. *IEEE Trans. on Evolutional Computation, 9*(3), 271-302.

Ginn, C. M. R., Willett, P., & Bradshaw, J. (2000). Combination of molecular similarity measures using data fusion [Perspectives]. *Drug Discovery and Design, 20*, 1-15.

Grammatikakis, M.D., Hsu, D.F., & Kraetzl, M. (2001). *Parallel system interconnections and communications*. Boca Raton, FL: CRC Press.

Heydemann, M.C. (1997). Cayley graphs and interconnection networks. In G. Hahn & G. Sabidussi (Eds.), *Graph symmetry* (pp. 161-224). Norwell, MA: Kluwer Academic Publishers.

Ho, T.K. (2002). Multiple classifier combination: Lessons and next steps. In H. Bunke & A. Kandel (Ed.), *Hybrid methods in pattern recognition* (pp. 171-198). Singapore: World Scientific.

Ho, T.K., Hull, J.J., & Srihari, S.N. (1992). Combination of decisions by multiple classifiers. In H.S. Baird, H. Burke, & K. Yamulmoto (Eds.), *Structured document image analysis* (pp. 188-202). Berlin: Springer-Verlag.

Ho, T.K., Hull, J.J., & Srihari, S.N. (1994). Decision combination in multiple classifier system. *IEEE Trans. on Pattern Analysis and Machine Intelligence, 16*(1), 66-75.

Hsu, D.F., & Palumbo, A. (2004). A study of data fusion in Cayley graphs $G(S_n, P_n)$. In: D.F. Hsu et al. (Ed.), *Proceedings of the 7th International Symposium on Parallel Architectures, Algorithms and Networks (I-SPAN'04)* (pp. 557-562). IEEE Computer Society.

Hsu, D.F., Shapiro, J., & Taksa, I. (2002). *Methods of data fusion in information retrieval: Rank vs. score combination* (Tech. Rep. 2002-58). Piscataway, NJ: DIMACS Center.

Hsu, D.F., & Taksa, I. (2005). Comparing rank and score combination methods for data fusion in information retrieval. *Information Retrieval, 8*(3), 449-480.

Hu, W., Tan, T., Wang, L., & Maybank, S. (2004). A survey on visual surveillance of object motion and behaviors. *IEEE Trans. on Systems, Man, and Cybernetics—Part C: Applications and Review, 34*(3), 334-352.

Huang, C.D., Lin, C.T., & Pal, N.R. (2003). Hierarchical learning architecture with automatic feature selection for multi-class protein fold classification. *IEEE Trans. on NanoBioscience, 2*(4), 503-517.

Ibraev, U., Ng, K.B., & Kantor, P.B. (2001). *Counter intuitive cases of data fusion in information retrieval* (Tech. Rep.). Rutgers University.

Jin, X., & Branke, J. (2005). Evolutional optimization in uncertain environments—a survey. *IEEE Trans. on Evolutional Computation, 9*(3), 303-317.

Kantor, P.B. (1998, Jan). *Semantic dimension: On the effectiveness of naïve data fusion methods in certain learning and detection problems.* Paper presented at the meeting of the Fifth International Symposium on Artificial Intelligence and Mathematics, Ft. Lauderdalee, FL.

Kendall, M., & Gibbons, J.D. (1990). *Rank correlation methods.* London: Edward Arnold.

Kittler, J., & Alkoot, F.M. (2003). Sum versus vote fusion in multiple classifier systems. *IEEE Transactions on Pattern Analysis and Machine Intelligence, 25*(1), 110-115.

Kuncheva, L.I. (2005). Diversity in multiple classifier systems [Guest editorial]. *Information Fusion, 6*, 3-4.

Kuriakose, M.A., Chen, W.T., He, Z.M., Sikora, A.G., Zhang, P., Zhang, Z.Y., Qiu, W.L., Hsu, D.F., McMunn-Coffran, C., Brown, S.M., Elango, E.M., Delacure, M.D., & Chen, F.A.. (2004). Selection and Validation of differentially expressed genes in head and neck cancer. *Cellular and Molecular Life Sciences, 61*, 1372-1383.

Lee, J.H. (1997). Analyses of multiple evidence combination. In N.J. Belkin, A.D. Narasimhalu, P. Willett, W. Hersh (Ed.), *Proceedings of the 20th Annual International ACM SIGIR Conference on Research and Development in Information Retrieval* (pp. 267-276). New York: ACM Press.

Lin, C.-Y., Lin, K.-L., Huang, C.-D., Chang, H.-M., Yang, C.-Y., Lin, C.-T., & Hsu, D.F. (2005). Feature selection and combination criteria for improving predictive accuracy in protein structure classification. In *Proceedings of IEEE BIBE'05* (pp. 311-315). IEEE Computer Society.

Lin, K.-L., Lin, C.-Y., Huang, C.-D., Chang, H.-M., Lin, C.-T., Tang, C.Y., & Hsu, D.F. (2005a). Improving prediction accuracy for protein structure classification by neural networks using feature combination. *Proceedings of the 5th WSEAS International Conference on Applied Informatics and Communications (AIC'05)*(pp. 313-318).

Lin, K.-L., Lin, C.Y., Huang, C.-D., Chang, H.-M., Lin, C.-T., Tang, C.Y., & Hsu, D.F. (2005b). Methods of improving protein structure prediction based on HLA neural networks and combinatorial fusion analysis. *WSEAS Trans. on Information Science and Application, 2*, 2146-2153.

Marden, J.I. (1995). *Analyzing and modeling rank data*. (Monographs on Statistics and Applied Probability, No. 64). London: Chapman & Hall.

McCullagh, P. (1992). Models on spheres and models for permutations. In *Probabality Models and Statistical Analysis for Ranking Data* (M.A. Fligner, & J.S. Verducci, Ed., pp. 278-283). (Lecture Notes in Statistics, No. 80). Berlin: Springer-Verlag.

Melnik, D., Vardi, Y., & Zhang, C.U. (2004). Mixed group ranks: preference and confidence in classifier combination. *IEEE Trans. on Pattern Analysis and Machine Intelligence, 26*(8), 973-981.

Ng, K.B., & Kantor, P.B. (1998). An investigation of the preconditions for effective data fusion in information retrieval: A pilot study. In C.M. Preston (Ed.), *Proceedings of the 61st Annual Meeting of the American Society for Information Science* (pp. 166-178). Medford, NJ: Information Today.

Ng, K.B., & Kantor, P.B. (2000). Predicting the effectiveness of naïve data fusion on the basis of system characteristics. *Journal of the American Society for Information Science. 51*(13), 1177-1189.

Paolucci, U., Vigneau-Callahan, K. E., Shi, H., Matson, W. R., & Kristal, B. S. (2004). Development of biomarkers based on diet-dependent metabolic serotypes: Characteristics of component-based models of metabolic serotype. *OMICS, 8*, 221-238.

Patil, G.P., & Taillie, C. (2004). Multiple indicators, partially ordered sets, and linear extensions: multi-criterion ranking and prioritization. *Environmental and Ecological Statistics, 11*, 199-288.

Perrone, M.P., & Cooper, L.N. (1992). *When networks disagree: Ensemble methods for hybrid neural networks* (Report AF-S260 045). U.S. Dept. of Commerce.

Shi, H., Paolucci, U., Vigneau-Callahan, K. E., Milbury, P. E., Matson, W. R., & Kristal, B. S. (2004). Development of biomarkers based on diet-dependent metabolic serotypes: Practical issues in development of expert system-based classification models in metabolomic studies. *OMICS, 8*, 197-208.

Shi, H., Vigneau-Callahan, K., Shestopalov, I., Milbury, P.E., Matson, W.R., & Kristal B.S. (2002a). Characterization of diet-dependent metabolic serotypes: Proof of principle in female and male rats. *The Journal of Nutrition, 132*, 1031-1038.

Shi, H., Vigneau-Callahan, K., Shestopalov, I., Milbury, P.E., Matson, W.R., & Kristal, B.S. (2002b). Characterization of diet-dependent metabolic serotypes: Primary validation of male and female serotypes in independent cohorts of rats. *The Journal of Nutrition, 132*, 1039-1046.

Shoichet, B.K (2004). Virtual screening of chemical libraries. *Nature, 432*, 862-865.

Thompson, G. L. (1992). Graphical techniques for ranked data. In *Probabality Models and Statistical Analysis for Ranking Data* (M.A. Fligner, & J.S. Verducci, Ed., pp. 294-298). (Lecture Notes in Statistics, No. 80). Berlin: Springer-Verlag.

Triesch, J., & von der Malsburg, C. (2001). Democratic integration: self-organized integration of adaptive cues. *Neural Computation, 13*, 2049-2074.

Tumer, K., & Ghosh, J. (1999). Linear and order statistics combinations for pattern classification. In: Amanda Sharkey (Ed.), *Combining artificial neural networks* (pp. 127-162). Berlin: Springer-Verlag.

Vogt, C. C., & Cottrell, G. W. (1999). Fusion via a linear combination of scores. *Information Retrieval, 1*(3), 151-173.

Xu, L., Krzyzak, A., & Suen, C.Y. (1992). Methods of combining multiple classifiers and their applications to handwriting recognition. *IEEE Transactions on Systems, Man, and Cybernetics, 22*(3), 418-435.

Yang, J.M., Chen, Y.F., Shen, T.W., Kristal, B.S., & Hsu, D.F. (2005). Consensus scoring for improving enrichment in virtual screening. *Journal of Chemical Information and Modeling, 45*, 1134-1146.

Young, H.P. (1975). Social choice scoring functions. *SIAM Journal on Applied Mathematics, 28*(4), 824-838.

Young, H. P., & Levenglick, A. (1978). A consistent extension of Condorcet's election principle. *SIAM Journal on Applied Mathematics, 35*(2), 285-300.

Chapter IV

DNA Sequence Visualization

Hsuan T. Chang, National Yunlin University of Science and Technology, Taiwan

Abstract

This chapter introduces various visualization (i.e., graphical representation) schemes of symbolic DNA sequences, which are basically represented by character strings in conventional sequence databases. Several visualization schemes are reviewed and their characterizations are summarized for comparison. Moreover, further potential applications based on the visualized sequences are discussed. By understanding the visualization process, the researchers will be able to analyze DNA sequences by designing signal processing algorithms for specific purposes such as sequence alignment, feature extraction, and sequence clustering, etc.

Introduction

Recently, the great progress of biotechnology makes the deoxyribonucleic acid (DNA) sequencing more efficient. Huge amounts of DNA sequences of various organisms have been successfully sequenced with higher accuracies. By analyzing DNA sequences, the biological relationships such as homologous and phylogeny of different species can be investigated. However, the analysis of DNA sequences by the use of biological methods is too slow for processing huge amount of DNA sequences. Therefore, the assistance of computers is necessary and thus bioinformatics is extensively developed. Efficient algorithms and implemented computer-based tools are desired to deal with the considerable and tedious biomolecular data.

In general, DNA sequences are stored in the computer database system in the form of character strings. In a human somatic cell, its haploid nuclear DNA contains 2.85 billion base pairs (bps), in which a tremendous wealth of genetic information resides (Collins et al., 2004). Distinguishing the differences and similarities among DNA sequences has been a major task for biologists. Most of the sequences in their character strings are too long to be displayed on the computer screen and, therefore, are very hard to be extracted for any feature or characteristic.

Development of visualization techniques for presenting biological sequences has been widely attempted (Roy, Raychaudhury, & Nandy, 1998; Loraine & Helt, 2002). Mehta & Sahni (1994) proposed some efficient algorithms that make use of the compact symmetric directed acyclic word graph (csdawg) data structure. Blumer, Blumer, Haussler, McConnell, and Ehrenfeucht (1987) proposed the analysis and visualization of patterns in long string. Some previous studies (Anastassiou, 2001; Berger, Mitra, Carli, & Neri, 2002; Wang & Johnson, 2002) have shown various methods (such as discrete Fourier transform or wavelet transform) of transforming the symbolic DNA sequences to numeric sequences for further processing. With the methods described above, the periodic patterns existed in DNA sequences can be observed from the determined scalograms or spectrograms. On the other hand, some methodologies (Cheever & Searls, 1989; Cork & Wu, 1993; Wu, Roberge, Cork, Nguyen, & Grace, 1993) were proposed to depict symbolic sequences by two-dimensional (2-D) images, three-dimensional (3-D) curves, or graphs. The calculation in some methods was troublesome and required intensive computation. Efficient and direct mapping methods are desired to convert the symbolic sequences into the numeric sequences, and have them displayed in graphs.

Visualization (i.e., graphical representation) of DNA sequences provides corresponding pseudo shapes in a global view, which makes the sorting, comparison, and feature extraction based on the pseudo shape available. Visual recognition of differences among related DNA sequences by inspection can be made through sequence visualization. The graphical form can be viewed on a computer display or be printed on a piece of paper. Therefore, global and local characterizations/features of sequences can be quickly grasped in a perceivable form. Moreover, numerical characterizations (or features) of sequences can be determined from the visualized data (Chang, Xiao, & Lo, 2005). The extracted characterizations or features make the sequence data in a much more manageable fashion. Visualization is an alternative for DNA sequence representations. In addition to complementing the limitations of conventional symbolic-based methods, more powerful application tools would be expected in the near future.

There have been many researchers around the world working on this topic. The related bioinformatics tools and databases of visualized DNA sequence also have been accessible over the Internet. Therefore, the objective of this chapter is a literature review and summarization of the visualization methods for DNA sequences. Further applications based on the visualization methods will also be mentioned.

Background

A DNA strand is a biomolecular polymer characteristic of four different bases, i.e., adenine (A), guanine (G), cytosine (C), and thymine (T). The length unit of a DNA sequence is in base pair (bp) for a double-stranded DNA, or in nucleotide (nt) for a single strand. With the rapid development and progress in bioinformatics and the completion of more genome sequencing projects, many sequence databases have been constructed and will be continuously constructed for sequence query.

It is not easy to directly access large amount of genomic DNA sequence data to perform mathematical analysis. That is, character-based representation of DNA sequences cannot provide immediately useful nor informative characterizations. To perform visualization, symbolic-based sequences should be transformed into numeric-based versions. Different transformation will provide various characterizations of the visualized sequences. Therefore, the transformation (or mapping, simply saying) is a critical issue in sequence visualization.

There are many transformation methods for mapping the symbolic characters A, T, C, G, to numerical values (Almeida & Vinga, 2002; Chen, Li, Sun, & Kim, 2003; Wang & Johnson, 2002; Zhang et al., 2002). Those methods transformed DNA sequences into one-dimensional sequence, 2-D, 3-D, higher dimensional, or some graphical/image-like data. The data values can be real, imaginary, complex, or quaternion numbers. Therefore, the resulting numerical characterizations of transformed sequences are model dependent. The simplest way to transform the symbolic sequence to a numerical one is to directly assign an integer (or real value) for each nucleotide. For example, one can select the integers 1, 2, 3, and 4 for representing A, T, C, and G, respectively. On the other hand, more complicated transformations such as a complex assignment or a statistical calculation were also proposed to derive different graphical representation of DNA sequences. In addition to providing a global view and to directly perform sequence comparison, further analyses based on the visualized data, such as feature extraction for concise data representation and data retrieval, and phylogenetic clustering, are all the possible objectives. Therefore, the significance of the studies on DNA sequence visualization is quite obvious.

Sequence Visualization Methods

The studies of DNA sequence visualization were initiated more than 20 years ago. From the literature survey, Hamori & Ruskin (1983), Gates (1986), Jeffery (1990), Peng et al. (1992), Zhang & Zhang (1994), and Roy & Nandy (1994) could be the pioneers in this research area during the 1980s and 1990s. Roy, Raychaudhury, and Nandy (1998) gave a comprehensive survey on the graphical representation methods and their applications in viewing and analyzing DNA sequences. Randic's group (Randic, 2000a; Randic & Vracko, 2000; Randic et al., 2000; Guo, Randic, & Basak, 2001; Randic & Basak, 2001; Randic & Balaban, 2003; Randic et al., 2003a; 2003b; 2003c; Randic, 2004; Randic, Zupan,

Figure 1. 3-D H-curves: Homo sapiens B-cell CLL/lymphoma 2 (BCL2) mRNA sequence (AC: NM_000657, 911 bp) and Human dihydrofolate reductase gene (AC: J00140, 979 bp) shown in blue/dark and green/light lines, respectively

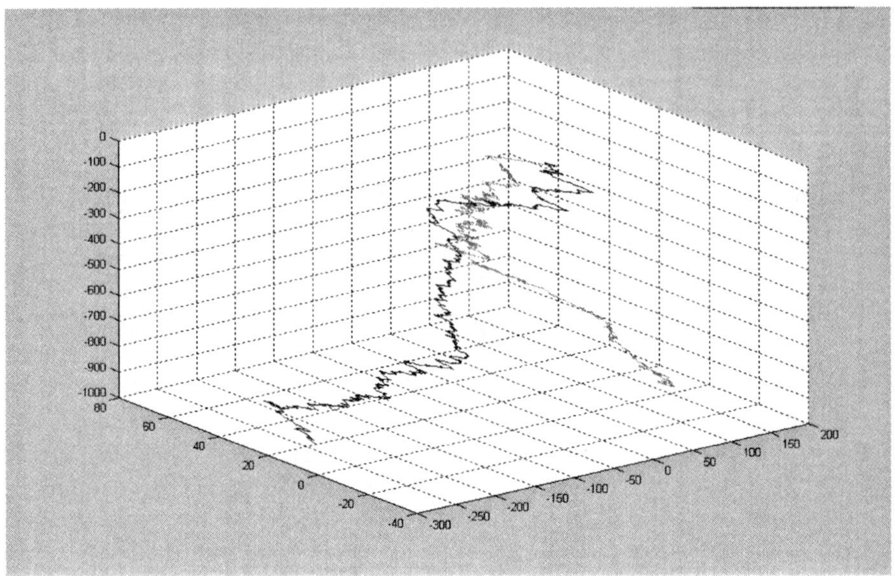

& Balaban, 2004; Zupan & Randic, 2005) proposed various visualization methods after 2000. Many new perspectives and practical applications based on the visualized DNA sequences are blossoming. In this section, we will briefly and historically review most of the published visualization methods. Their methodologies, characterizations, and their applications on bioinformatics will be discussed as well.

H-Curves

Most recent studies have referred to the first visualization method for DNA sequences as the *H*-curves proposed by Hamori and Ruskin (1983). The authors claimed that the important global features of long symbolic sequences, such as characteristic nucleotide ratios of certain sections, repetition of longer fragments, sudden changes in prevailing nucleotide compositions at certain locations, are not readily conveyed. The analysis, recognition, comparison, mental recollection, and so forth, of long sequences without long range features serving as visual clues become nearly impossible tasks for most people.

In this method, a vector function $g(z)$ is defined as:

$$g(z) = \begin{cases} i + j - k, & \text{if } z = \text{A} \\ i - j - k, & \text{if } z = \text{T} \\ -i - j - k, & \text{if } z = \text{C} \\ -i + j - k, & \text{if } z = \text{G} \end{cases} \tag{1}$$

where $i, j,$ and k are unit vectors pointing in the direction of the Cartesian $x, y,$ and z axes, respectively. The 3-D curve (H-curve) consisting of a series of n joined base vectors is defined as

$$H_{1,n} = h(z) = \sum_{1}^{n} g(z). \tag{2}$$

As shown in Eq. (1), four bases are represented by four directions in 2-D x-y plane. The H-curve is constructed by moving one unit in the corresponding direction and another for each unit in z direction. H-curve can uniquely represent a DNA sequence. In addition, more advantages H-curves offer include: (1) H-curve preserves all the important characteristics of the long DNA sequence; (2) The average nucleotide composition of a certain fragment is an important sequence property, which is not directly evident from the

Figure 2(a). 2-D DNA walks: Homo sapiens B-cell CLL/lymphoma 2 (BCL2) mRNA sequence (AC: NM_000657, 911 bp)

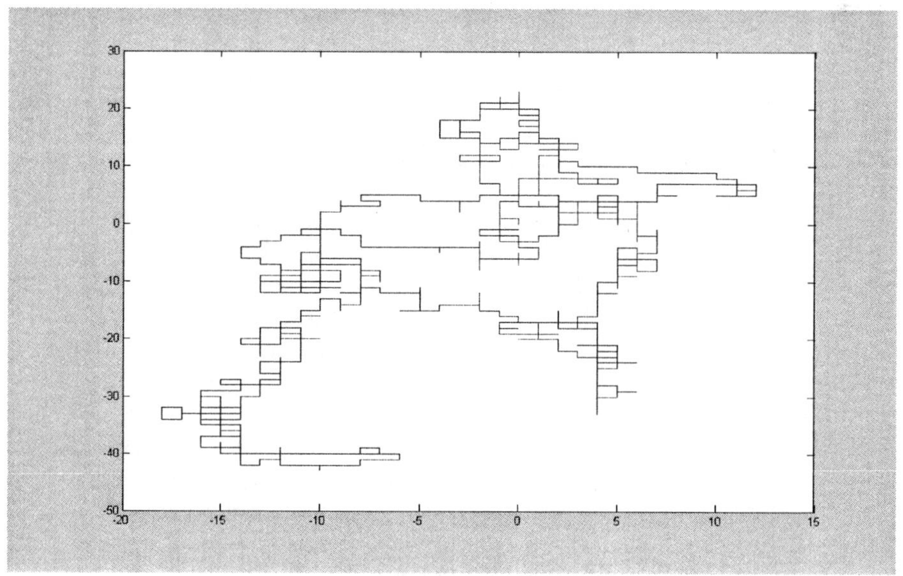

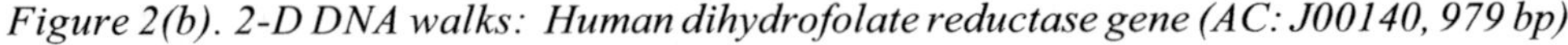

Figure 2(b). 2-D DNA walks: Human dihydrofolate reductase gene (AC: J00140, 979 bp)

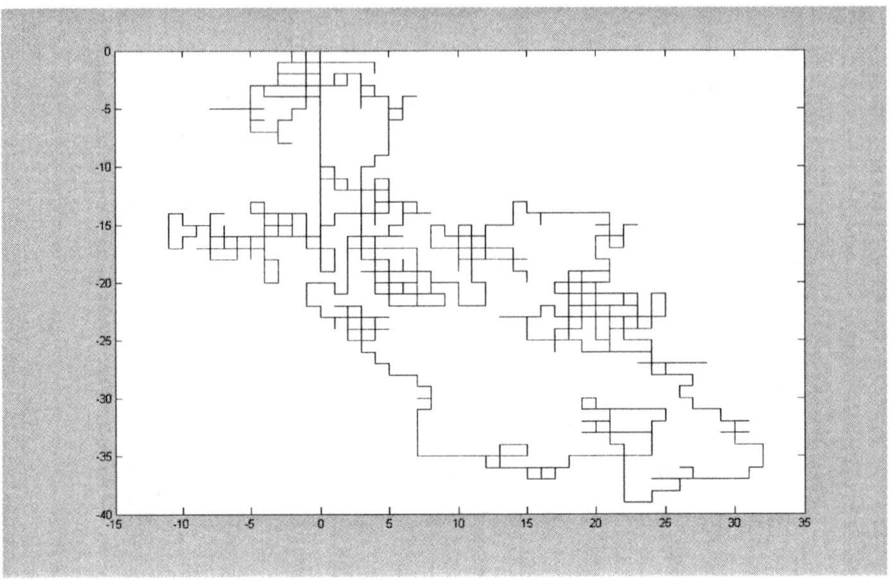

symbolic sequence; (3) Visual comparisons and matching between different DNA sequences are facilitated; (4) The information associated with the location of the end point of an *H*-curve is provided. Figure 1 shows the *H*-curves of Homo sapiens B-cell CLL/lymphoma 2 (BCL2) mRNA sequence (AC: NM_000657, 911 bp) and Human dihydrofolate reductase gene (AC: J00140, 979 bp) shown in blue/dark and green/light lines, respectively. The characterizations of two curves corresponding to the two sequences are obviously quite different.

DNA Walks

DNA walks for representing the DNA sequence have been widely utilized since 1992 (Peng et al., 1992). The first work is the 1-D DNA walk, which uses binary alphabets. Let a DNA sequence of length N_s be denoted as $\{x[i], i=1,2,3...,N_s.\}$. The presences of purines (A, G) and pyrimidines (C, T) in DNA sequences correspond to the $x[k]$ values +1 and -1, respectively, where k denotes the position in the sequence. Then a DNA walk sequence $s[k]$ is defined as the cumulative sum of $x[i]$, i.e.,

$$s[k] = \sum_{i=1}^{k} x[i]. \tag{3}$$

Similarly, other two letter alphabets such as *S-W* and *K-M* can be employed to obtain the 1-D DNA walks. Note that the symbols *S* (strong), *W* (weak), *K* (keto), and *M* (amino) are defined as follow: $S \in \{C,G\}, W \in \{A,T\}, R \in \{A,G\}, Y \in \{C,T\}$.

The dimensionality of the numerical DNA sequence can be more than one for a better representation. For example, the 3-D vector notation for mapping A, T, C, and G, is similar to Eq. (1). Furthermore, the complex representation for DNA walk can be defined as follows:

$$x(k) = \begin{cases} 1, & \text{for A} \\ -1, & \text{for T} \\ j, & \text{for C} \\ -j, & \text{for G} \end{cases} . \quad (4)$$

This complex DNA walk can be plotted in a 3-D Cartesian coordinate system by treating the accumulated values of the real part and imaginary part, and the variable *k* as the values for *x, y,* and *z* axes, respectively. Some repeat sequence can be found by projecting the DNA walk down to the *x-y* plane (Berger et al., 2004). DNA walk representations can be used to extract useful information, such as the long-range correlation information and sequence periodicities, from DNA sequences. To elucidate further information, the

Figure 3. 3-D DNA walks: Homo sapiens B-cell CLL/lymphoma 2 (BCL2) mRNA sequence (AC: NM_000657, 911 bp) and Human dihydrofolate reductase gene (AC: J00140, 979 bp) shown in blue/dark and green/light lines, respectively

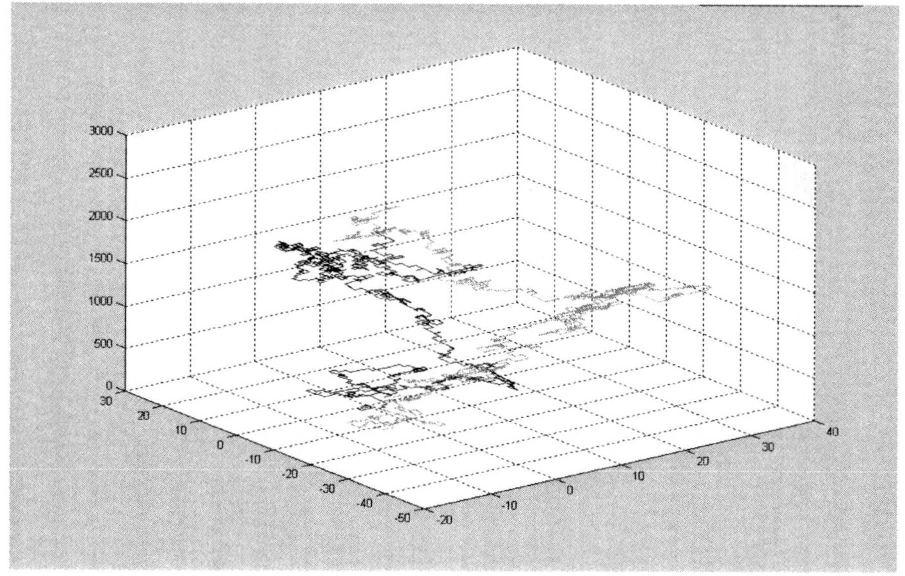

wavelet transform analysis can be applied to 1-D or 2-D walks for observing dependencies of nucleotide arrangement in long sequences (Arneodo, Bacry, Graves, & Muzy, 1995; Berger et al., 2004). Figures 2(a) and 2(b) show the 2-D DNA walks of Homo sapiens B-cell CLL/lymphoma 2 (BCL2) mRNA sequence (AC: NM_000657, 911 bp) and Human dihydrofolate reductase gene (AC: J00140, 979 bp), respectively. On the other hand, Figure 3 shows the corresponding 3-D DNA walks of two sequences.

Z-Curves

Zhang and Zhang (1994) proposed the *Z*-curve representation to visualize DNA sequences in 1994. Consider a DNA sequence of N bases and let the accumulated numbers of the bases A, T, C, and G be four positive integers A_n, T_n, C_n, and G_n, respectively. The *Z*-curve consists of a series of nodes P_n ($n = 0, 1, 2, \ldots, N$), whose coordinates are denoted by x_n, y_n, and z_n. Then these three coordinates are determined by the use of the four integers A_n, T_n, C_n, and G_n. Their relationships are called the *Z* transform and are expressed as below:

$$\begin{cases} x_n = (A_n + G_n) - (C_n + T_n) \\ y_n = (A_n + C_n) - (C_n + T_n), \quad x_n, y_n, z_n \in [-N, N], \quad n = 0, 1, \ldots, N, \\ z_n = (A_n + T_n) - (C_n + G_n) \end{cases} \tag{5}$$

Figure 4. 3-D Z-curves: Homo sapiens B-cell CLL/lymphoma 2 (BCL2) mRNA sequence (AC: NM_000657, 911 bp) and Human dihydrofolate reductase gene (AC: J00140, 979 bp) shown in blue/dark and green/light lines, respectively

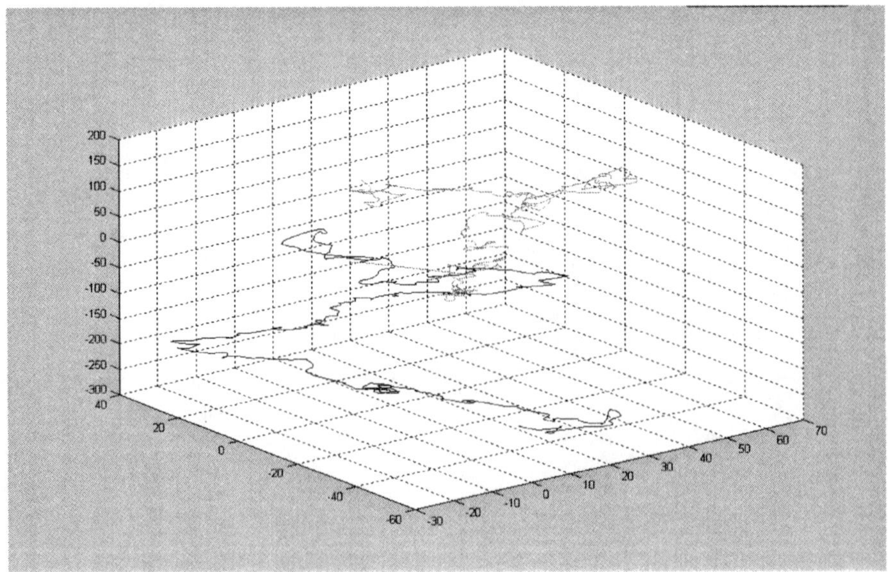

where $A_0 = T_0 = C_0 = G_0 = 0$ and thus $x_n = y_n = z_n = 0$. Given the coordinates of a *Z*-curve, the corresponding DNA sequence can be reconstructed by the use of inverse *Z*-transform, which is expressed as

$$\begin{pmatrix} A_n \\ C_n \\ G_n \\ T_n \end{pmatrix} = \frac{n}{4}\begin{pmatrix} 1 \\ 1 \\ 1 \\ 1 \end{pmatrix} + \frac{1}{4}\begin{pmatrix} 1 & 1 & 1 \\ -1 & 1 & -1 \\ 1 & -1 & -1 \\ -1 & -1 & 1 \end{pmatrix}\begin{pmatrix} x_n \\ y_n \\ z_n \end{pmatrix}, \quad n = 0,1,\ldots,N, \tag{6}$$

where the relation $A_n + T_n + C_n + G_n = n$ is used. Figure 4 shows the smoothed *Z*-curves of Homo sapiens B-cell CLL/lymphoma 2 (BCL2) mRNA sequence (AC: NM_000657, 911 bp) and Human dihydrofolate reductase gene (AC: J00140, 979 bp).

As shown in Eq. (5), some statistical characterizations of the DNA sequences can be obtained directly from the *Z*-curves. For example, the distributions of purine/pyrimidine, amino/keto, and strong-H bond/weak-H bond bases along the sequences correspond to the three components of the *Z*-curve: x_n, y_n, and z_n, respectively. Moreover, algorithms for exons identification, gene finding, prediction, or recognition using the Fourier transform performed on x_n, y_n, and z_n components of the *Z*-curve are also possible (Guo,

Figure 5. Randic's 3-D curves: Homo sapiens B-cell CLL/lymphoma 2 (BCL2) mRNA sequence (AC: NM_000657, 911 bp) and Human dihydrofolate reductase gene (AC: J00140, 979 bp) shown in blue/dark and green/light lines, respectively

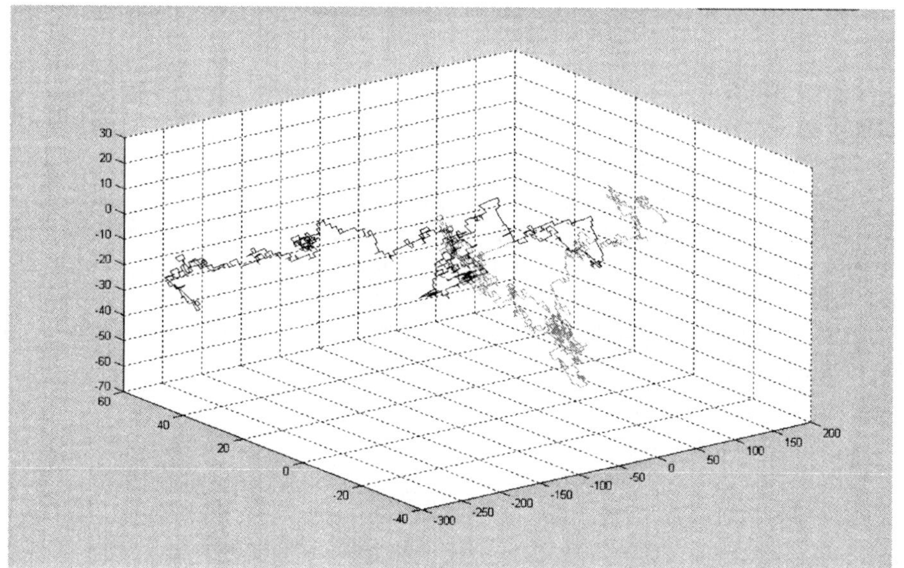

Ou, & Zhang, 2003; Zhang & Zhang, 2004). Identification of replication origins in genomes based on the *Z*-curve method is also reported by Zhang and Zhang (2005).

The *Z*-curve database now is also available on the Internet (Zhang, Zhang, & Ou, 2003). In additional to draw and manipulate the *Z*-curve of a user's input sequence online, several software services and archives are accessible in the Web site: http://tubic.tju.edu.cn/zcurve/.

Randic's Visualization Methods

Since 2000, Randic's research group had proposed several visualization schemes for DNA sequences (Randic, 2000a; Randic & Vracko, 2000; Randic et al., 2000; Guo, Randic, & Basak, 2001; Randic & Basak, 2001; Randic & Balaban, 2003; Randic et al., 2003a; 2003b; 2003c; Randic 2004; Randic, Zupan, & Balaban, 2004; Zupan & Randic, 2005). A 3-D graphical representation method was first proposed (Randic et al., 2000). Then the sequences can be analyzed based on the matrix characteristics (Randic, Kleiner, & DeAlba, 1994; Randic 2000; Randic & Basak, 2001) extracted from the original DNA sequences. In this method, the four vertices associated with a regular tetrahedron are assigned to four nucleotides. The mapping between four nucleotides and corresponding 3-D coordinates is shown below:

$$\begin{aligned} (+1,\ -1,\ -1) &\rightarrow \mathrm{A} \\ (-1,\ +1,\ -1) &\rightarrow \mathrm{G} \\ (-1,\ -1,\ +1) &\rightarrow \mathrm{C} \\ (+1,\ +1,\ +1) &\rightarrow \mathrm{T} \end{aligned} \tag{7}$$

For a sequence, a 3-D curve is plotted by connecting all the points determined by sequentially adding the (x, y, z) coordinates assigned for the nucleotides. Figure 5 shows the 3-D curves of Homo sapiens B-cell CLL/lymphoma 2 (BCL2) mRNA sequence (AC: NM_000657, 911 bp) and Human dihydrofolate reductase gene (AC: J00140, 979 bp). Note that the 2-D representations in Nandy's (1994) and Leong's (1995) methods can be considered as the projections on (x, y) and (x, z) plane, respectively.

The problems of degeneracy and overlap in graphical representation of DNA sequence may result in identical graphs for different sequences. Therefore, a 2-D graphical representation of low degeneracy was proposed to solve the mentioned problems (Guo & Basak, 2001). The basic idea is to introduce a positive integer d while assigning the four special vectors in 2-D Cartesian (x, y) coordinates for nucleotides. One of the possible axes systems for this 2-D graphical representation can be

$$\begin{aligned} &(-1, +\frac{1}{d}) \rightarrow \text{A} \\ &(+\frac{1}{d}, -1) \rightarrow \text{T} \\ &(+1, +\frac{1}{d}) \rightarrow \text{G} \\ &(+\frac{1}{d}, +1) \rightarrow \text{C} \end{aligned} \qquad (8)$$

The mathematical derivation for calculating the minimum length of the DNA sequence that can form a loop in the graphical representation was performed. It shows that if d is a larger even number, the corresponding graphical representation has a lower degeneracy. Figure 6 shows the 2-D curves of Homo sapiens B-cell CLL/lymphoma 2 (BCL2) mRNA sequence (AC: NM_000657, 911 bp) and Human dihydrofolate reductase gene (AC: J00140, 979 bp). The degeneracy problem is obviously reduced and much less than that shown in Figure 2. Liu, Guo, Xu, Pan, and Wang (2002) gave some notes on 2-D graphical representation with lower or no degeneracy.

A 2-D graphical representation (Randic et al., 2003a), which does not consider four directions along the Cartesian coordinate axes, was then proposed in 2003. Four

Figure 6. 2-D curves with low degeneracy: Homo sapiens B-cell CLL/lymphoma 2 (BCL2) mRNA sequence (AC: NM_000657, 911 bp) and Human dihydrofolate reductase gene (AC: J00140, 979 bp) shown in blue/dark and green/light lines, respectively

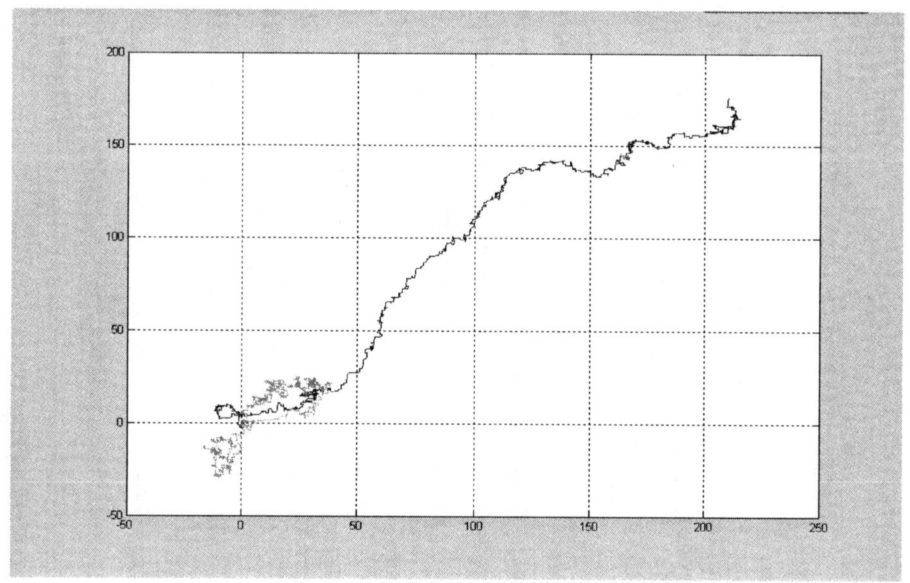

horizontal lines separated by unit distance are used to represent four nucleotides and the nucleotides constituting the considered sequence are denoted as dots and placed on the corresponding lines. By connecting the adjacent dots, we can obtain a zigzag like curve for sequence visualization. Figure 7 shows the zigzag-like curve of a given sequence:

"aggctggagagcctgctgcccgcccgcccgtaaaatggtcccctcggctggacagctcgccctgttcgct ctgggtattgtgttggctgcg."

There is no degeneracy problem with this representation.

Another very compact 2-D visualization scheme (Randic et al., 2003c; Randic, 2004) based on a template curve was proposed to inspect lengthy DNA sequences without requiring large space. A zigzag spiral representing the worm curve is used as the template for constructing compact graphical representation. Then four binary codes for the four nucleotides are assigned as A=00, G=01, C=10, and T=11. Starting from the center of the template, a binary code sequence is drawn by putting dark spots for ones and nothing for zeros on the vertices of template. There is no degeneracy problem from this graphical representation. On the other hand, only a square space with side approximately $\sqrt{n}$ is required to visualize a DNA sequence of length n. Figures 8(a) and 8(b) show the 2-D spiral worm curves of Homo sapiens B-cell CLL/lymphoma 2 (BCL2) mRNA sequence

Figure 7. 2-D zigzag like curve

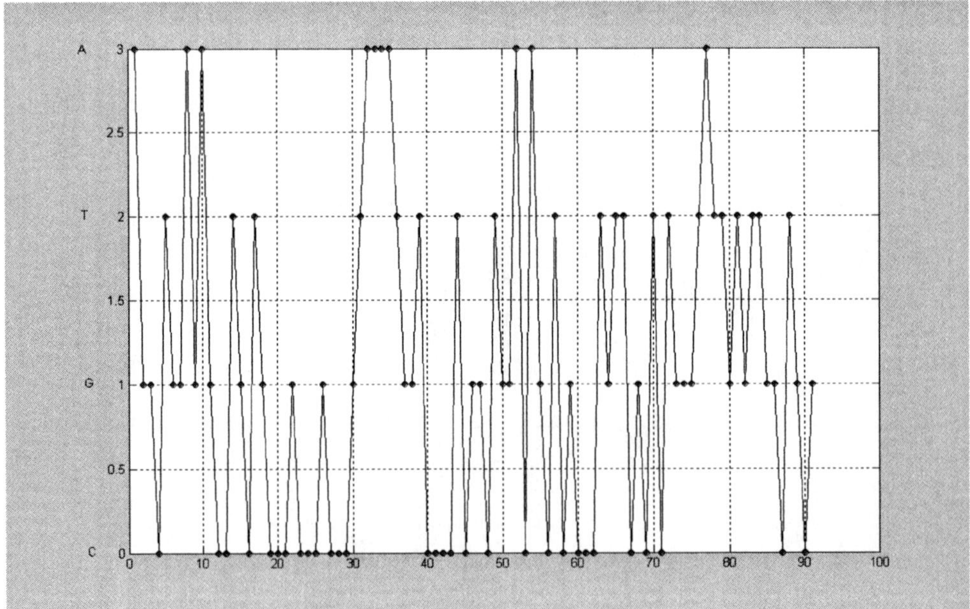

Figure 8(a). The 2-D spiral worm curves: Homo sapiens B-cell CLL/lymphoma 2 (BCL2) mRNA sequence (AC: NM_000657, 911 bp)

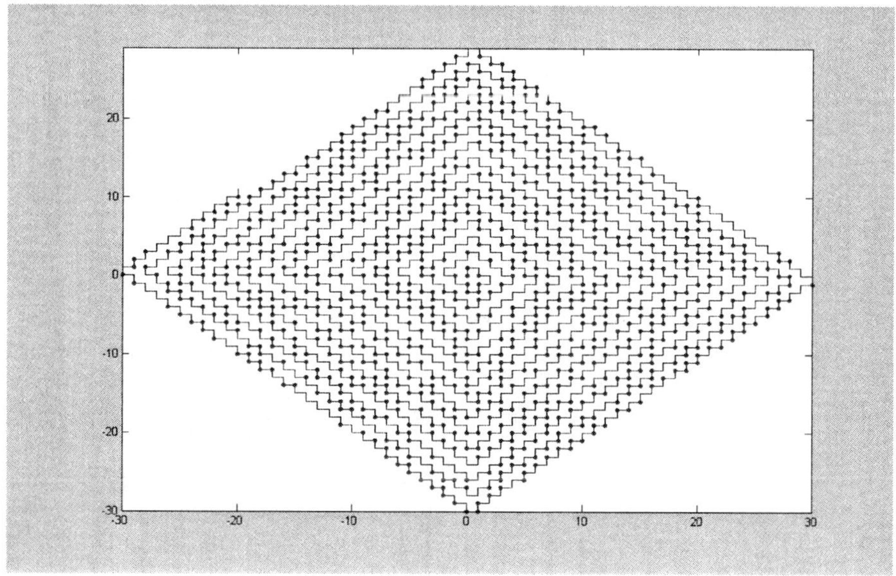

Figure 8(b). The 2-D spiral worm curves: Human dihydrofolate reductase gene (AC: J00140, 979 bp)

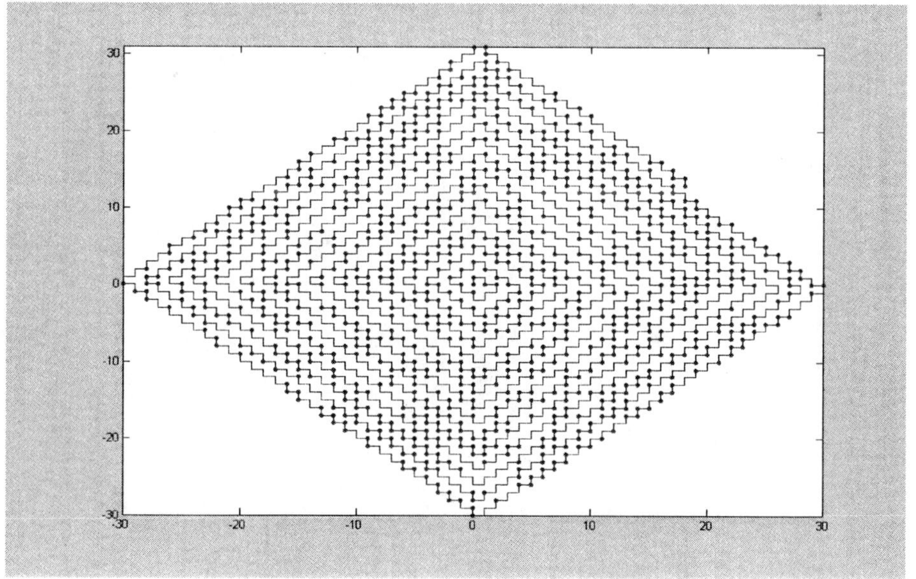

(AC: NM_000657, 911 bp) and Human dihydrofolate reductase gene (AC: J00140, 979 bp), respectively. They demonstrate a quite different visualization form from those demonstrated in Figures 1-7.

Chaos Game Representation and Fractals

In addition to the curve form representation, more complicated patterns can be utilized for DNA sequence visualization. Jeffrey (1990) developed the chaos game representation (CGR) method, which uses an iterated function system (IFS) (Barnsley, 1988) to map the character-based DNA sequence to a set of points in $\boldsymbol{R}^2$. The result points are not connected by line segments but rather are displayed in a 2-D pattern image, which is a picture of fractal structure. On the other hand, the fractal dimension of a DNA sequence can be measured as one of sequence characterizations (Berthelsen, Glazier, & Skolnik, 1992). Sequence visualization through the iterated function systems (Tino, 1999) and fractals (Hao, Lee, & Zhang, 2000; Ichinose, Yada, & Takagi, 2001) are also available. The properties of the corresponding DNA sequence can be revealed because the sequence basically is non-random and thus an attractor is visually observable. Figures 9 and 10 show the CGR and fractals (square), respectively, of Homo sapiens B-cell CLL/lymphoma 2 (BCL2) mRNA sequence (AC: NM_000657, 911 bp) and Human dihydrofolate reductase gene (AC: J00140, 979 bp).

With the CGR, different fractal patterns can be plotted from groups of genes. The most significant property of the CGR different from the curve form is that the adjacent bases

Figure 9(a). Chaos Game Representation (CGR): Homo sapiens B-cell CLL/lymphoma 2 (BCL2) mRNA sequence (AC: NM_000657, 911 bp)

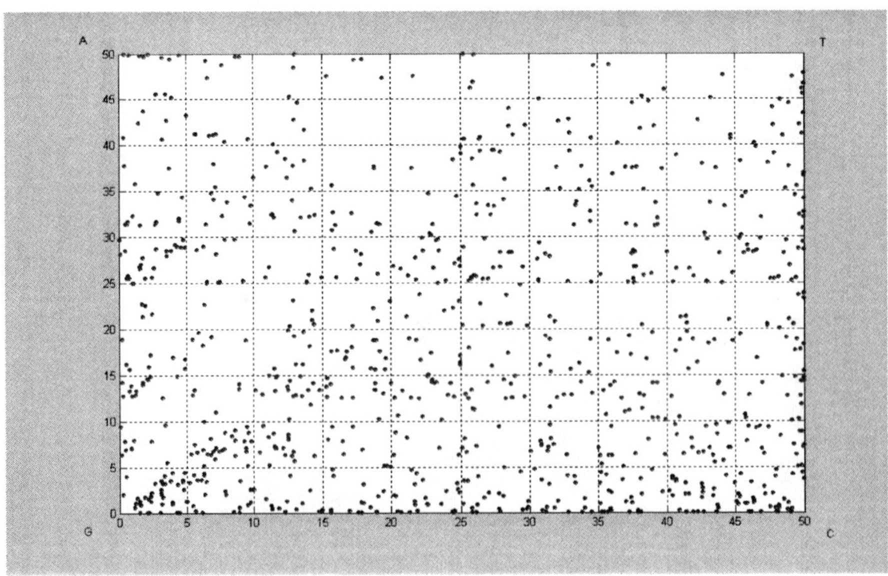

Figure 9(b). Chaos Game Representation (CGR):Human dihydrofolate reductase gene (AC: J00140, 979 bp)

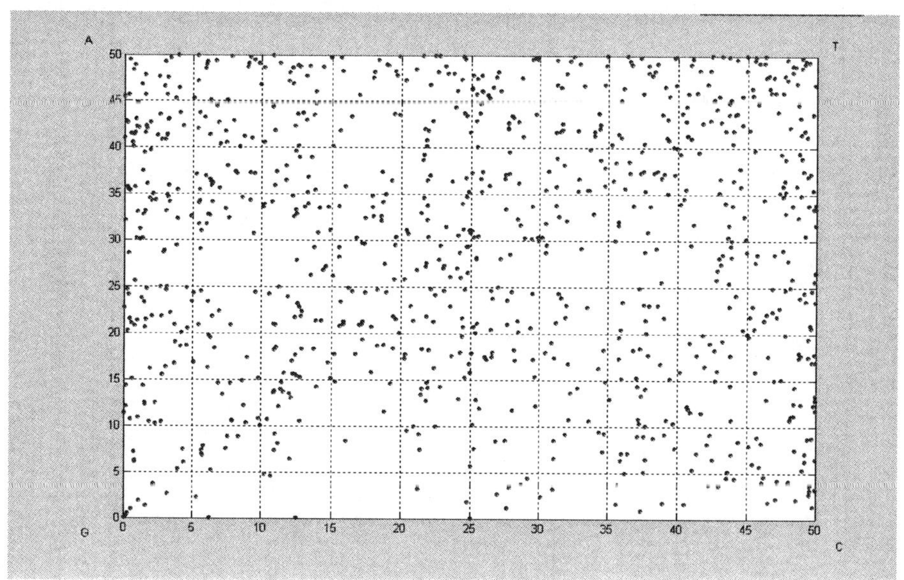

Figure 10(a). Fractals: Homo sapiens B-cell CLL/lymphoma 2 (BCL2) mRNA sequence (AC: NM_000657, 911 bp)

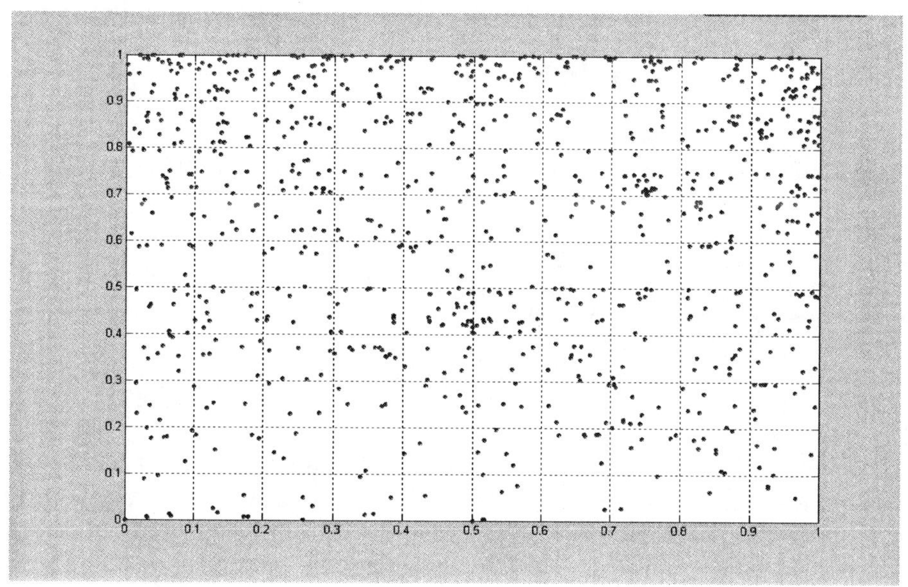

Figure 10(b). Fractals: Human dihydrofolate reductase gene (AC: J00140, 979 bp)

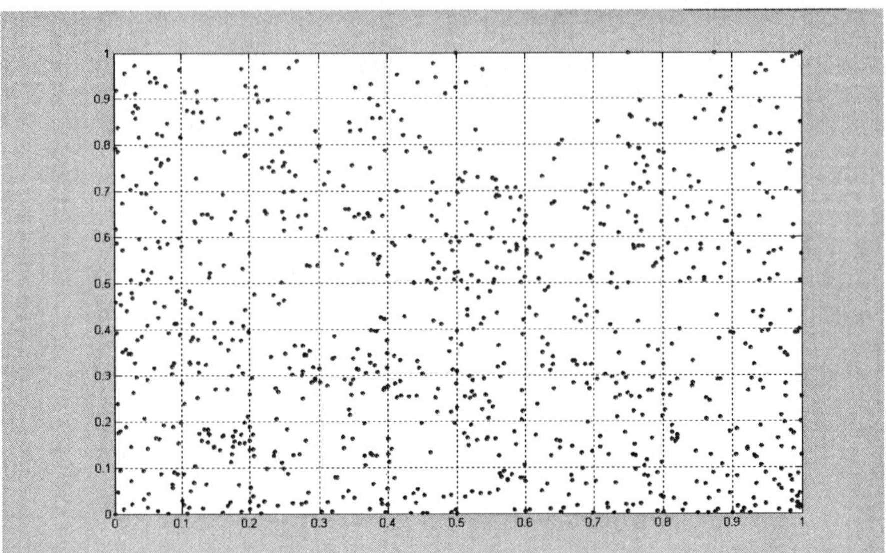

in the sequence are not always plotted neighboring to each other. Therefore, a new metric should be developed for measuring the similarity between two sequences.

One of the applications of the CRG for a DNA sequence is to discriminate the random and natural DNA sequences (Jeffery, 1990). By measuring the histogram entropy of a CRG image, the entropic profiles of DNA sequences can be constructed. On the other hand, a more generalized visualization model based on IFS was proposed by Wu et al. (1993). Three visual models based on the IFS were introduced: (1) *H*-curve; (2) Chaos game representation; (3) *W*-curve. By controlling two parameters in the proposed model, one of the three visual models can be selected.

Other Visualization Methods

A more detailed discussion on the 2-D graphical representation shown in Figure 8 is given by Randic et al. (2003b). Yau et al. (2003) proposed a visualization method in which the 2-D graphical representation is shown without degeneracy. Li and Wang (2004) proposed a 3-D representation of DNA sequences. The elements in the 3-D vectors assigned for the four bases are all positive such that no overlap and intersection appear. Yuan, Liao, and Wang (2003) proposed the 3-D representation of DNA sequences that can avoid overlaps and intersect. The basic concept is that the *z*-axis coordinates are the accumulated length of the bases. Chang et al. (2003) proposed a 3-D trajectory method based on a regular tetrahedron, which is similar to Randic's 3-D graphical representation method. Liao, Tan, and Ding (2005) proposed a unique 4-D representation of DNA sequences, which is different with Randic's 4-D representation method (Randic & Balaban, 2003). The studies on DNA sequence visualization are obviously flourishing.

Discussion and Future Trends

The most important issue in investigating various visualization methods is "what the direct applications are." In addition to some direct curve features, such as the geometrical center and end points (Liao, Tan, & Ding, 2005), many methods employ the distance matrix and its invariants (Randic, Kleiner, & DeAlba, 1994; Randic & Basak, 2001; Yuan, Liao, & Wang, 2003) to extract the eigenvalues as the condensed sequences for comparison. The protein coding measure, identification of replication origins and termination for some bacterial and archaeal genomes based on the *Z*-curve method was reported (Yan, Lin, & Zhang, 1998; Zhang & Zhang, 2005). As for the chaos game and fractal representation, global distance and local similarity can be determined for sequence comparison (Almeida, Carrico, Maretzek, Noble, & Fletcher, 2001) and for measuring phylogenetic proximity (Deschavanne, Giron, Vilain, Fagot, & Fertil, 1999).

Finally, the possible further studies and future work based on the visualized DNA sequences are summarized as follows:

1. Apply mathematical analysis or signal-processing algorithms such as the Fourier and Wavelet transforms to the visualized sequence data for further sequence analysis, clustering, and annotation (Bernaola-Galvan, Carpena, Roman-Roldan, & Oliver, 2002; Dodin, Vandergheynst, Levoir, Cordier, & Marcourt, 2000; Issac et al., 2002).
2. Develop better human machine interface for scientists in operating the bioinformatics tools based on graphical representation of sequences (Ovcharenko, Boffelli, & Loots, 2004).
3. Cooperate with computer graphics techniques for better graphical representation of DNA sequences (Ghai, Hain, & Chakraborty, 2004; Montgomery et al., 2004).
4. Extract the features from the visualized/numerical sequences and then develop feature databases for fast and efficient database retrieval. (Chang et al., 2005)
5. The investigation of the effects of insertion, deletion, substitution, reverse, and complementary in sequences (Chang et al., 2003). Find more biological applications and develop various visualization tools such as characterization and classifications of species, visualizing regular pattern in sequences (Loraine & Helt, 2002).
6. Adopt the existing visualization methods to protein sequences (Basu, Pan, Dutta, & Das, 1997; Bayley, Gardiner, Willett, & Artymiuk, 2005; Randic & Krilov, 1997; Randic, Zupan, & Balaban, 2004).

Summary

The aids of computer and network technologies provide better data access, acquisition, update, and management on huge genomic sequences over the worldwide Internet databases. Through the data visualization methods mentioned in this chapter, various graphical representations of DNA sequences provide alternative observations, which can derivate certain sequence characteristics that cannot be directly obtained from the original character-based sequences. With the rapid increase of genomic sequences and databases, the development of new techniques for sequence visualization, analysis, characterization, comparison, and annotation is more and more important in a bioinformatics scenario. Most of recent researches on DNA sequence visualization/ graphical representation have been briefly reviewed in this chapter. Based on the visualized data formats, well known graphic/image processing and pattern recognition techniques for exploring the intra- and inter-sequence relationships in genomic research would be a significant trend.

References

Almeida, J. S., Carrico, J. A., Maretzek, A., Noble, P. A., & Fletcher, M. (2001). Analysis of genomic sequences by Chaos Game Representation. *Bioinformatics, 17*(5), 429-437.

Almeida, J. S., & Vinga, S. (2002). Universal sequence map (USM) of arbitrary discrete sequences. *BMC Bioinformatics, 3(6).*

Anastassiou, D. (2001). Genomic signal processing. *IEEE Signal Processing Magazine, 18*, 8-20.

Arneodo, A., Bacry, E., Graves, P. V., & Muzy, J. F. (1995). Characterizing long-range correlations in DNA sequences from wavelet analysis. *Physics Review Letters, 74*(16), 3293-3296.

Barnsley, M. F. (1988). *Fractals everywhere.* New York: Springer-Verlag.

Basu, S., Pan, A., Dutta, C., & Das, J. (1997). Chaos game representation of proteins. *Journal of Molecular Graphics and Modelling, 15,* 279-289.

Bayley, M. J., Gardiner, E. J., Willett, P., & Artymiuk, P. J. (2005). A Fourier fingerprint-based method for protein surface representation. *Journal of Chemical Information Model, 45*(3), 696-707.

Bernaola-Galvan, P., Carpena, P., Roman-Roldan, R., & Oliver, J. L. (2002). Study of statistical correlations in DNA sequences. *Gene, 300*, 105-115.

Berger, J., Mitra, S., Carli, M., & Neri, A. (2002). New approaches to genome sequence analysis based on digital signal processing. *Proceedings of IEEE Workshop on Genomic Signal Processing and Statistics (GENSIPS)*, CP2-08.

Berger, J., Mitra, S., Carli, M., & Neri, A. (2004). Visualization and analysis of DNA sequences using DNA walks. *Journal of the Franklin Institute, 341*(1-2), 37-53.

Berthelsen, C. L., Glazier, J. A., & Skolnik, M. H. (1992). Global fractal dimension of human DNA sequences treated as pseudorandom walks. *Physics Review A, 45*, 8902-8913.

Blumer, A., Blumer, J., Haussler, D., McConnell, R., & Ehrenfeucht, A. (1987). Complete inverted files for efficient text retrieval and analysis. *Journal of ACM, 34*, 578-595

Chang, H. T., Lo, N. W., Lu, W. C., & Kuo, C. J. (2003). Visualization and comparison of DNA sequences by use of three-dimensional trajectory. *Proceedings of The First Asia-Pacific Bioinformatics Conference* (pp. 81-85).

Chang, H. T., Xiao, S. W., & Lo, N. W. (2005). Feature extraction and comparison of TDTs: an efficient sequence retrieval algorithm for genomic databases. *In Abstract Book of The 3rd Asia Pacific Bioinformatics Conference* (APBC2005) (pp. 86).

Cheever, E. A., & Searls, D. B. (1989). Using signal processing techniques for DNA sequence comparison. *Proceedings of the 1989 Fifteenth Annual Northwest Bioengineering Conference* (pp. 173-174).

Chen, J., Li, H., Sun, K., & Kim, B. (2003). Signal processing applications — how will bioinformatics impact signal processing research? *IEEE Signal Processing Magazine, 20*(6), 16-26.

Collins, F. S., Lander, E. S., Rogers, J., Waterston, R. H., et al. (2004). Finishing the euchromatic sequence of the human genome. *Nature, 431*, 931-945.

Cork, D. J., & Wu, D. (1993). Computer visualization of long genomic sequences. *Proceedings of the 4th Conference on Visualization,* (pp. 308-315).

Deschavanne, P. J., Giron, A., Vilain, J., Fagot, G., & Fertil, B. (1999). Genomic signature: Characterization and classification of species assessed by Chaos Game Representation of sequences. *Molecular Biology Evolution, 16*(10), 1391-1399.

Dodin, G., Vandergheynst, P., Levoir, P., Cordier, C., & Marcourt, L. (2000). Fourier and wavelet transform analysis, a tool for visualizaing regular patterns in DNA sequences. *Journal of theoretical Biology, 206,* 323-326.

Gates, M. (1986). A simple way to look at DNA. *Journal of Theoretical Biology, 119*(3), 319-328.

Ghai, R., Hain, H., & Chakraborty, T. (2004). GenomeViz: visualization microvial genomes. *BMC Bioinformatics, 5*, 198.

Guo, F.-B., Ou, H.-Y., & Zhang, C.-T. (2003). Z-CURVE: A new system for recognizing protein-coding genes in bacterial and archaeal genomes. *Nucleic Acids Research, 31*(6), 1780-1789.

Guo, X., Randic, M., & Basak, S. C. (2001). A novel 2-D graphical representation of DNA sequences of low degeneracy. *Chemical Physics Letters, 350,* 106-112.

Hamori, E., & Ruskin, J. (1983). H curves, A novel method of representation of nucleotide series especially suited for long DNA sequences. *The Journal of Biological Chemistry, 258*(2), 1318-1327.

Hao, B.-L., Lee, H. C., & Zhang, S.-Y. (2000). Fractals related to long DNA sequences and complete genomes. *Chaos, Solitons and Fractals, 11*, 825-836.

Ichinose, N., Yada, T., & Takagi, T. (2001). Quadtree Representation of DNA Sequences. *Genome Informatics, 12*, 510-511.

Issac, B., Singh, H., Kaur, H., & Raghave, G.P.S. (2002). Locating probable genes using Fourier transform approach. *Bioinformatics, 18*(1), 196-197.

Jeffrey, H. J. (1990). Chaos game representation of gene structure. *Nucleic Acids Research, 18*(8), 2163-2170.

Leong, P. M., & Morgenthaler, S. (1995). Random walk and gap plots of DNA sequence. *Cabios, 11*, 503-507.

Li, C., & Wang, J. (2004). On a 3-D representation of DNA primary sequences. *Combinatorial Chemistry and High Throughput Screening, 7*, 23-27.

Liao, B., Tan, M., & Ding, K. (2005). A 4D representation of DNA sequences and its application. *Chemical Physics Letters, 402*, 380-383.

Liu, Y., Guo, X., Xu, J., Pan, L., & Wang, S. (2002). Some notes on 2-D graphical representation of DNA sequence. *Journal of Chemical Information and Computer Sciences, 42*(3), 529-533.

Loraine, A. E., & Helt, G. A. (2002). Visualizing the genome: techniques for presenting human genome data and annotations. *BMC Bioinformatics, 3*(19).

Mehta, D. P., & Sahni, S. (1994). Computing display conflicts in string visualization, *IEEE Transaction on Computers, 43,* 350-361.

Montgomery, S. B., Astakhova, T., Bilenky, M, Birney, E., Fu, T., Hassel, M., et al. (2004). Sockeye: a 3D environment for comparative genomics. *Genome Research, 14*, 956-962.

Nandy, A. (1994). A new graphical representation and analysis of DNA sequence structure: I. Methodology and application to globin genes. *Current Science, 66*(4), 309-314.

Ovcharenko, I., Boffelli, D., & Loots, G. G. (2004). eShadow: A tool for comparing closely related sequences. *Genome Research,* 1191-1198.

Peng, C. K., Buldyrev, S. V., Goldberger, A. L., Havlin, S., Sciortino, F., Simons, M., et al. (1992). Longrange correlations in nucleotide sequences. *Nature, 356*, 168-170.

Randic, M. (2000) On characterization of DNA primary sequences by a condensed matrix. *Chemical Physics Letters, 317*, 29-34.

Randic, M. (2004) Graphical representations of DNA as 2-D map. *Chemical Physics Letters, 386*, 468-471.

Randic, M., & Balaban, A. T. (2003). On a four-dimensional representation of DNA primary sequences. *Journal of Chemical Information and Computer Sciences, 43*(2), 532-539.

Randic, M., & Basak, S.C. (2001). Characterization of DNA primary sequences based on the average distances between bases. *Journal of Chemical Information and Computer Sciences, 41*(3), 561-568.

Randic, M., & Krilov, G. (1997). Characterization of 3-D sequence of proteins. *Chemical Physics Letters, 272*, 115-119.

Randic, M., & Vracko, M. (2000). On the similarity of DNA primary sequences. *Journal of Chemical Information and Computer Sciences, 40*(3), 599-606.

Randic, M., Kleiner, A. F., & DeAlba L. M. (1994). Distance/Distance matrices. *Journal of Chemical Information and Computer Science, 34*, 277-286.

Randic, M., Vracko, M., Lers, N., & Plavsic, D. (2003a). Novel 2-D graphical representation of DNA sequences and their numerical characterization. *Chemical Physics Letters, 368*, 1-6.

Randic, M., Vracko, M., Lers, N., & Plavsic, D. (2003b). Analysis of similarity/dissimilarity of DNA sequences based on novel 2-D graphical representation. *Chemical Physics Letters, 371*, 202-207.

Randic, M., Vracko, M., Nandy, A., & Basak, S. C. (2000). On 3-D graphical representation of DNA primary sequences and their numerical characterization. *Journal of Chemical Information and Computer Science, 40*(5), 1235-1244.

Randic, M., Vracko, M., Zupan, J., & Novic, M. (2003c) Compact 2-D graphical representation of DNA. *Chemical Physics Letters, 373*, 558-562.

Randic, M., Zupan, J., & Balaban, A. T. (2004) Unique graphical representation of protein sequences based on nucleotide triplet codons. *Chemical Physics Letters, 397*, 247-252.

Roy, A., Raychaudhury, C., & Nandy, A. (1998). Novel techniques of graphical representation and analysis of DNA sequences — a review. *Journal of Bioscience, 23*(1), 55-71.

Tino, P. (1999). Spatial representation of symbolic sequences through iterative function systems. *IEEE Transactions on Signal Processing, 29*, 386-393.

Wang, W., & Johnson, D. H. (2002). Computing linear transforms of symbolic signals. *IEEE Transaction on Signal Processing, 50*, 628-634.

Wu, D., Roberge, J., Cork, D. J., Nguyen, B. G., & Grace, T. (1993). Computer visualization of long genomic sequence. *Proceedings of IEEE Conference on Visualization* (pp. 308-315).

Yan, M., Lin, Z. S., & Zhang, C. T. (1998). A new Fourier transform approach for protein coding measure based on the format of the Z curve. *Bioinformatics, 14*(8), 685-690.

Yau, S. T., Wang, J., Niknejad, A., Lu, C., Jin, N., & Ho, Y. K. (2003). DNA sequence representation without degeneracy. *Nucleic Acids Research, 31*(12), 3078-3080.

Yuan, C., Liao, B., & Wang, T.M. (2003). New 3D graphical representation of DNA sequences and their numerical characterization. *Chemical Physics Letters, 379*, 412-417.

Zhang, C. T., Zhang, R., & Ou, H. Y., (2003). The Z curve database: a graphic representation of genome sequences. *Bioinformatics, 19*(5), 593–599.

Zhang, R., & Zhang, C. T. (1994). Z curves, an intuitive tool for visualizing and analyzing DNA sequences. *Journal of Biomolecular Structure Dynamics, 11*, 767-782.

Zhang, R., & Zhang, C. T. (2005). Identification of replication origins in archaeal genomes based on the Z-curve method. *Archaea, 1*, 335-346.

Zhang, X. Y., Chen, F., Zhang, Y. T., Agner, S. C., Akay, M., Lu, Z. H., et al. (2002). Signal processing techniques in genomic engineering. *Proceedings of The IEEE, 90*(12), 1822-1833.

Zupan, J., & Randic, M. (2005). Algorithm for coding DNA sequences into "spectrum-like" and "zigzag" representations. *Journal of Chemical Information Model, 45*(2), 309-313.

Chapter V

Proteomics with Mass Spectrometry

Simon Lin, Northwestern University, USA
Salvatore Mungal, Duke University Medical Center, USA
Richard Haney, Duke University Medical Center, USA
Edward F. Patz, Jr., Duke University Medical Center, USA
Patrick McConnell, Duke University Medical Center, USA

Abstract

This chapter provides a rudimentary review of the field of proteomics as it applies to mass spectrometry, data handling, and analysis. It points out the potential significance of the field suggesting that the study of nuclei acids has its limitations and that the progressive field of proteomics with spectrometry in tandem with transcription studies could potentially elucidate the link between RNA transcription and concomitant protein expression. Furthermore, we describe the fundamentals of proteomics with mass spectrometry and expound the methodology necessary to manage the vast amounts of data generated in order to facilitate statistical analysis. We explore the underlying technologies with the intention to demystify the complexities of the nascent field and to fuel interest by readers in the near future.

Introduction

The science of proteomics seeks to identify and characterize protein expression in biological systems. It leverages technologies that have been around for decades, such as mass spectrometry, liquid chromatography, and electrophoresis. Though the basic science is mature, proteomics technology has made much progress over the last decade (Aebersold & Mann, 2003). Proteomics has been gaining momentum as the limitations of studying DNA and RNA alone have been documented (Greenbaum, Colangelo, Williams, & Gerstein, 2003). Gene sequences themselves yield little information about how much of its transcribed protein will be expressed and in what cellular states. Furthermore, the nature and prevalence of alternative splicing has shown that studying gene expression at the protein level can yield complementary knowledge at the nuclei acid level.

Proteomics is a vast and complex field consisting of a variety of platforms. In this chapter, we focus on mass spectrometry technology and data because of its wide use in the field for protein identification and profiling experiments. The goal of identification is to find out the amino acid sequence of extracted proteins; whereas the goal of profiling is to quantify the expression level of proteins.

Proteomics data is measured in mass-to-charge ratios, termed m/z values. Peaks can be identified when we plot measured intensities against m/z values. The location of a peak corresponds to the chemical composition of the protein and thus can be used for protein identification, and the amplitude of a peak carries information on the protein abundance, which is the basis of profiling. Such data provides a unique challenge both to encode and to analyze. Firstly, even modest-sized experiments provide gigabytes worth of data. Encoding and exchanging data of this size is quite a technical challenge without mention of the problem of analyzing such a large dataset. Secondly, the data itself has intrinsic properties that make it particularly challenging to analyze and interpret. For example, one common task in analyzing spectra data is to locate data peaks. However, what characteristics of the peaks are most important: height, area under the peak, or some ratio thereof? Furthermore, noise in a single spectrum and aggregating multiple spectra provide unique statistical challenges.

There has been some work towards standardization of proteomics data. The Proteomics Standards Initiative (PSI) has been created under the auspices of the Human Proteome Organization (HUPO). PSI has working groups to develop data standards for protein-protein interaction and mass spectrometry (Orchard et al., 2004). Though good intentioned, the PSI mass spectrometry data standards are still in an abstract form. The Minimum Information About Proteomics Experiments is a document describing the types of data items needed to fully described proteomics experiments, but no data standard exists. The PSI Object Model (PSI-OM) and PSI Markup Language (PSI-ML) have defined a set of concepts and the relationships of those concepts, but not concrete data formats. Finally, the PSI Ontology (PSI-Ont) is being developed as an extension to the Microarray and Gene Expression Data Ontology (MGED Ontology) with the 'PSI' namespace, but is not yet complete. Proteomics data standardization is far from complete, and there is much work to be done to define data formats and ontologies.

Data

Production of Proteomics Data

Proteomics data is produced from mass spectrometry experiments, which measures the atomic mass of peptides (see Figure 1). Classical proteomics study begins with 2D-gel for protein separation prior to mass spectrometry identification of gel spots. First, a biologic sample containing some proteins of interest is prepared. The proteins are then digested, or split, into peptide fragments. These fragments are ionized just before they enter the mass analyzer and detector, where their masses are measured. There are several techniques for measuring the mass. One popular approach, termed Time-of-Flight or just TOF, measures the amount of time an ion takes to reach a detector after it enters the mass spectrometer. Currently, the most accurate method, Fourier-transform ion cyclotron resonance (FTICR), involves measuring the frequency that ions circle the high magnetic field of a super-conducting magnet. The resulting data from the mass spectrometer is a list of masses and the relative abundance of the ions for that mass, termed the intensity. A secondary, or tandem, mass spectrometry reading is sometimes taken in which the peptide itself is then broken into its components and mass readings are taken. This allows for a much more accurate reading (Steen and Mann, 2004). The fingerprint generated by the mass spectrometry is then matched against a database of known proteins. The peak height information is usually discarded during the database search.

Although mass spectrometry is often used to identify proteins first separated by 2D-gel, several recent studies have suggests that direct analysis of intact protein (without gel separation and digestion) with MALDI-TOF or SELDI-TOF can provide a measurement of the protein expression level, as indicated by the peak heights (Rosenblatt et al., 2004).

The goal of a proteomics experiment is usually one of two objectives: accurately identify proteins in a sample or create a predictive model from a set of proteomics spectra. These are the protein identification and protein profiling problems, respectively. In protein identification, the goal is to as accurately as possible measure the peptide fragments of proteins of interest. The masses of the high intensity peptides are searched against a

Figure 1. Process flow for a typical proteomics experiment: Different options/ technologies are listed below each step (Steen and Mann, 2004, reprinted with copyright permission from Nature Reviews Molecular Cell Biology)

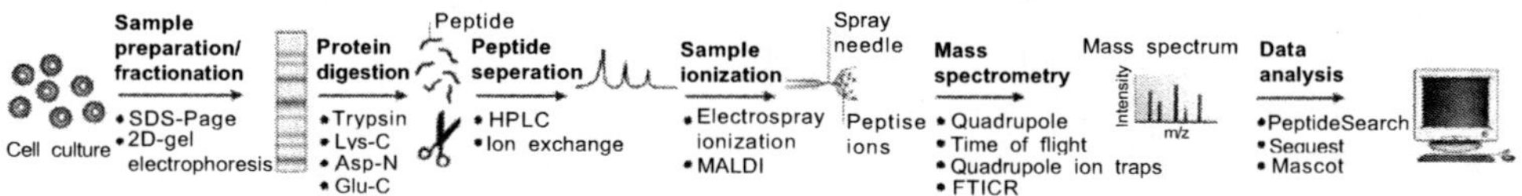

database to identify which proteins exist in the sample. In proteomics profiling studies, the mass/intensity spectra of different samples are used to create a predictive model. The data is fed into a machine learning algorithm which can later be used as a model to predict what class a particular spectra falls into, for example diseased and healthy.

Overview of m/z Data

A spectrum collected from a proteomics experiment is a list of peaks that consist of a pair of values. The first value is the mass-to-charge ratio, often referred to as the m/z value. The charge is included because sometimes larger fragments can gain greater than a single charge during the ionization step. The second value is the intensity, or abundance of the peak. That intensity value is relative to a particular run of the mass spectrometer and typically requires statistical normalization to be comparable to other runs. This full list of m/z-intensity values makes up a proteomics spectra (see Figure 2).

As noted previously in Figure 1, there can be a third dimension to proteomics experiments, such as when a mass spectrometer is used to analyze data that comes out of a liquid chromatography (LC) data. In that case, each spectrum is associated with a retention time, that is, the time that a sample takes to go through the chromatography column. Successive scans come out at the intervals of several seconds. So, the data has three dimensions overall, involving m/z values, LC retention times, and finally the MS intensity values. In addition to making 2-dimensional (2-D) plots such as the one depicted in Figure 2, systems analyzing this data also make plots in 3-D.

Figure 2. A proteomics spectra has the mass values running along the x-axis and the intensities along the y axis; the mass is actually the mass-to-charge ratio, as peptides can gain more than a single charge when ionized. the intensity is typically measure by counts of ions for each mass value (Steen and Mann, 2004, reprinted with copyright permission from Nature Reviews Molecular Cell Biology)

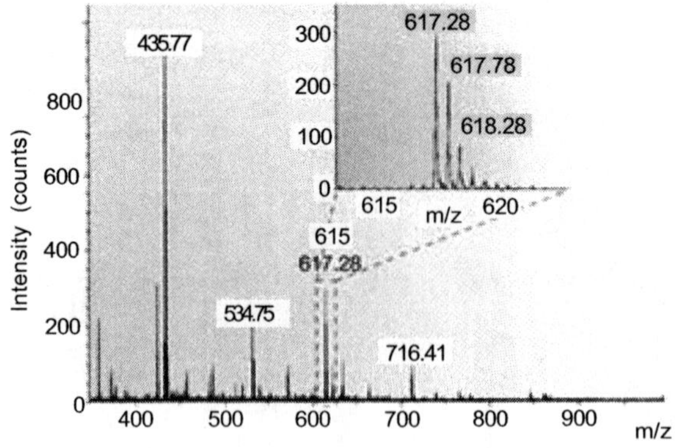

Important Metadata Considerations

In addition to the m/z-intensity values, it is important to keep track of other experiment-related metadata. In the data modeling world, metadata is simply data about the data of interest. For proteomics, there are some important categories of metadata.

Data Acquisition

It is important to keep track of how the data was acquired, from the biological sample through any lab procedures used, and up to the ionization procedure. There is a class of software called Laboratory Information Management Systems (LIMS) for tracking these sorts of data. LIMS are usually quite large, complex, and expensive systems, and when it comes to analysis, the information stored in a LIMS must be attached to the data for processing. Data acquisition considerations can impact how data is processed, analyzed, and interpreted.

Instrumentation

The equipment used to ionize the peptides and the mass-spectrometry equipment should be tracked with the m/z-intensity data. This instrumentation can be an important consideration when performing data processing and analysis, as the mass spectrometer can have a large impact on the sensitivity and accuracy of the data.

Data Processing

As you will read in the Analysis section of this chapter, there is a full pipeline of processes that can be run on the data. Many of these statistical analyses simply modify or reduce the m/z-intensity values. This is important to keep track of because the data processing can have a large impact on how the data should be interpreted.

The Proteomic Standard Initiative (PSI) is a group that is attempting to define the Minimum Information About a Proteomics Experiment (MIAPE) that should be collected (Orchard et al., 2004). At the time of writing, the MIAPE standard is only a draft document outlining some metadata guidelines. However, when the document reaches maturity, it will be an invaluable resource for proteomics metadata standards.

Available Data Standards

As of the writing of this chapter, there are two major efforts underway to provide a standard encoding of proteomics data. The first effort called mzXML is by the Institute for System Biology (ISB) (Pedrioli et al., 2004). The second effort called mzData is by the

Proteomics Standards Initiative (PSI) (Orchard et al., 2004). Both these standards are very similar and can be used interchangeably with some translation of formats, so we will focus on mzXML because it was the first and is the most widely used. However, it there may be convergence on mzData sometime in the future because it has the backing of a standards group.

mzXML is a standard for encoding proteomics m/z-intensity data. All data is encoded in the Extensible Markup Language (XML), which is a textual format for encoding arbitrary data. The m/z-intensity values are encoded as in Base64, which is an efficient encoding of binary data into text. Although Base64 is unreadable by humans, it provides the relatively low overhead of encoding three bytes of data in four textual characters (Table 1).

Table 1. Example mzXML file (Notice the instrumentation and dataprocessing metadata associated with the msRun and the metadata attributes associated with the scan; the second row shows a standard SAX XML parser written in Java, and characters between start and end elements are stored in a StringBuffer, attributes are parsed in the startElement method, and the actual peak values should be parsed in the endElement method.)

```
<?xml version="1.0" encoding="ISO-8859-1"?>
<msRun
  xmlns="http://sashimi.sourceforge.net/schema/"
  xmlns:xsi="http://www.w3.org/2001/XMLSchema-instance"
  xsi:schemaLocation="http://sashimi.sourceforge.net/schema/
    http://sashimi.sourceforge.net/schema/MsXML.xsd"
  scanCount="7198"
  startTime="PT0S"
  endTime="PT120S"
>
  <parentFile fileName="myData.wiff" fileType="RAWData"
    fileSha1="068adb0e388196bc03ad95cd7140c4cce233a181"
  />

  <instrument
    manufacturer="ABI" model="QSTAR"
    ionisation="ESI" msType="Q-TOF"
    cid="LE"
  >
    <software type="acquisition" name="Analyst" version="1.0"/>
  </instrument>

  <dataProcessing intensityCutoff="0">
    <software type="conversion" name="mzStar" version="1.0.6"/>
    <processingOperation name="min_peaks_per_spectra" value="1"/>
  </dataProcessing>

  <scan num="1" msLevel="1" peaksCount="4246" polarity="+"
    retentionTime="PT1.00499999523163S" startMz="400" endMz="3500"
    basePeakMz="447.352228504701" basePeakIntensity="8"
totIonCurrent="9016"
  >
    <peaks precision="32">Q8gbrEAAAABDyJx6QAAAAEPJF6 . . .</peaks>
  </scan>

</msRun>
```

Table 1. continued

```
private class MzXmlHandler extends DefaultHandler
{
  private StringBuffer chars = new StringBuffer(); // chars between els
  private int peakCount = 0; // number of value pairs to decode
  private int precision = 0; // 32 or 64
  private int scanNumber = 0; // scan number

  public void startElement(
    String namespaceURI, String localName, String qName,
    Attributes atts
  )
    throws SAXException
  {
    chars.delete(0, chars.length());

    // parse out attribute values
    if (qName.equals("scan")) {
      peakCount = Integer.parseInt(atts.getValue("peaksCount"));
      scanNumber = Integer.parseInt(atts.getValue("num"));
    } else if (qName.equals("peaks")) {
      precision = Integer.parseInt(atts.getValue("precision"));
    }
  }

  public void characters(char buf [], int offset, int len)
    throws SAXException
  {
    // store values between start and end tags
    chars.append(buf, offset, len);
  }

  public void endElement(
    String namespaceURI, String localName, String qName
  )
    throws SAXException
  {
    if (qName.equals("peaks")) {
      // decode base64 values based on peakCount and precision
      // handle base64 values based on scanNumber
    }

    chars.delete(0, chars.length());
  }
}
```

mzXML also encodes some of the necessary metadata associated with instrumentation and data processing. For each mass spectrometry run, there is an instrumentation element that encodes the mass spectrometer manufacturer, model, ionization method, and mass spectrometry type. Additionally, there is a dataProcessing element that provides a description of the software employed, and a generic description of the operation utilized (Table 1).

Data Parsing

At this time, most proteomics data is exchanged as comma delimited (CSV) text files (see Table 2). A drawback of this approach is that it takes upwards of 12 bytes to encode four bytes of data. The benefits of CSV files are that the files are convenient for people to read and the ease of integration with existing software. For example, CSV files can be read into statistical packages such as R or Mathematica with one line of code and can be imported into Microsoft Excel with ease (see Table 2).

To be sure, CSV files of m/z and intensity values contain no metadata. Hence it is still necessary to provide an additional file containing that data. It is possible to have this file of attributes and attribute values also be in a CSV form. However, as of this writing, when proteomics data is exchange using CSV files, most of the time, such files of metadata are not included.

The other main approach involves encoding proteomics data using an XML-based format. One major issue with encoding proteomics data in XML is that XML is more technically difficult to parse than other text formats (see Table 1). This problem is compounded by the fact that the peak values are stored in base64, which is not immediately useful to the programmer. An extra step of decoding the base64 values is needed. Luckily, there are numerous free base64 parsers available on the Web. And, to ease the use of mzXML, ISB has created a suite of tools for converting and parsing mzXML data. They have written converters from the binary output of many popular mass spectrometers, libraries in C++ and Java for parsing mzXML files, and graphical viewers

Table 2. Example of comma-delimited proteomics data (the first column contains m/z values and the second column contains intensity values; unlike mzXML, there is not metadata associated with the data, the second row shows R code to read the m/z-intensity data into a data frame for easy manipulation)

```
M/Z,Intensity
1999.883455, 1207.3
2000.178622, 1279.94
2000.473811, 1324.59
2000.769021, 1360.99
2001.064254, 1412.31
2001.359508, 1486.62
2001.654784, 1457.04
2001.950082, 1442.05
2002.245402, 1376.06
2002.540743, 1331.33
2002.836107, 1344.02
2003.131492, 1410.01
2003.426899, 1427.38
2003.722328, 1461.4
2004.017778, 1514.69
2004.313251, 1557.36
2004.608745, 1567.99
2004.904261, 1597.97
2005.199799, 1593.68
2005.495359, 1585.36
2005.790940, 1527.7
2006.086544, 1474.01
2006.382169, 1448.39
```

```
# R code to read a CSV file into a data frame
data = read.csv(file, header=TRUE)
```

for inspecting mzXML. However, one of ISB's most interesting contributions is the ability to index an mzXML file and access scans in a non-sequential way.

It should be noted that the above two formats (CSV and XML-based formats) most often represent data messaging formats. In practice, pipelines used for high throughput proteomics frequently take data directly from the raw vendor formats from manufacturers and convert that to binary files or store it in databases, without going through the intermediary step of converting to CSV or XML form.

Publicly Available Data Sources

As of the writing of this chapter, there are few large-scale proteomics efforts or data repositories. A good starting point for finding public data is the Open Proteomics Database, which is a repository containing four organisms and 1.2 million spectra (Prince, Carlson, Wang, Lu, & Marcotte, 2004). The mzXML Web site (http://sashimi.sourceforge.net/repository.html) has a small repository of useful designed and other experiments. Finally, the PeptideAtlas site has proteomics data from four organisms and hundreds of thousands of spectra (Desiere et al., 2004).

Open Source and Commercial Databases

Currently, there are no open-source or commercial databases for proteomics data available. The Fox Chase Cancer Center is currently working on an open-source implementation of a proteomics LIMS for managing proteomics data from the point-of-view of the lab. This project is slated to be completed in year 2006. Many mass spectrometer vendors provide software for a fee that will manage proteomics data as it is produced by their instrumentation. The fee is usually nominal in comparison to the price of the mass spectrometer itself, being 1%-5% of the cost of the machine, so many labs choose this option.

Analysis

We will use the data set in the Duke Proteomics Data Mining Conference (September 23, 2002) to illustrate the data analysis challenges. A collection of 41 clinical specimens were measured by MALDI-TOF, of which 17 were normal and 24 were lung cancer. The first goal is to find out whether we can classify the patients based on proteomics measurements. If so, the second goal is to find out protein biomarkers that can differentiate cancer versus normal. Results from this conference were compiled into a special issue of Proteomics, a journal by Wiley-VCH (volume 3, No. 9, 2003).

Raw Data and Preprocessing

Without appropriate preprocessing, some systematic bias (irrelevant information) can drive the data mining process to get superficial results. This argument is supported by observed differences in replicated measurements of the same biosample. The goal of data preprocessing is to reduce the variance seen in replicates of the same biosample, but also to preserve the difference among different biosamples. Variance caused by systematic bias can be corrected with baseline removal and normalization.

Raw Data

Ideally, a protein corresponds to a vertical line on the spectrum, and its abundance is represented by the height of the line. In reality, due to heterogeneous molecular weight caused by isotopic elements, and stochastic ionic movements during the measurement

Figure 3. Part of spectra showing the difference between cancer and normal: Ten replicate runs of each biosample are shown superimposed (left) and stacked (right)

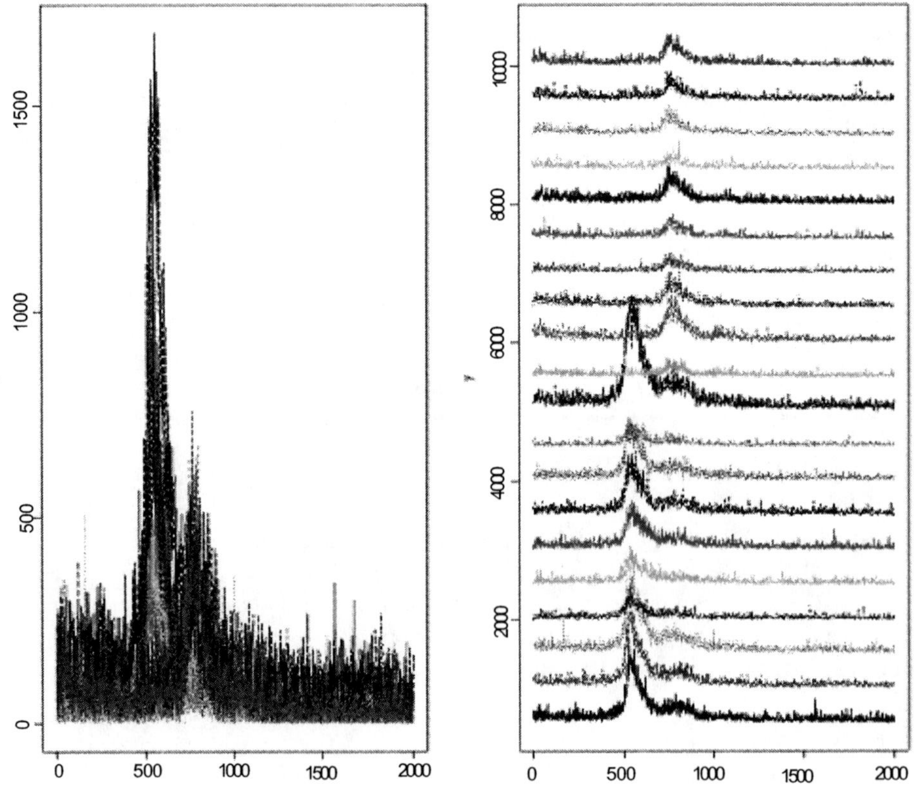

process, the vertical line is broadened into a peak. A total of 20 raw spectra (10 replicates each from a cancer and a normal specimen) are shown in Figure 3.

Baseline Correction

With both MALDI-TOF and SELDI-TOF data, we usually observe an elevated baseline at the low m/z range. Figure 4 illustrates two scans of the same biological sample. Elevated baseline of scan 3 can be removed by subtracting the local minima smoothed by a cubic spline (Berndt, Hobohm, & Langen, 1999). From this example, we clearly see that without appropriate baseline correction, it is pointless to compare the intensity difference between normal and cancer samples.

Similar to background removal in microarray image analysis, baseline correction of spectrum data have the following implications:

1. Some negative intensities values may be generated after baseline removal. These values can be reset to zero if necessary.
2. Incorrect choice of algorithm and parameters can results in erroneous results: to one extreme, excessive correction will remove all the peaks and result in a flat spectrum with noise only; to the other extreme, inadequate correction will not remove the systematic bias.

Figure 4. The effect of baseline removal on spectra

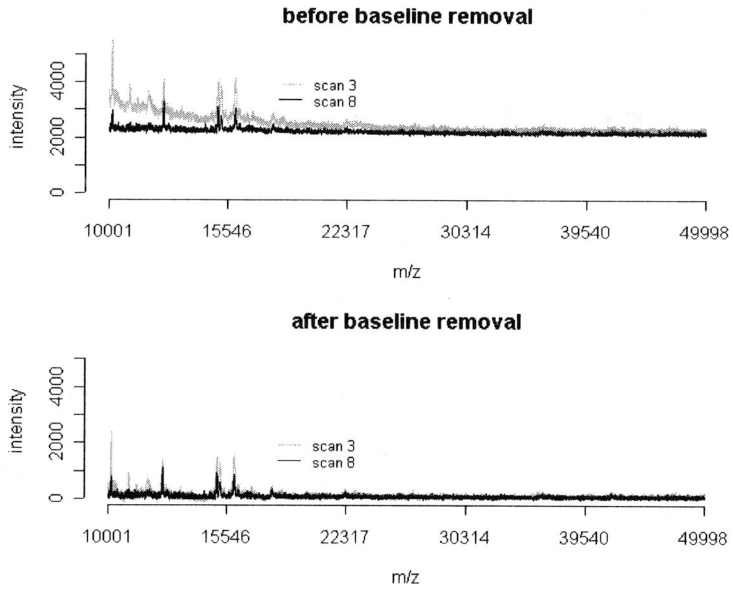

Normalization

As seen in Figure 5, the maximum amplitude of scan 3 and scan 8, which are from the same biosample, are different even after baseline correction. This difference can be corrected by a multiplication factor. This simple operation corresponds to scale-normalization of microarray data processing. The utilities of other more complex normalization methods are yet to be evaluated.

Smoothing

As with time series data, the spectrum data contains a lot of noise. A smoothing process can help removing some superficial spikes (see more discussion below in the peak finding section), and thus help the peak finding process (Figure 6). It can be achieved by moving average, smoothing splines (Gans & Gill, 1984), Fourier transform filtering (Cameron & Moffatt, 1984), or wavelet filtering (Cancino-De-Greiff, Ramos-Garcia, & Lorenzo-Ginori, 2002).

Notice that excessive smoothing can change the shape of the peak: the peak will be shorter and broader with moving average operation. Extreme smoothing can destroy the peak features by generating a flat line.

Figure 5. The effect of scale normalization on spectra

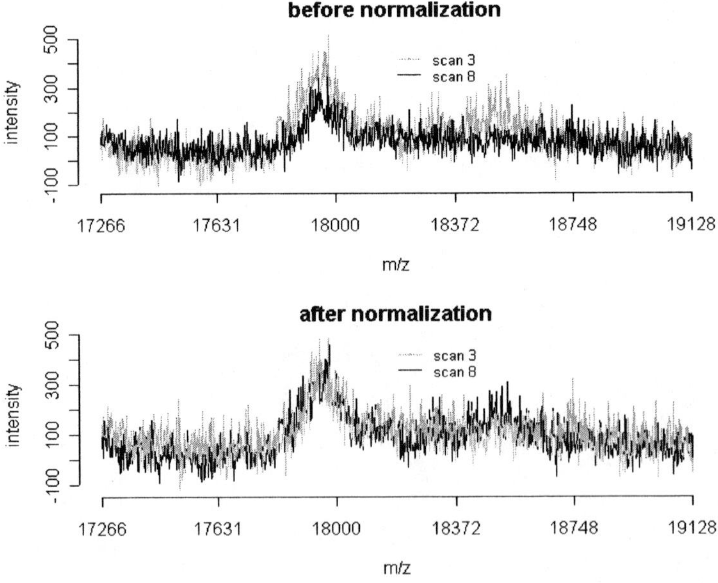

Figure 6. Smoothing (black line) the raw data (grey line) helps the identification of the peaks (dots)

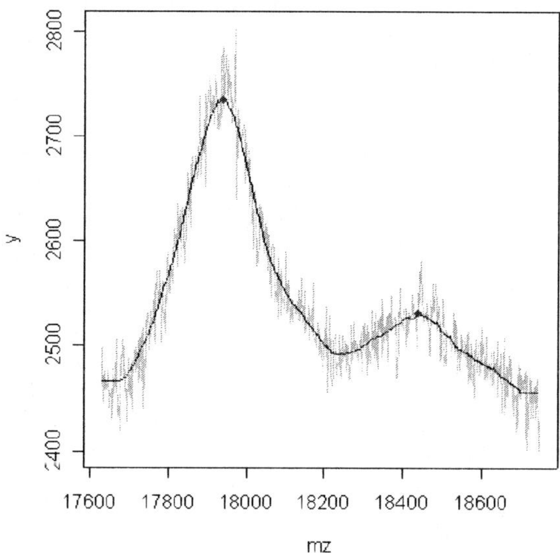

Feature Extraction

Each patient is described by a collection of m/z values. These raw m/z data can be used directly to drive the pattern analysis (Sorace & Zhan, 2003). However, since some spectra contain from one hundred thousand to one million data points, it is common to reduce the amount of data, and use more interesting spectrum features. Here we briefly discuss two widely used feature extraction methods in analytical chemistry for spectrum data.

Binning

The simplest possible method for reducing the information used in subsequent analyses involves binning the data. In this approach, all m/z values within a given window are condensed to one point. The window size used can be either fixed or varied. Domain-specific knowledge of the x-axis measurement accuracy can guide the selection the windowsize (Petricoin, Rajapaske, Herman, Arekani, Ross, & Johann, et al., 2004). For example, by a variable-length window, 350 000 raw data features were reduced to 7 084 features (Petricoin, Rajapaske, et al., 2004), before they were analyzed with genetic algorithm-based feature selection and self-organizing map (SOM) classification.

Notice that the discussion of binning here is in the context of feature reduction of x-axis, rather than the usual discussion of binning continuous input into ordinary variable of the y-axis.

Peak Finding

Theoretically, a protein is observed as a peak on the spectrum. Usually, peak height is used to drive the down-stream analysis. Other measurements of a peak include full width at half maximum (FWHM), peak location, and area under the curve (Figure 7).

Peaks can be found by the local maximum or by modeling peaks using a parametric model. In this context, however, the presence of noise can interfere with such algorithms for finding local maxima. Thus, it is often used together with adequate smoothing (Morrey, 1968). Parametric models can handle noise and outliers more elegantly with a model fitting process. However, their parametric assumptions can also in practice limit their utility (Scheeren, Barna, & Smit, 1985).

Data Mining and Analysis

After appropriate data preprocessing and feature extraction, proteomics data mining shares a lot of similarity with microarrays data mining. For preprocessing steps, general signal processing techniques are usually relevant as we have already discussed.

Depending on whether the patient outcomes are available at the model-building process, proteomics data mining can be classified as either a supervised learning (outcome is known) or an unsupervised learning (outcome is unknown) process (Figure 8).

Figure 7. A peak can be characterized with the following parameters: Peak location (m), peak height (h), full width at half maximum (FWHM), and area under the curve (AUC)

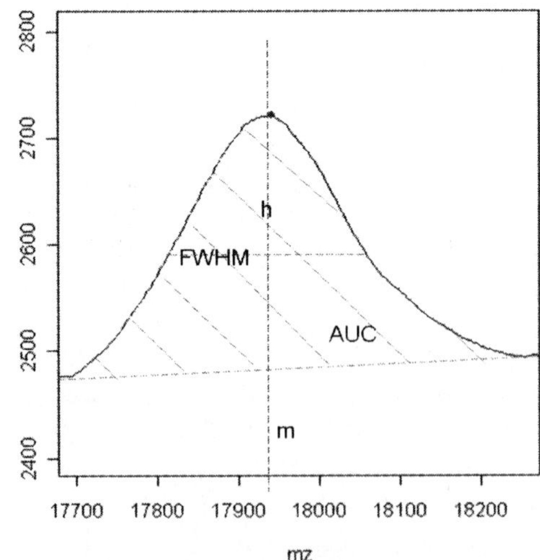

Figure 8. Predictive data mining: Proteomics measurements (features) of patients are linked to outcomes, such as cancer or norma; if the outcomes are not available at the time of the modeling process, it becomes an unsupervised machine learning problem

	Feature 1	Feature 2	...	Feature p
Patient 1				
Patient 2				
...				
Patient n				

→

Outcomes
Outcome 1
Outcome 2
...
Outcome n

These data mining methods are related to clinical applications in the following ways:

- Clustering can provide new insights into disease classification and can discover new subtypes of a disease.
- Supervised machine learning methods can automatically classify new measurements into predefined disease categories, such as cancer or normal. Survival modeling is a special kind of supervised analysis used in cancer prognostics, where the clinical outcome variable is time to death (Shoemaker & Lin, 2005).

A majority of data mining techniques have been applied to proteomics, we list some representative papers from 2002-2004 in Table 3.

Open Source and Commercial Tools

A commercial tool of ProteomeQuest (Correlogic Inc., Bethesda, MD) was used in (Petricoin, Rajapaske, et al., 2004c). It is a combination of genetic-algorithm-based feature selection and self-organizing-map-based pattern recognition. Excellent classification result was reported by using this tool (Petricoin, Ardekani, Hitt, Levine, Fusaro, & Steinberg, et al., 2002), although there are later controversies of this study (Baggerly, Morris, & Coombes, 2004; Petricoin, Fishman, Conrads, Veenstra, & Liotta, 2004).

Table 3. Representative data mining methods used in proteomics

Proteomics Study	Data Mining Methodology	Citation
Breast cancer	multivariate logistic regression	(Li et al., 2002)
Drug toxicity	genetic algorithm and SOM	(Petricoin, Rajapaske, et al., 2004)
Taxol resistance	artificial neural network	(Mian et al., 2003)
Renal carcinoma	decision tree	(Won et al., 2003)
Prostate cancer	boosted decision tree	(Qu et al., 2002)
Breast cancer	unified maximum separation analysis	(Li et al., 2002)
Colorectal cancer	support vector machine	(Yu et al., 2004)

There are several open source tools written in the R statistical language for analyzing proteomics data, including PROcess in the Bioconductor package (http://www.bioconductor.org), the mscalib at the CRAN distribution site (http://cran.r-project.org/), and the Rproteomics under the caBIG project (http://cabig.nci.nih.gov). Since these tools are all on the R computing platform, many modules can be mixed-and-matched for practical uses.

Summary

The State-of-the-Art

The expression level of a single protein, such as PSA or Ca125, has been used for clinical diagnostics of prostate and ovarian cancer. From a data mining point of view, these tests are univariate classifiers. With the advance of proteomics, we are able to measure the expression level of many proteins in one run. Thus, a more powerful multivariate pattern can be used for clinical diagnostics. However, this approach faces typical data mining challenges:

1. Accuracy and reproducibility of high-throughput measurements
2. Curse of high dimensionality: very large number of features vs. very small number of training samples
3. Concerns of over-fitting the model
4. Interpretability of the potential "black-box" models

Even with these obstacles, current researches have shown the early promises of clinical proteomics application (Carr, Rosenblatt, Petricoin, & Liotta, 2004; Petricoin, Wulfkuhle, Espina, & Liotta, 2004).

The Future of Proteomics

Experimental studies have suggested there is not necessarily a good correlation between the mRNA expression and the protein expression level (Greenbaum et al., 2003; Greenbaum, Luscombe, Jansen, Qian, & Gerstein, 2001). Thus, proteomics measurements provide additional information to microarray measurements. Combining proteomics with transcriptomics seems logical but aggravates the high-dimensionality problem of data mining and data analysis.

Integrating multiple 'omics' data— which include transcriptome, proteome, glycome, interactome, and metabolome— to study 'systems biology' is a major challenge in bioinformatics. Usually, information on different omics is available from separate data sources. Data fusion provides a way to combine information for data mining. Proteome adds a new dimension to this already complex problem (Yates, 2004). As the measurement platforms of proteomics getting mature and becoming more accurate, we will expect to see more discussion of data mining and data fusion challenges in system biology.

References

Aebersold, R., & Mann, M. (2003). Mass spectrometry-based proteomics. *Nature, 422*, 198-207.

Baggerly, K. A., Morris, J. S., & Coombes, K. R. (2004). Reproducibility of SELDI-TOF protein patterns in serum: Comparing datasets from different experiments. *Bioinformatics, 20*, 777-785.

Berndt, P., Hobohm, U., & Langen, H. (1999). Reliable automatic protein identification from matrix-assisted laser desorption/ionization mass spectrometric peptide fingerprints. *Electrophoresis, 20*, 3521-3526.

Cameron, D. G., & Moffatt, D. J. (1984). Deconvolution, derivation, and smoothing of spectra using Fourier-transforms. *Journal of Testing and Evaluation, 12*, 78-85.

Cancino-De-Greiff, H. F., Ramos-Garcia, R., & Lorenzo-Ginori, J. V. (2002). Signal denoising in magnetic resonance spectroscopy using wavelet transforms. *Concepts in Magnetic Resonance, 14*, 388-401.

Carr, K. M., Rosenblatt, K., Petricoin, E. F., & Liotta, L.A. (2004). Genomic and proteomic approaches for studying human cancer: Prospects for true patient-tailored therapy. *Human Genomics, 1*, 134-140.

Desiere, F., Deutsch, E. W., Nesvizhskii, A. I., Mallick, P., King, N. L., Eng, J. K., et al. (2005). Integration with the human genome of peptide sequences obtained by high-throughput mass spectrometry. *Genome Biology, 6*(1), R9.

Gans, P., & Gill, J.B. (1984). Smoothing and differentiation of spectroscopic curves using spline functions. *Applied Spectroscopy, 38*, 370-376.

Greenbaum, D., Colangelo, C., Williams, K., & Gerstein, M. (2003). Comparing protein abundance and mRNA expression levels on a genomic scale. *Genome Biology, 4*, 117.

Greenbaum, D., Luscombe, N. M., Jansen, R., Qian, J., & Gerstein, M. (2001). Interrelating different types of genomic data, from proteome to secretome: 'oming in on function. *Genome Research, 11*, 1463-1468.

Li, J., Zhang, Z., Rosenzweig, J., Wang, Y. Y., & Chan, D. W. (2002). Proteomics and bioinformatics approaches for identification of serum biomarkers to detect breast cancer. *Clinical Chemistry, 48*, 1296-1304.

Mian, S., Ball, G., Hornbuckle, J., Holding, F., Carmichael, J., Ellis, I., et al. (2003). A prototype methodology combining surface-enhanced laser desorption/ionization protein chip technology and artificial neural network algorithms to predict the chemoresponsiveness of breast cancer cell lines exposed to Paclitaxel and Doxorubicin under in vitro conditions. *Proteomics, 3*, 1725-1737.

Morrey, J. R. (1968). On determining spectral peak positions from composite spectra with a digital computer. *Analytical Chemistry, 40*, 905-914.

Orchard, S., Hermjakob, H., Julian, R. K., Jr., Runte, K., Sherman, D., Wojcik, J., Zhu, W., & Apweiler, R. (2004). Common interchange standards for proteomics data: Public availability of tools and schema. *Proteomics*, 4, 490-491.

Pedrioli, P. G., Eng, J. K., Hubley, R., Vogelzang, M., Deutsch, E. W., Raught, B., et al. (2004). A common open representation of mass spectrometry data and its application to proteomics research. *Nature Biotechnology, 22*(11), 1459-1466.

Petricoin, E., Wulfkuhle, J., Espina, V. & Liotta, L.A. (2004). Clinical proteomics: revolutionizing disease detection and patient tailoring therapy. *Journal of Proteome Research, 3*, 209-217.

Petricoin, E. F., Ardekani, A. M., Hitt, B. A., Levine, P. J., Fusaro, V. A., Steinberg, S. M., et al. (2002). Use of proteomic patterns in serum to identify ovarian cancer. *Lancet, 359*, 572-577.

Petricoin, E. F., Fishman, D. A., Conrads, T. P., Veenstra, T. D., & Liotta, L. A. (2004). Proteomic pattern diagnostics: Producers and consumers in the era of correlative science. *BMC Bioinformatics*. Retrieved February 1, 2006, from http://www.biomedcentral.com/1471-2105/4/24/comments#14454

Petricoin, E. F., Rajapaske, V., Herman, E. H., Arekani, A. M., Ross, S., Johann, D., et al. (2004). Toxicoproteomics: Serum proteomic pattern diagnostics for early detection of drug induced cardiac toxicities and cardioprotection. *Toxicologic Pathology, 32*, 122-130.

Prince, J. T., Carlson, M. W., Wang, R., Lu, P., & Marcotte, E. M. (2004). The need for a public proteomics repository. *Nature Biotechnology, 22*, 471-472.

Qu, Y., Adam, B. L., Yasui, Y., Ward, M. D., Cazares, L. H., Schellhammer, P. F., et al. (2002). Boosted decision tree analysis of surface-enhanced laser desorption/ionization mass spectral serum profiles discriminates prostate cancer from noncancer patients. *Clinical Chemistry, 48*, 1835-1843.

Rosenblatt, K. P., Bryant-Greenwood, P., Killian, J. K., Mehta, A., Geho, D., Espina, V., et al. (2004). Serum proteomics in cancer diagnosis and management. *Annual Review of Medicine, 55*, 97-112.

Scheeren, P. J. H., Barna, P., & Smit, H. C. (1985). A Software package for the evaluation of peak parameters in an analytical signal based on a non-linear regression method. *Analytica Chimica Acta, 167*, 65-80.

Shoemaker, J. S., & Lin, S. M. (2005). *Methods of microarray data analysis IV*. New York: Springer.

Sorace, J. M., & Zhan, M. (2003). A data review and re-assessment of ovarian cancer serum proteomic profiling. *BMC Bioinformatics, 4*, 24.

Won, Y., Song, H. J., Kang, T. W., Kim, J. J., Han, B. D., & Lee, S. W. (2003). Pattern analysis of serum proteome distinguishes renal cell carcinoma from other urologic diseases and healthy persons. *Proteomics, 3*, 2310-6.

Yates, J. R., 3rd. (2004). Mass spectrometry as an emerging tool for systems biology. *Biotechniques, 36*, 917-9.

Yu, J. K., Chen, Y. D., & Zheng, S. (2004). An integrated approach to the detection of colorectal cancer utilizing proteomics and bioinformatics. *World Journal of Gastroenterology, 10*, 3127-3131.

Chapter VI

Efficient and Robust Analysis of Large Phylogenetic Datasets

Sven Rahmann, Bielefeld University, Germany
Tobias Müller, University of Würzburg, Germany
Thomas Dandekar, University of Würzburg, Germany
Matthias Wolf, University of Würzburg, Germany

Abstract

The goal of phylogenetics is to reconstruct ancestral relationships between different taxa, e.g., different species in the tree of life, by means of certain characters, such as genomic sequences. We consider the prominent problem of reconstructing the basal phylogenetic tree topology when several subclades have already been identified or are well known by other means, such as morphological characteristics. Whereas most available tools attempt to estimate a fully resolved tree from scratch, the profile neighbor-joining (PNJ) method focuses directly on the mentioned problem and has proven a robust and efficient method for large-scale datasets, especially when used in an iterative way. We describe an implementation of this idea, the ProfDist software package, which is freely available, and apply the method to estimate the phylogeny of the eukaryotes. Overall, the PNJ approach provides a novel effective way to mine large sequence datasets for relevant phylogenetic information.

Introduction

Basics

Phylogenetic analyses aim to reconstruct ancestral relationships (the *phylogeny*) between different taxa, e.g., different species in the tree of life, by means of certain morphological or molecular *characters*. The latter ones are available from the increasing number of sequencing projects. Because of the wealth of data generated in these projects, we need efficient methods to mine the sequence databases for relevant phylogenetic information if we want to reconstruct the most likely phylogeny.

Over the last 50 years, researchers have proposed many methods for phylogenetic tree reconstruction, based on different concepts and models. Each method has its strengths and weaknesses. Distance-based methods, such as Neighbor-joining (Saitou and Nei, 1987) or improved variants thereof, e.g., WEIGHBOR (Bruno, Socci, & Halpern, 2000), BIONJ (Gascuel, 1997), FASTME (Desper & Gascuel, 2002), are relatively fast (the running times generally grows as a cubic polynomial of the number of taxa), but first reduce the information contained in the characters to a matrix of distances. Character-based methods, on the other hand, such as maximum parsimony (Camin & Sokal, 1965; Fitch, 1971), maximum-likelihood (Felsenstein, 1981), or Bayesian methods (e.g., MrBayes; Huelsenbeck & Ronquist, 2001; see Holder & Lewis, 2003, for a review) work directly with character data, but usually require an evaluation of super exponentially many tree topologies; therefore one reverts to heuristics. There seems to be no universally accepted best method, especially for large datasets.

All of the above-mentioned methods aim to estimate a fully resolved tree from scratch. In some cases, this results in more computational work than needs to be done, or even than the data can robustly support, as usually evidenced by low bootstrap values in the basal branching pattern of the tree. In several analyses, however, we are mainly interested in the basal branching pattern of known or clearly separated and fully supported subclades. In other words, given families of closely related sequences, what is the topology showing the relationships between these families? The problem arises for many large phylogenetic datasets. Müller, Rahmann, Dandekar, and Wolf (2004) recently published a case study on the Chlorophyceae (green algae). Further prominent large phylogenetic datasets in question are the given families of related sequences of, e.g., the Viridiplantae, the Metazoa, the eukaryotes, or even of all the species classified within the tree of life.

As the field of phylogenetics involves both the biological and the mathematical community, the language can sometimes be confusing. For the purpose of this article, we use the terms "subtree" and "subclade" synonymously; each subtree defines a monophyletic group of taxa, i.e., a grouping in which all species share a common ancestor, and all species derived from that common ancestor are included. We usually work with unrooted trees, i.e., connected undirected acyclic graphs, in which every internal node has exactly three neighboring nodes. The taxa are the leaves of the tree.

It is well known that there exist super-exponentially many unrooted tree topologies for N taxa, the precise number is $(2N-5)!! := (2N-5) \times (2N-7) \times \ldots \times 3$; see the book by Felsenstein (2003) for more details. For the purposes of illustration, we shall assume that there are five taxa A, B, C and D, and E, for which there are 15 unrooted tree topologies. If we already know that A and B form a subclade S, the number of possibilities reduces to 3 unrooted trees (e.g., C and S could be neighbors, along with the other neighboring pair D and E; we denote this tree by CS|DE. The other two trees would be DS|CE and ES|CD). So even a little additional knowledge leads to a significant reduction in the number of possible tree topologies, an important aspect when using cubic-time tree reconstruction algorithms. As an additional benefit, we observe below that the resulting basal trees tend to be more robust, i.e., are supported by higher bootstrap values, than they would be when reconstructed from scratch.

Seen from a broader perspective, the knowledge of a single split (i.e., a single edge of the tree that separates the taxa into two disjoint nonempty subsets) allows us to divide the large problem of reconstructing the tree on n taxa into two smaller problems, namely reconstructing trees on n_1 and n_2 taxa, respectively, with $n_1+n_2=n$.

The Challenge: Integrating Additional Knowledge

Interest in large phylogenies, e.g., the whole tree of life, or "only" the ancestral relationships among Eukaryotes, has been and still is increasing, but none of the standard methods mentioned above is well suited for datasets of this scale, at least for reconstructing a fully resolved tree from scratch. However, we often do have access to additional information that allows the definition of subclades. Thus, we face the challenge of integrating additional knowledge efficiently into the established tree reconstruction algorithms. A solution to the problem can be formulated in several ways; we shall now compare some of them to give the reader an overview.

The first possibility consists of keeping the whole set of taxa, but to restrict the set of allowable tree topologies to those that are consistent with the known monophyletic groups. Such approaches work best when used with character-based methods, e.g., in a maximum likelihood or in a Bayesian framework, since the constraints can be integrated in the heuristic topology search that has to be performed anyway. The PAUP* package (Swofford, 2003) implements such constraints, for example. This type of method works less efficiently with distance-based methods, since we do not save any distance computations.

A second possibility consists of solving many small problems, each with a reduced set of taxa, and to assemble the resulting (sub-)trees into a consistent supertree. There are two scenarios.

1. The subtrees are overlapping, i.e., they may share taxa, but propose inconsistent neighbor relationships, and the problem becomes to find a supertree that is maximally consistent with the set of subtrees for some measure of consistency. Such general supertree methods are reviewed by Bininda-Emonds (2004); the TREE-PUZZLE software (Schmidt & von Haeseler, 2003) that assembles a phylo-

genetic tree from its quartets (subtrees with four taxa) is an example. These methods do not require previous knowledge of monophyletic groups, but can benefit from it.

2. In the second scenario, all subtrees are disjoint and form a partition of the whole set of taxa. Such approaches are also called divide-and-conquer approaches; here knowledge of the partition of the taxa is required. The problem reduces to (a) representing the subtree by a new type of characters (it is not always represented as a simple molecular sequence), and (b) to find a method that allows tree reconstruction on this new set of characters.

To solve problem (a), it has been proposed to use consensus sequences, in other words, a majority vote of all sequences in the subtree, or the most parsimonious ancestral sequence, which is the sequence that requires the smallest total number of mutations in the subtree, or to derive the most likely ancestor in a likelihood setting, or to use a probabilistic model for the ancestral sequence. To solve problem (b) in the case of more complex characters than molecular sequences, one has the possibility to extend existing character-based methods or to first compute distances between the subtrees and then use an existing distance-based method.

The approach taken in this chapter consists of replacing the taxa in a known subclade by a single supertaxon, which is represented by a "probabilistic" sequence, also called a sequence profile (Gribskov, McLachlan, & Eisenberg, 1987). Then evolutionary distances can be estimated by a generalization of the standard maximum likelihood distance estimator — this is explained in detail below and represents our solution to problem (b) above. Subsequently, any distance-based reconstruction method can be used. In our implementation, we have chosen the Neighbor Joining algorithm; hence we call the resulting methop *profile neighbor-joining* (PNJ).

Naturally, other combinations of approaches would also work. Currently there is relatively little practical experience with supertree methods of the PNJ type, but we expect to see more research on the topic in the future.

The remainder of this chapter is organized as follows: After discussing the ideas behind PNJ in detail, as well as an automatic iterative variant, we present a freely available implementation, ProfDist (Friedrich, Wolf, Dandekar, & Müller, 2005), and as a biological application, a reconstructed tree of the eukaryotes. A discussion concludes this chapter.

Profile Neighbor-Joining

Profiles

A sequence *profile* is a stochastic model of a sequence family. It can also be pictured as a fuzzy or "smeared-out" sequence. Formally, a profile consists of a number of probability distribution vectors, one for each position. So each position k specifies its own nucleotide distribution $\alpha_k = (\alpha_{k,\mathbf{A}}, \alpha_{k,\mathbf{C}}, \alpha_{k,\mathbf{G}}, \alpha_{k,\mathbf{T}})$. Nucleotide sequences are thus special profiles (singleton families), where each α_k is given by one of the unit vectors $\mathbf{A} = (1, 0, 0, 0)$, $\mathbf{C} = (0, 1, 0, 0)$, and so on.

There exist different methods for estimating profiles from sequence families. For example, given a phylogenetic tree of sequence families, we could estimate the most likely profile at the root, i.e., the profile that maximizes the joint probability of all descendant sequences under a given evolutionary model. This idea is used, for example, in Felsenstein's maximum likelihood tree reconstruction method (Felsenstein, 1981). In the context of this chapter, however, we are more interested in the "center of gravity" of the sequence family distribution. Therefore, we use the position-specific relative character frequencies over all sequence family members. This results in a robust estimate that is independent of estimated subclade topologies; in fact, it corresponds to a star tree topology in the subclade, giving the same weight to all sequences. Naturally, sequence weighting schemes (e.g., see Krogh & Mitchison, 1995) can be used to obtain a different behavior and simulate different subclade topologies. Our choice is motivated by the attempt to find a "middle road" between simplicity, robustness, and accuracy; this will be discussed further below.

The profiles can now be interpreted as the taxa's "characters", and any method that supports "characters" with a continuous value range could directly be used to build a tree. For example, Felsenstein's maximum likelihood (ML) method easily accommodates profiles. For efficiency reasons, however, we consider it advantageous to use a distance-based method and thus require a way to estimate distances between sequence profiles.

Evolutionary Markov Processes

Maximum likelihood methods for distance estimation (also called correction formulas, because they map or "correct" the number of observed mutations to an evolutionary distance measured in units of percent of expected mutations) rely on a model of molecular substitution. Substitutions are modeled by an evolutionary Markov process (EMP) acting independently on each site of the sequence. Different models have been proposed and estimated for nucleotide and amino acid sequences (Lanave, Preparata, Saccone, & Serio, 1984; Müller & Vingron, 2000).

An EMP is uniquely described by its *starting distribution* π_0 and its rate matrix Q, called the *substitution model*. In the nucleotide case, π_0 is a row vector of nucleotide frequencies summing to 1, and Q is a 4x4 matrix, where for $i \neq j$, the entry Q_{ij} specifies the rate at which nucleotide i mutates into nucleotide j. We assume that all rates are positive, i.e., that all substitutions have nonzero probability in any time interval. The diagonal entry Q_{ii} is set to the negative sum of the other rates in row i, so each row sums to zero. We assume that Q has a unique stationary distribution π, i.e., a distribution that remains unchanged under the evolutionary dynamics defined by Q; it is the only probability row vector π that satisfies $\pi \cdot Q = 0$. Further, we assume that the process starts in its stationary distribution, so $\pi_0 = \pi$. We also need to fix a time unit: The rates are calibrated in such a way that per 100 time units, one substitution is expected to occur:

$$\Sigma_i \pi_i \Sigma_{j \neq i} Q_{ij} = -\Sigma_i \pi_i Q_{ii} = 1/100.$$

This condition can always be satisfied by re-scaling Q by a constant factor, if necessary.

Depending on its parameterization, the rate matrix is called Jukes-Cantor model (1 parameter, all rates are equal; Jukes & Cantor, 1969), Kimura model (2 parameters, with different rates for transitions and transversions; Kimura, 1980), or general time reversible (GTR) model (6 free parameters; Lanave et al., 1984). The parameters (such as the transition-transversion ratio in the Kimura model) must be estimated from the data. Since our main aim is to reconstruct the basal branching topology, we can use sequence data from all subclades to estimate a single overall rate matrix; different methods have been described in the literature (e.g., Müller & Vingron, 2000) or integrated into phylogenetic software (e.g., MrBayes; Huelsenbeck & Ronquist, 2001).

Distance Estimation with EMPs

We symbolically write $i \rightarrow j$ for the event that character i has been substituted by nucleotide j after some unknown time t. The probability of this substitution is given by the (i,j) entry of the time-t transition matrix P^t, which is related to the rate matrix Q via the matrix exponential $P^t = \exp(tQ)$. The entry P^t_{ij} is equal to the conditional probability that what is now character i will have become character j after time t.

To estimate the evolutionary distance between two sequences, we first compute a pairwise alignment (or take a 2-dimensional projection of a multiple alignment) and count the number of all nucleotide pair types in the alignment. Let N_{ij} be the number of observed events $i \rightarrow j$. In general, the higher the off-diagonal entries (mismatch counts) are in comparison to the diagonal entries (match counts), the larger is the evolutionary distance. As we are working with unrooted trees where we initially cannot infer the direction of time, we work with time-reversible models. In particular, we count substitutions in both directions, so seeing i aligned with j, we add a counter to both N_{ij} and N_{ji}, thus obtaining a symmetric count matrix.

A well-founded framework is given by the maximum likelihood principle. If the distance is t time units, the joint probability of all events is given by $\Pi_{i,j} (P^t_{ij})^{N_{ij}}$. We seek the value t maximizing this probability, or equivalently, the *log-likelihood* function,

$$L(t) = \Sigma_{i,j} N_{ij} \log(P^t_{ij}).$$

Note that $L(t)$ is a sum of log-probabilities $\log(P^t_{ij})$, weighted by their observed counts N_{ij}, the counts summing up to the total number $2n$ of sites in the alignment (the total number is $2n$ instead of n because we count reversibly). For general models Q, the solution of this one-dimensional maximization problem cannot be given in closed form, but is easily obtained by numerical methods.

We now generalize this estimator from sequences to profiles. A site need not contribute an integer count of 1 to a single substitution category, but fractional counts summing to 1 can be spread out over all 16 categories. Assume that the profiles at a particular site of the two families are given by $\alpha=(\alpha_A, \alpha_C, \alpha_G, \alpha_T)$ and $\beta=(\beta_A, \beta_C, \beta_G, \beta_T)$, respectively. Intuitively, if the sequence families consist of m_1 and m_2 sequences and we independently draw a sequence from each family, we observe the nucleotide pair (i, j) on average $m_1\alpha_i \cdot m_2\beta_j$ times, corresponding to a relative frequency of $\alpha_i\beta_j$. The product distribution mirrors the conditionally independent drawing from within the sequence families and does not imply that the families might be unrelated. Thus the total counts N_{ij} for event $i \rightarrow j$ are given by the sum over all n sites. Again, we enforce symmetry by counting again with exchanged α and α:

$$N_{ij} = \Sigma^{n}_{k=1} \ (\alpha_{k,i} \ \beta_{k,j} + \alpha_{k,j} \ \beta_{k,i})$$

Except for the computation of N_{ij}, the likelihood function $L(t)$ remains unchanged, and so does the numerical maximization over t. In this way, we obtain a distance matrix of evolutionary times between each pair of (super-)taxa, to which the neighbor-joining method can then be applied.

As is evident from the description above, all of the sequence profiles do need to be of the same length. This is a general requirement for molecular sequence data in phylogenetics; usually one computes a multiple sequence alignment of all sequences and for distance evaluation considers only columns without gap characters (some methods treat a gap like an ordinary fifth character besides the nucleotides). Of course, if multiple sequence alignments are computed separately for each subclade and then converted into profiles, the resulting profiles can be of different length. In such a case, we may apply evolutionary models that do take gaps into account to derive a maximum likelihood distance estimator, such as the Thorne-Kishino-Felsenstein model (see Steel & Hein, 2000). In our experiments, we have so far not used this option, but instead worked with a single multiple alignment of all taxa sequences, which automatically results in all profiles being of the same length.

Bootstrapping

In phylogenetics, bootstrapping is a widely accepted method to assess the robustness of an inferred tree topology. The topology is robust if it remains invariant after slight perturbations of the input data (i.e., the alignments and distances). Note that robustness is in no obvious way related to accuracy (see Chapter 14.9 in Ewens & Grant, 2001, for more careful distinctions). The following procedure has become the standard one (Felsenstein, 1985; Hillis & Bull, 1993; Efron, Halloran, & Holmes, 1996): From the initial sequence alignment A, several so-called *bootstrap replicates* $A^*_1,\ldots, A^*_B$, with B » 1000 to 10000, are created by randomly resampling entire columns of A. That is, each A^* has the same length as A, but contains each column of A zero times, once, or several times. Each A^* is now considered a new input alignment to the tree reconstruction algorithm. From the B estimated bootstrap trees a so-called consensus tree T is derived (e.g., a majority rule consensus tree; see, e.g., Felsenstein, 2003).

Each edge of T is then annotated with its bootstrap value, i.e., the percentage of bootstrap trees that share the same edge. Edges with bootstrap values above 75% are generally considered as reliable. If low bootstrap values are encountered, it probably makes sense to aim for a less resolved tree, i.e., to allow multifurcating inner nodes.

We observe that when we estimate a tree directly from all taxa, some bootstrap values in the basal branching pattern tend to drop, possibly because of a single taxon "jumping" from one family into a different one in several bootstrap replicates, even if it is initially obvious (e.g., from additional data) where the taxon belongs to within the phylogeny. For the same reason, we may expect that using family profiles would make the basal branching pattern more robust; and this expectation is confirmed by our experiments.

Iterative Profile Neighbor-Joining

The basic PNJ method requires defining monophyletic groups in the initial set of taxa using any kind of additional knowledge; the sequences within a group are then converted into profiles, and finally distances between pairs of profiles can be computed. It would be even more advantageous, however, to have a fully automatic approach. We therefore apply the PNJ method iteratively. In the first step, all taxa's sequences are used to produce a tree. Then the algorithm finds all maximal subtrees, in which all bootstrap values exceed a pre-defined threshold (as mentioned above, a good choice is 75%). The sequences of each so defined subtree are then converted into a profile, and the next iteration starts. This procedure continues until no further profiles are created. To save time in the first iteration, almost identical sequences are initially grouped together and immediately converted into a profile; this behavior can be tuned by specifying a so-called identity threshold.

Software

A user-friendly software package called ProfDist (Friedrich et al., 2005) implements the PNJ and iterative PNJ methods on nucleotide sequence data. Windows, Linux, and Macintosh versions of ProfDist are available at http://profdist.bioapps.biozentrum.uni-wuerzburg.de and come with a graphical user interface.

Subclades for profile formation can be chosen manually or automatically based on bootstrap confidence values. ProfDist implements the BIONJ variant (Gascuel, 1997) of the neighbor-joining algorithm and uses the iterative version of the PNJ method to automatically build a tree if no subclades are chosen manually. Standard tree viewers, such as TREEVIEW (Page, 1996) or ATV (Zmasek & Eddy, 2001), can be used to visualize the output.

The main feature of ProfDist is its efficient and one-stop-shop implementation of the PNJ method. The different steps (drawing of bootstrap samples, distance computation, tree estimation via BIONJ, consensus tree formation) can be started separately using the buttons in the ProfDist main window (see Figure 1).

Figure 1. ProfDist user interface: Sequences can be loaded in FASTA or EMBL format; trees are output in NEWICK format

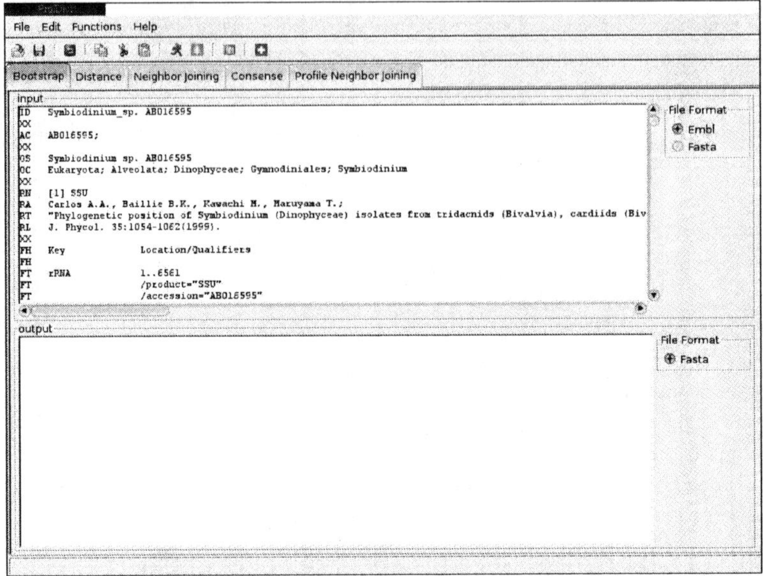

ProfDist supports different file formats for both trees and sequences (e.g., NEWICK, FASTA, EMBL and PHYLIP). The essential parameters of ProfDist are (1) the number of bootstrap replicates, (2) the distance estimation method, possibly supplemented with a user-defined substitution model, and (4) the PNJ agglomeration procedure. ProfDist implements the JC (Jukes & Cantor, 1969), K2P (Kimura, 1980), GTR (Lanave et al., 1984) distance estimation methods, and the Log-Det transformation (Barry & Hartigan, 1987).

For the iterative PNJ variant, there are two further parameters: (1) the minimal bootstrap value that an estimated monophyletic group has to achieve before it is considered trustworthy and represented as a profile in the next iteration, and (2) an identity threshold for immediate profile formation in the first iteration.

Nucleotide ProfDist is very time and space efficient: Nucleotide alignments are handled in an encoded diff-like format that makes it easy to count the number of nucleotide substitutions between any pair of sequences in a multiple alignment.

A Tree of the Eukaryotes Based on 18S rRNA

From the European ribosomal rRNA database (Wuyts, Perriere, & Van de Peer, 2004), we obtained all available eukaryotic 18S SSU (small subunit) rRNA sequences in aligned

Figure 2. A tree of the eukaryotes computed with ProfDist, based on SSU rRNA data: The upper part shows the unrooted tree; the lower part shows a rooted tree with bootstrap support values (numbers in square brackets denote the number of taxa in the respective subtree)

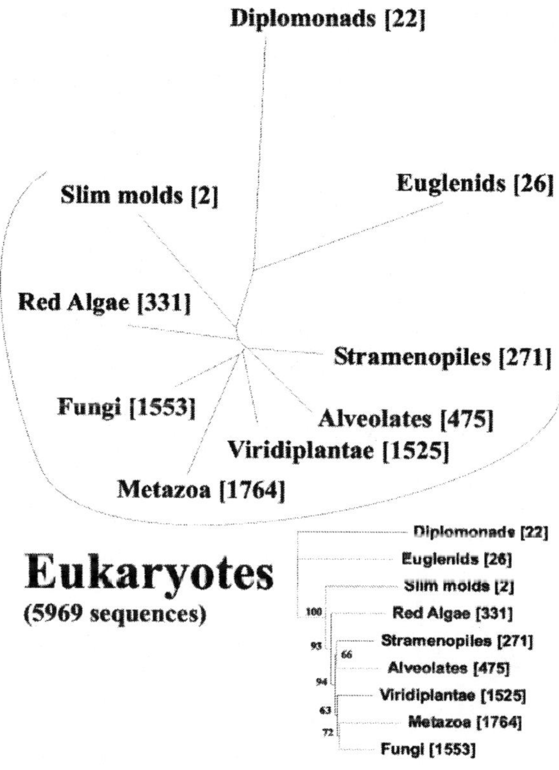

format. Using the Jukes-Cantor substitution model, an initial identity threshold of 90%, and a bootstrap confidence threshold of 50%, we ran ProfDist with the iterative version of the PNJ algorithm, after we had generated profiles on predefined monophyletic groups (Metazoa, Fungi, Viridiplantae, Alveolates, Stramenopilates, Red algae, Slim molds, Euglenids and Diplomonads). The resulting tree is shown in Figure 2. It corresponds to trees computed from SSU rRNA with other methods (van de Peer, Baldauf, Doolittle, & Meyer, 2000), but has been computed within only a few minutes and has high bootstrap support values.

Summary

We have presented the profile neighbor-joining (PNJ) approach, in which known or previously inferred subclades are converted into sequence profiles. As discussed in the introduction, other approaches are possible: For example, we might use one representative sequence from each subclade, or estimate the most likely sequence in the subclade root. The profile-based approach appears preferable because it integrates information from all family members; each member sequence is weighted equally, i.e., the subtree topology is ignored in this approach. This could be changed through sequence weighting schemes, but we may question whether this is desirable: We have to take into account the possibility that our sequence alignment contains fast evolving species or strong variation over sites. Then it could happen that one or more sequences in the subtree are connected accidentally to the rest by a Long Branch Attraction (LBA) phenomenon. This would result in a very strong signal of this sequence in the weighted profile generated from these sequences. In this sense, our approach is more robust. Also, knowledge of the intra-subclade topologies is not necessary to construct the profile. However, this robustness and simplicity come at a price: We must expect a drop in accuracy, i.e., in simulated experiments, the true basal branching pattern may be reconstructed less often if subclade topologies differ extremely from star trees, because of the resulting bias in the profiles. Interestingly, though, we could not observe such an accuracy drop in our simulations (Müller et al., 2004), indicating that even this simple profile construction method works well in practice. Further studies, especially comparing the relative efficiency of different tree estimators, would be an interesting research topic for the future.

Profile distances could be viewed more or less as averaged sequence distances, but there is one difference between profile distances and average distances: Average distances are a mean of maximum-likelihood distances of sequences between sequences of both groups, whereas the profile distance is a maximum-likelihood estimation between the "mean" of the sequences in both groups. In this sense, using profile distances results in a more computationally efficient estimator and leads to a more robust formulation, compared to consensus methods.

As is well known, if the distances between a set of objects (taxa or subclades) are additive (i.e., if they fulfill the four-point condition, see Felsenstein (2003) for an in-depth discussion), there exists a unique unrooted tree representing these distance, and the NJ method will recover it exactly. Even if the distance data is not tree-like (i.e., far from being additive), the NJ method always reconstructs a tree, but the meaning of this tree is not obvious. Such trees will tend to have relatively low bootstrap support since their topology is not well supported by the data.

We observe that trees constructed with PNJ tend to have much better bootstrap support than trees constructed by other methods. However, this by itself is not necessarily of biological significance. The central question is whether the PNJ method is more accurate than other methods, i.e., whether it reconstructs the true tree topology more often. While this question is difficult to answer for real large datasets, where the true topology is usually unknown, it can be answered for simulated datasets: Based on the Robinson-Foulds distance measure (Robinson & Foulds, 1981) between true and estimated topology, Müller et al. (2004) have shown that PNJ indeed outperforms NJ in terms of accuracy.

Many questions remain unanswered about PNJ and related methods at this point. Is it statistically consistent like maximum likelihood methods in the case of tree-like data, or how do we need to modify the profile computation procedure to make it consistent? How long do the sequences need to be to recover the true tree with high probability, in other words, what is the convergence rate? Can it be boosted by general principles like the disc-covering method (DCM; Huson, Nettles, & Warnow, 1999)? How will improving the convergence rate affect the robustness against single outlier (e.g., fast-evolving) sequences or errors in the sequence alignments? These are questions of general interest that are increasingly investigated by the scientific community.

In the future, we plan to extend the PNJ software to handle protein alignment data, to automatically estimate the substitution model from the data (up to now it has to be specified by the user), and to include the possibility of rate variation over sites into the distance estimators. We cordially invite the reader to try out the latest version at http://profdist.bioapps.biozentrum.uni-wuerzburg.de.

References

Barry, D., & Hartigan, J. (1987). Asynchronous distance between homologous DNA sequences. *Biometrics, 43*, 261-276.

Bininda-Emonds, O. R. P. (2004). The evolution of supertrees. *Trends in Ecology and Evolution, 19*(6), 315-322.

Bruno, W. J., Socci, N. D., & Halpern, A. L. (2000). Weighted neighbor joining: A likelihood-based approach to distance-based phylogeny reconstruction. *Molecular Biology and Evolution, 17*, 189-197.

Camin, J., & Sokal, R. (1965). A method for deducing branching sequences in phylogeny. *Evolution, 19*, 311-326.

Desper, R., & Gascuel, O. (2002). Fast and accurate phylogeny reconstruction algorithms based on the minimum-evolution principle. *Journal of Computational Biology, 19*, 687-705.

Efron, B., Halloran, E., & Holmes, S. (1996). Bootstrap confidence levels for phylogenetic trees. *PNAS, 93*, 7085-7090.

Ewens, W. J., & Grant, G. R. (2001). *Statistical Methods in Bioinformatics*. New York: Springer-Verlag

Felsenstein, J. (1981). Evolutionary trees from DNA sequences: A maximum likelihood approach. *Journal of Molecular Evolution, 17*, 368-376.

Felsenstein, J. (1985). Confidence limits on phylogenies: An approach using the bootstrap. *Evolution, 39*, 783-791.

Felsenstein, J. (2003). *Inferring Phylogenies*. Sunderland, MA: Sinauer Associates.

Fitch, W.M. (1971). Toward defining the course of evolution: Minimum change for a specified tree topology. *Systematic Zoology, 20*, 406-416

Friedrich, J., Wolf, M., Dandekar, T., & Müller, T. (2005): ProfDist: A tool for the construction of large phylogenetic trees based on profile distances. *Bioinformatics, 21*, 2108-2109.

Gascuel, O. (1997). BIONJ: An improved version of the NJ algorithm based on a simple model of sequence data. *Molecular Biology and Evolution, 14*, 685-695.

Gribskov, M., McLachlan, A., & Eisenberg, D. (1987). Profile analysis: Detection of distantly related proteins. *PNAS, 84*, 4355-4358.

Hillis, D., & Bull, J. (1993). An empirical test of bootstrapping as a method for assessing confidence in phylogenetic analysis. *Systematic Biology, 42*, 182-192.

Holder, M., & Lewis, P. (2003). Phylogeny estimation: Traditional and Bayesian approaches. *Nature Reviews, 4*, 275-284.

Huelsenbeck, J., & Ronquist, F. (2001). MRBAYES: Bayesian inference of phylogeny. *Bioinformatics, 17*, 754-755.

Huson, D. H., Nettles, S. M., & Warnow, T. J. (1999). Disk-covering, a fast-converging method for phylogenetic tree reconstruction. *Journal of Computational Biology, 6*, 369-386.

Jukes, T., & Cantor, C. R. (1969). Evolution of protein molecules. In H. Munro (Ed.), *Mammalian protein metabolism* (pp. 21-132). New York: Academic Press.

Kimura, M. (1980). A simple method for estimating evolutionary rate of base substitutions through comparative studies of nucleotide sequences. *Journal of Molecular Evolution, 16*, 111-120.

Krogh, A., & Mitchison, G. (1995). Maximum entropy weighting of aligned sequences of proteins or DNA. In C. Rawlings, D. Clark, R. Altman, L. Hunter, T. Lengauer, & S. Wodak (Eds.), *Proceedings of the Third International Conference on Intelligent Systems for Molecular Biology* (pp. 215-221). Menlo Park, CA: AAAI Press.

Lanave, C., Preparata, G., Saccone, C., & Serio, G. (1984). A new method for calculating evolutionary substitution rates. *Journal of Molecular Evolution, 20*, 86-93.

Müller, T., Rahmann, S., Dandekar, T., & Wolf, M. (2004). Accurate and robust phylogeny estimation based on profile distances: a study of the Chlorophyceae (Chlorophyta). *BMC Evolutionary Biology, 4*(20).

Müller, T., & Vingron, M. (2000). Modeling amino acid replacement. *Journal of Computational Biology, 7*, 761-776.

Page, R. D. M. (1996). TREEVIEW: An application to display phylogenetic trees on personal computers. *Computer Applications in the Biosciences, 12*, 357-358.

Robinson, D., & Foulds, L. (1981). Comparison of phylogenetic trees. *Mathematical Biosciences, 53*, 131-147.

Saitou, N., & Nei, M. (1987). The neighbor-joining method: A new method for reconstructing phylogenetic trees. *Journal of Molecular Evolution, 4*, 406-425.

Schmidt, H. A., & von Haeseler, A. (2003). Maximum-likelihood analysis using TREE-PUZZLE. In A. D. Baxevanis, D. B. Davison, R. D. M. Page, G. Stormo, & L. Stein (Eds.), *Current protocols in bioinformatics* (Unit 6.6). New York: Wiley and Sons.

Steel, M., & Hein, J. (2001). Applying the Thorne-Kishino-Felsenstein model to sequence evolution on a star-shaped tree. *Applied Mathematics Letters, 14*, 679-684.

Swofford, D. L. (2003). *PAUP*: Phylogenetic Analysis Using Parsimony (*and other methods), Version 4*. Sunderland, MA: Sinauer Associates.

van de Peer, Y., Baldauf, S. L., Doolittle, W. F., & Meyer, A. (2000). An updated and comprehensive rRNA phylogeny of (crown) eukaryotes based on rate-calibrated evolutionary distances. *Journal of Molecular Evolution, 51*, 565-76.

Wuyts, J., Perriere, G., & Van de Peer, Y. (2004). The European ribosomal RNA database. *Nucleic Acids Research, 32,* 101-103. Retrieved February, 2006, from http://www.psb.ugent.be/rRNA

Zmasek, C., & Eddy, S. (2001). ATV: display and manipulation of annotated phylogenetic trees. *Bioinformatics, 17*, 383-384.

Chapter VII

Algorithmic Aspects of Protein Threading

Tatsuya Akutsu, Kyoto University, Japan

Abstract

This chapter provides an overview of computational problems and techniques for protein threading. Protein threading is one of the most powerful approaches to protein structure prediction, where protein structure prediction is to infer three-dimensional (3-D) protein structure for a given protein sequence. Protein threading can be modeled as an optimization problem. Optimal solutions can be obtained in polynomial time using simple dynamic programming algorithms if profile type score functions are employed. However, this problem is computationally hard (NP-hard) if score functions include pairwise interaction preferences between amino acid residues. Therefore, various algorithms have been developed for finding optimal or near-optimal solutions. This chapter explains the ideas employed in these algorithms. This chapter also gives brief explanations of related problems: protein threading with constraints, comparison of RNA secondary structures and protein structure alignment.

Introduction

Inference and mining of functions of genes is one of the main topics in bioinformatics. Protein structure prediction provides useful information for that purpose because it is known that there exists close relationship between structure and function of a protein, where protein structure prediction is a problem of inferring three-dimensional structure of a given protein sequence. Computational inference of protein structure is important since determination of three-dimensional structure of a protein is much harder than determination of its sequence.

There exist various kinds of approaches for protein structure prediction (Clote & Backofen, 2000; Lattman, 2001; Lattman, 2003). Ab initio approach tries to infer structure of a protein based on the basic principles (e.g., energy minimization) in physics. In this approach, such techniques as molecular dynamics and Monte Carlo simulations have been employed. Homology modeling approach tries to infer structure of a protein using structure of a homologous protein (i.e., a protein whose structure is already known and whose sequence is very similar to the target protein sequence). In this approach, backbone structure of a protein is first computed from structure of a homologous protein and then the whole structure is computed by using molecular dynamics and/or some optimization methods. Secondary structure prediction approach does not aim to infer three-dimensional structure. Instead, it tries to infer which structural class (α, β, others) each residue belongs to. Such information is believed to be useful for inference of three-dimensional structure and/or function of a protein. In secondary structure prediction approach, various machine learning methods have been employed, which include neural networks and support vector machines.

Protein threading is another major approach for protein structure prediction. In this approach, given an amino acid sequence and a set of protein structures (structural templates), a structure into which the given sequence is most likely to fold is computed. In order to test whether or not a sequence is likely to fold into a structure, an alignment (i.e., correspondence) between spatial positions of a 3-D structure and amino acids of a sequence is computed using a suitable score function. That is, an alignment which minimizes the total score (corresponding to the potential energy) is computed. This minimization problem is called the protein threading problem. Though there exists some similarity between protein threading and homology modeling, these are usually considered to be different: alignment between a target sequence and a template structure is computed in protein threading whereas alignment between two sequences is computed in homology modeling.

Many studies have been done on protein threading. Most of them focus on improvement of practical performances using heuristic and/or statistical techniques, and few attentions had been paid to algorithmic aspects of protein threading. Since protein threading is NP-hard in general, heuristic algorithms have been widely employed, which do not necessarily guarantee optimal solutions. However, recent studies (Xu, Xu, Crawford, & Einstein, 2000; Xu, Li, Kim, & Xu, 2003) suggested that it is possible to compute optimal solutions in reasonable CPU time for most proteins and computation of optimal threadings is useful for improving practical performances of protein threading. Furthermore, there exist several important problems in bioinformatics, which are closely related to protein threading. Therefore, in this chapter, we overview algorithmic aspects of protein threading and related problems.

This chapter is organized as follows. First, we formally define the protein threading problem and show NP-hardness of protein threading. Then, after briefly reviewing heuristic methods, we describe three exact approaches for computing optimal threading: branch-and-bound approach (Lathrop & Smith, 1996), divide-and-conquer approach (Xu, Xu, & Uberbacher, 1998) and linear programming approach (Xu et al., 2003). Next, we briefly explain a variant of protein threading (protein threading with constraints) and related problems (comparison of RNA secondary structures and protein structure alignment). Finally, we conclude with future directions.

Protein Threading Alignment Between Sequence and Structure

As mentioned in the introduction, protein threading is a problem of computing an alignment between a target sequence and a template structure. First we define a usual alignment for two sequences. Let Σ be the set of amino acids (i.e., $|\Sigma|=20$). Let $s=s_1 \ldots s_m$ and $t=t_1 \ldots t_n$ be strings over Σ. An alignment between s and t is obtained by inserting gap symbols ('-') into or at either end of s and t such that the resulting sequences s' and t' are of the same length l, where it is not allowed for each i that both s_i' and t_i' are gap symbols.

For example, consider two sequences $s=$ LDVQWAVDEGDKVV, $t=$ DVQWSVEKRHGDKLVLT. Then, the followings are examples of alignments:

AL_1	AL_2	AL_3
LDVQWAVDE---GDKVV--	LDVQWAVDE---GDK-VV-	LDVQWAVDE---GDKVV
-DVQWSV-EKRHGDKLVLT	-DVQWSV-EKRHGDKLVLT	VDQWSVEKRHGDKLVLT

Let $f(x,y)$ be a score function from $\Sigma \times \Sigma$ to the set of real numbers that satisfies $f(x,y)=f(y,x)$. We extend $f(x,y)$ to include gap symbols by defining $f(x,-)=f(-,y)=-d$ for all x, y in Σ, where $-d$ ($d>0$) is a penalty per gap. The score of alignment (s',t') is defined by:

$$score(s',t') = \sum_{i=1}^{l} f(s'_i, t'_i).$$

An optimal alignment is an alignment having the maximum score. Suppose that $f(x,x)=10$ and $f(x,y)=-10$ for all $x \neq y$, and $d=10$. Then, the scores of AL_1, AL_2 and AL_3 are 10, 10 and

-150, respectively. In this case, both AL_1 and AL_2 are optimal alignments whereas AL_3 is a non-optimal alignment.

In this example, we assumed linear gap cost. Though affine gap cost should be used in practice, we use linear gap cost in this chapter for simplicity. However, all the results in this chapter can be extended for affine gap cost.

It is well-known that optimal alignments can be computed in O(mn) time using a simple dynamic programming algorithm. The following procedure computes the score of an optimal alignment:

$D[i,0] \leftarrow i,$

$D[0,j] \leftarrow j,$

$D[i,j] \leftarrow \max(D[i-1,j]-d, D[i,j-1]-d, D[i-1,j-1]+f(s_i,t_j)),$

where $D[i,j]$ corresponds to the optimal score between $s_1s_2...s_i$ and $t_1t_2...t_j$. An optimal alignment can also be obtained from this matrix by using the traceback technique (Clote & Backofen, 2000).

Next, we define the protein threading problem (see Figure.1). Recall that in protein threading, we compute an alignment between a target sequence and a template structure. Let $s = s_1 ... s_m$ be a target sequence over Σ. Let $t = t_1...t_n$ be a template protein structure, where t_i is the i-th residue in t. We can regard t as as a sequence of Cα(or Cβ) atoms of the protein structure. Then, a threading between s and t is an alignment between s and t.

Various score functions have been used in protein threading. One of the simplest score functions is a profile. A profile f_t is defined for a template structure t, and is a function

Figure 1. Threading between a target sequence s *and a template structure t. Shaded residues are aligned with gap symbols. In this case, s'*=MTLC-VERLD-E *and* $t' = t_1\, t_2$ - $t_3\, t_4\, t_5\, t_6$ - - $t_7\, t_8\, t_9$.

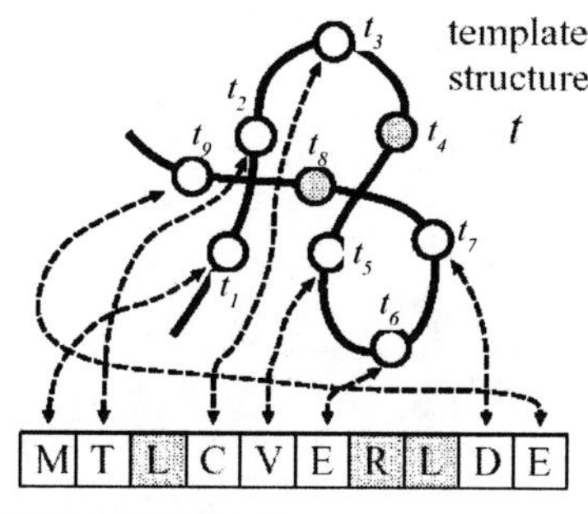

from $(\Sigma \cup \{-\}) \times \{t_1,...,t_n,-\}$ to the set of reals. Then, the score of a threading (s',t') is defined by:

$$score(s',t') = \sum_{i=1}^{l} f_t(s'_i, t'_i).$$

It should be noted that s'_i is aligned to t'_i where t'_i denotes some residue in t or a gap. For a profile type score function, an optimal threading can be computed in O(mn) time as in the case of sequence alignment. Various types of profile-type score functions have been proposed. The 3-D profile proposed by Bowie, Luthy, & Eisenberg (1991) is one of the earliest profiles. A well-known sequence search software PSI-BLAST (Altschul et al., 1997) employs a kind of profiles. Sequence alignment using an HMM (Hidden Markov Model) can also be considered as alignment with profiles (Durbin,Eddy, Krogh, & Mitchison , 1998). Recently, profile-profile alignment is regarded as a powerful method for protein threading, in which a kind of profile is used also for s instead of a sequence. An optimal alignment can still be computed in O(mn) time in this case. Edgar and Sjolander (2004) compared 23 types of score functions for profile-profile alignment.

In the above, score functions do not include pairwise intersection preferences. However, it is well-known that pairwise interactions between amino acid residues play a very important role in the process of protein folding (i.e., folding into a three-dimensional protein structure). Therefore, various types of score functions have been proposed in which pairwise interaction preferences are taken into account. In this case, the score function basically consists of two terms: $f_t(x,y)$ and $g_t(x,y,t_i,t_j)$, where $g_t(x,y,t_i,t_j)$ is a function from $\Sigma \times \Sigma \times \{t_1,...,t_n\} \times \{t_1,...,t_n\}$ to the set of reals. In a simplest case, g_t can be a contact potential defined by

$$g_t(x,y,t_i,t_j) = \begin{cases} 0, & \text{if } dist(t_i,t_j) > \Theta, \\ g_0(x,y), & \text{otherwise,} \end{cases}$$

where $dist(t_i,t_j)$ is the Euclidean distance between Cα atoms of t_i and t_j, Θ is a threshold, and $g_0(x,y)$ is a function from $\Sigma \times \Sigma$ to the set of reals. Then, the score of a threading (s',t') is defined by

$$score(s',t') = \sum_{i=1}^{l} f_t(s'_i, t'_i) + \sum_{i<j} g_t(s'_i, s'_j, t'_i, t'_j),$$

where the sum in the second term is taken over all pairs of i,j such that none of s'_i, s'_j, t'_i, t'_j is a gap symbol. It is worthy to note that alignment problems (including threading with profiles) are usually defined as maximization problems whereas threading problems with pair score functions are usually defined as minimization problems. In this chapter, we basically follow this tradition.

Hardness of Threading with Pair Score Functions

As previously shown, protein threading with profiles can be solved in O(mn) time. On the other hand, it is known that protein threading with pair score functions is NP-hard (Lathrop, 1994). In the following, we present a simple proof based on Akutsu and Miyano (1999).

Theorem 1.

Protein threading with pair score functions is NP-hard.

Proof. We use a polynomial time reduction from MAX CUT, where MAX CUT is a well-known NP-hard problem (Garey & Johnson, 1979). Recall that MAX CUT is, given an undirected graph $G(V,E)$, to find a subset V' of V maximizing the cardinality of the cut (i.e., the number of edges between V' and $V-V'$). From $G(V,E)$, we construct an instance of protein threading in the following way (see also Fig. 2) ,where we only consider two types of amino acids (Alanine (A) and Leucine (L)). Let $V=\{v_1,v_2,\ldots,v_n\}$. Then, we let $t=t_1\ldots t_n$ (we identify v_i with t_i) and let s be n repetitions of substring AL. Next, we define $f_t(x,y)=0$ for all x, y, and we define $g_t(x,y,t_i,t_j)$ by

$$g_t(x,y,t_i,t_j) = \begin{cases} -1, & \text{if } i<j,\ \{v_i,v_j\}\in E, \text{ and } x\neq y, \\ 0, & \text{otherwise.} \end{cases}$$

Then, each threading corresponds to a cut by letting $V'=\{v_i|$ A is assigned to $t_i\}$. Moreover, the score of the threading corresponds to the cardinality of the

Figure 2. Reduction from MAXCUT *to protein threading. A set of vertices aligned to* A (Alanine) *corresponds to* V' *and a set of vertices aligned to* L (Leucine) *corresponds to* $V-V'$.

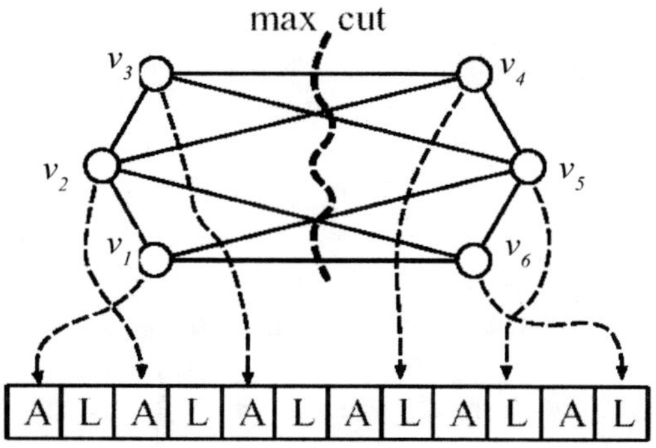

cut where the sign of the score should be inverted. Since the reduction can be clearly done in polynomial time, protein threading with pair score functions is NP-hard. Q.E.D.

Akutsu and Miyano (1999) extended this proof for showing strong hardness results on finding approximate solutions. They also developed approximation algorithms for special cases.

Algorithms for Threading with Pair Score Functions

Although NP-hardness of protein threading was proven, several approaches can be considered. One approach is to develop exact algorithms that guarantee optimal solutions using such techniques as branch-and-bound. Though such an approach is desirable, all of existing algorithms are somewhat complicated, to be explained later. Another approach is to develop heuristic algorithms. Though optimal solutions are not necessarily guaranteed in this approach, some of existing algorithms are very simple and have reasonable performances. Since the second approach is much simpler, we begin with the second one.

Heuristic Algorithms

Here, we briefly overview two well-known heuristic algorithms for protein threading with pair score functions: frozen approximation (Godzik & Skolnick, 1992) and double dynamic programming (Jones, Taylor, & Thornton, 1992).

The idea of frozen approximation is very simple. It replaces $g_t(x,y,t_i,t_j)$ by $g_t(x,l(t_j),t_i,t_j)$ where $l(t_j)$ is the amino acid type of the j-th residue in the template structure. That is, we use amino acid type of the template structure for y instead of using amino acid type of the target sequence. If we use $g_t(x,l(t_j),t_i,t_j)$ in place of $g_t(x,y,t_i,t_j)$, the optimal solution can be computed by using a dynamic programming algorithm similar to that for protein threading with profiles. It should be noted that this optimal solution is different from that of the original threading problem, and there is no theoretical guarantee on the accuracy of the solution obtained by using frozen approximation. However, frozen approximation has been adopted by many protein threading programs because it is very easy to implement.

The double dynamic programming technique was originally developed for computing structure alignment between two protein structures. It uses two levels of dynamic programming: upper level dynamic programming and lower level dynamic programming. Lower level dynamic programming is performed to compute the score $F[i,j]$ which indicates how likely it is that the pair (s_i,t_j) is on an optimal alignment. Then, upper level dynamic programming is performed as in sequence alignment using $F[i,j]$ as a score function.

Exact Algorithms

Though protein threading is NP-hard in general, the sizes of real proteins is limited (usually each protein consists of less than several hundred residues) and thus there exist chances that practically efficient exact algorithms can be developed. Indeed, several algorithms have been developed which can compute optimal threadings for proteins consisting of up to several hundred residues. Here, we briefly review important three of such algorithms: a branch-and-bound algorithm (Lathrop & Smith, 1996), PROSPECT (Xu et al., 2000), and RAPTOR (Xu et al., 2003).

Before presenting algorithms, we simplify the threading problem (see also Figure. 3). It is known that gaps are seldom inserted in the core regions of proteins, where a core region is either an α–helix or a β-strand (Lathrop & Smith, 1996). Therefore, we assume that gaps are not inserted into core regions (either in the target sequence or in the template structure). Let $C_1,\ldots,C_M$ be the core regions of the template structure. Since it is not difficult to modify the algorithms so that terms of $f_t(x,y)$ can be taken into account, we assume that the score consists of terms given by $g_t(x,y,t_i,t_j)$. Moreover, we consider pairwise interaction terms among residues only in core regions because it is considered that pairwise interactions among core regions are much more important than the other pairwise interactions. Let p_i denote the position of s to which the first position of C_i is aligned. Let c_i denote the length of C_i. It is easy to pre-compute the score between any pair of core regions because no gaps are inserted into any core region. Thus, we denote by $g(i,j,p_i,p_j)$ the score between C_i and C_j when the first positions of C_i and C_j are aligned to the p_i-th and p_j-th positions of s respectively. That is:

$$g(i,j,p_i,p_j) = \sum_{0 \le i' < c_i} \sum_{0 \le j' < c_j} g_t(s_{p_i+i'}, s_{p_j+j'}, t_{q_i+i'}, t_{q_j+j'}),$$

where q_i and q_j are the beginning positions of cores C_i and C_j, respectively.

Figure 3. In exact algorithms, we assume that gaps are not inserted into core regions. The first residue position of core C_i is aligned to the p_i-th residue in a target sequence.

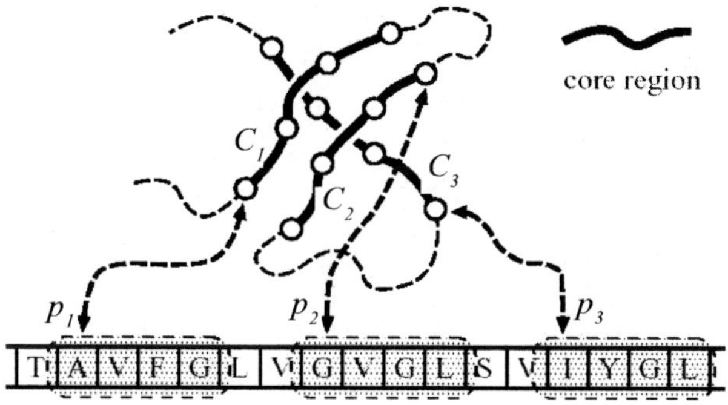

Since we do not consider scores for non-core regions and we assume that gaps are not inserted into core regions, it is enough to represent a threading by means of the first positions of cores. Therefore, each threading is represented by M-tuple $(p_1,\ldots,p_M)$. It should be noted that core regions must not overlap in a feasible threading. That is, $p_i + c_i \leq p_{i+1}$ must hold for all $i<M$. Then, we re-formulate the threading as: given a target sequence $s=s_1\ldots s_m$, core lengths $c_1, \ldots, c_M$, and a score function $g(i,j,p_i,p_j)$, to find a feasible threading $(p_1,\ldots,p_M)$ minimizing the score defined by:

$$score(p_1,\ldots,p_M) = \sum_{i<j} g(i,j,p_i,p_j).$$

Lathrop and Smith (1996) first proposed an efficient and exact algorithm based on a branch-and-bound technique. They defined the set of allowed ranges $([b_1,d_1],[b_2,d_2],\ldots,[b_M,d_M])$, which means that $b_i \leq p_i \leq d_i$ must hold for each i. Their algorithm begins with the set of ranges corresponding to all possible threadings and iteratively subdivides each set of ranges into smaller sets of ranges. For a simplest example, $P=([b_1,d_1],[b_2,d_2],\ldots,[b_M,d_M])$ can be subdivided into three sets of ranges:

$$P_1 = ([b_1,d_1],\ldots,[b_{i-1},d_{i-1}],[b_i,e_i-1],[b_{i+1},d_{i+1}],\ldots,[b_M,d_M]),$$
$$P_2 = ([b_1,d_1],\ldots,[b_{i-1},d_{i-1}],[e_i,e_i],[b_{i+1},d_{i+1}],\ldots,[b_M,d_M]),$$
$$P_3 = ([b_1,d_1],\ldots,[b_{i-1},d_{i-1}],[e_i+1,d_i],[b_{i+1},d_{i+1}],\ldots,[b_M,d_M]),$$

where choice of i and e_i is done based on some heuristic criteria. It should be noted that union of three sets of ranges is equal to the original set of ranges. If we repeat this subdivision process, we finally obtain a set of all individual threadings. However, an exponential number of threadings would be generated. Therefore, we should adopt some technique to limit the subdivision process. For that purpose, Lathrop and Smith proposed a few kinds of lower bounds of the scores attained by any threading from P. For example, the following is the simplest one:

$$\min_{(p_1,\ldots,p_M)\in P} score(p_1,\ldots,p_M) = \min_{(p_1,\ldots,p_M)\in P} \sum_{i<j} g(i,j,p_i,p_j) \geq \sum_{i<j} \min_{b_i \leq x \leq d_i, b_j \leq y \leq d_j} g(i,j,x,y)$$

Let $LB(P)$ denote the rightmost term in the above inequality. Let S_{opt} denote the minimum score among threadings found so far. Then, we need not search for subdivisions of P if $LB(P)$ is greater than S_{opt} because any threading from P can not attain the score lower than S_{opt}. Therefore, we can reduce the search space. In order to improve the efficiency, Lathrop and Smith proposed stronger bounds and succeeded to compute optimal threadings for many medium-size proteins in reasonable CPU time.

Figure 4. A schematic of the divide-and-conquer procedure of PROSPECT

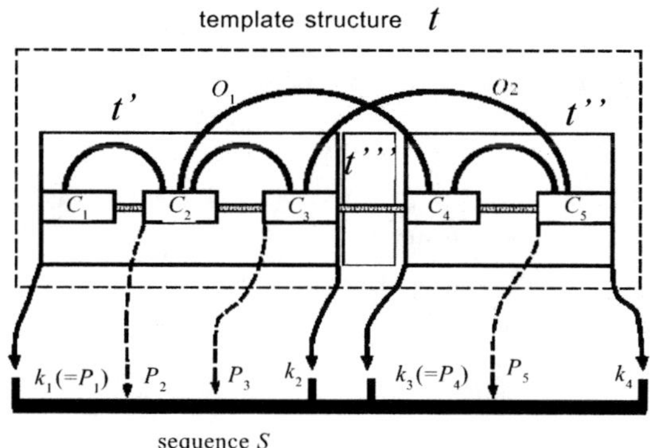

Xu et al. (1998) developed the PROSPECT algorithm/system based on a divide-and-conquer strategy. PROSPECT repeatedly subdivides the template structure into substructures until each substructure contains one core region, where each (sub)structure t is divided into two smaller substructures t' and t'' and one non-core region t'''. Then, PROPECT finds optimal threadings for each substructure from smaller ones to larger ones. In order to compute optimal threadings for t, PROSPECT combines optimal threadings for t' and t''. PROSPECT also utilizes a property that only a few cores strongly interact with each core. Thus, PROSPECT considers interactions for pairs of strongly interacting cores, where this simplification is also employed by RAPTOR. We denote by E the set of such pairs. For example:

$$E = \{(C_1, C_2), (C_2, C_3), (C_2, C_4), (C_3, C_5), (C_4, C_5)\}$$

in Figure 4. We explain the key idea of PROSPECT using this figure. Suppose that we need to compute optimal threadings between template structure t and all possible subsequences . We call a pair of aligned positions a link. For example, link o_1 is (p_2, p_4) and o_2 is (p_3, p_5). Suppose that optimal threadings for t' are already computed for all combinations of k_1, k_2, o_1, o_2. Suppose also that optimal threadings for t'' are already computed for all combinations of k_3, k_4, o_1, o_2. Then, an optimal threading between t and $s_{k_1} \ldots s_{k_4}$ can be obtained by taking the minimum of the sum of threadings for t' and t'' where the minimum is taken over all combinations of k_2, k_3, o_1, o_2. Though PROSPECT takes exponential time in general, it works efficiently if the number of links between t' and t'' is small for each substructure t.

Xu et al. (2003) developed a novel algorithm/system named RAPTOR by formulating protein threading as an integer program (IP). Let a variable $x_{i,j}$ mean that core C_i is aligned to the j-th position of the target sequence s (if $x_{i,j} = 1$). Let a variable $y_{(i,l),(j,k)}$ mean that cores C_i and C_j are aligned to the l-th and k-th positions of the target sequence s, respectively. Then, the objective function of the threading problem can be represented as a linear combination of $y_{(i,l),(j,k)}$'s:

$$\min \sum_{(C_i,C_j)\in E} \sum_{l\in D[i]} \sum_{k\in R[i,j,l]} y_{(i,l)(j,k)} g(i,j,l,k)$$

where $D[i]$ denotes all valid target sequence positions that C_i could be aligned to, and $R[i,j,l]$ denotes all valid alignment positions of C_j given C_i is aligned to the l-th position of s. In order to guarantee that a set of variables represents to a valid (feasible) threading, we should add constraints. For that purpose, it is shown in Xu et al. (2003) that the following linear constraints are enough:

$$\sum_{j\in D[i]} x_{i,j} = 1, \qquad i = 1,2,\ldots,M,$$

$$x_{i,j} \geq 0, \qquad j \in D[i],\ i = 1,2,\ldots,M,$$

$$y_{(i,l)(j,k)} \geq 0, \qquad l \in D[i],\ k \in D[j],\ i,j = 1,2,\ldots,M,$$

$$\sum_{k\in R[i,j,l]} y_{(i,l)(j,k)} = x_{i,l}, \qquad (C_i, C_j) \in E,$$

$$\sum_{l\in R[j,i,k]} y_{(i,l)(j,k)} = x_{j,k}, \qquad (C_i, C_j) \in E.$$

For example, the first constraint means that each core can be aligned to a unique sequence position. The last two constraints mean that if two interacted cores C_i and C_j are aligned to two sequence positions l and k respectively, then the interaction score between these two sequence segments should be counted in the threading score. Since integer programming is known to be NP-hard, Xu et al. (2003) relaxed this integer program to a linear program (LP) by removing integral constraints on all variables. In that case, each variable may take non-integer value. Therefore, they employed some branch-and-bound procedure for cases where non-integral solutions are obtained by an LP-solver. However, they obtained a surprising result that the relaxed linear programs generated integral solutions directly in most cases.

Threading with Constraints

As seen above, extensive studies have been done for protein threading. However, predictive accuracy is not yet satisfactory for practical use. Recently, some advances were made by the significant utilization of distance restraints, which can be obtained from some biological experiments. Xu et al. (2000) showed that (partial) information obtained from NMR experiments is useful to improve the accuracy of the protein threading method. For that purpose, they modified PROSPECT so that constraints given from NMR experiments were taken into account. Young et al. (2000) developed a novel experimental method to aid in construction of a homology model by using chemical cross-linking and time-of-flight (TOF) mass spectrometry to identify LYS-LYS cross-links. They also suggested that distance restraints on Lysine residues are useful to improve accuracy of structure prediction (fold recognition) based on protein threading. Some other heuristic approaches have also been proposed (Albrecht, Hanisch, Zimmer, & Lengauer, 2002; Li.

Zhang, & Skolnick, 2004; Meiler & Baker, 2003). Based on these, Akutsu, Hayashida, Tomita, Suzuki, and Horimoto (2004) considered the protein threading with profiles and constraints. In this section, we briefly explain their approach.

In order to cope with constraints, we modify protein threading with profiles as follows (see Figure 5). Let A_s be the set of pairs of residue positions in s for each of which the distance is known (i.e., the distance is known for each pair $(i, j) \in A_s$ where $i<j$). We define a function $IC(i,j,i',j')$, where $(i, j) \in A_s$ and i',j' ($i'<j'$) denote positions of residues in the template structure. Then, $IC(i,j,i',j')=1$ if the distance between the i-th and j-th residues and the distance between the i'-th and j'-th residues satisfy some constraints (e.g., the difference between these distances is less than some threshold). Otherwise, $IC(i,j,i',j')=0$. It should be noted that $IC(i,j,i',j')$ is not defined if $(i, j) \notin A_s$. Then, profile threading with strict constraints is defined as a protein threading with profiles under the following conditions: (i) any residue included in A_s must not be aligned with a gap symbol, (ii) for any pairs $(i, j) \in A_s$, $IC(i,j,i',j')=0$ must hold, where the i'-th and j'-th residues of t are aligned with the i-th and j-th residues of s in alignment (s',t'), respectively. These conditions mean that distance constraints must be satisfied for all known distances in the target sequence.

However, it may not be possible in some cases to satisfy all distance constraints. Therefore, we consider profile threading with non-strict constraints. In this problem, an optimal (maximum) threading is computed under the following conditions: (i) any residue included in A_s must not be aligned with a gap symbol, (ii) $\sum_{(i,j)\in A_s} IC(i,j,i',j')$ is the minimum.

Though profile threading can be solved in $O(mn)$ time, profile threading with (both strict and non-strict) constraints is proven to be NP-hard (Akutsu et al., 2004). They developed two exact algorithms for profile threading with strict constraints: CLIQUETHREAD and BBDPTHREAD. CLIQUETHREAD reduces the threading problem to the maximum edge-weight clique problem whereas BBDPTHREAD combines dynamic programming and

Figure 5. Protein threading with constraints. $IC(i,j,i',j')=0$ if $|dist(s_i,s_j) - dist(t_{i'},t_{j'})|$ is less than a threshold, otherwise $IC(i,j,i',j')=1$.

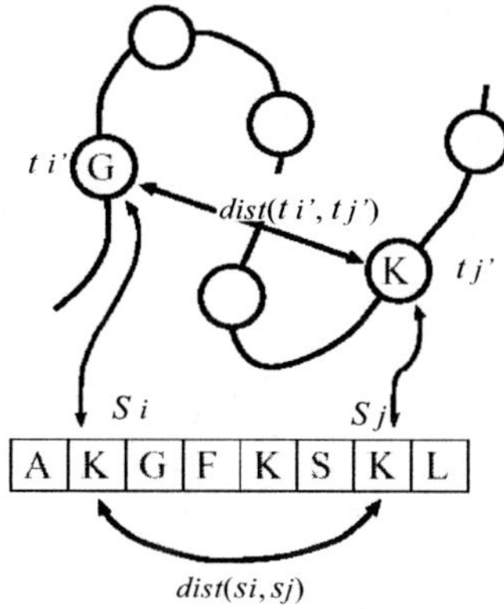

branch-and-bound techniques. Though BBDPTHREAD is faster than CLIQUETHREAD, variants of CLIQUETHREAD can be applied to profile threading with non-strict constraints.

Related Problems

There exist several problems in bioinformatics that are closely related to protein threading. Among these, comparison of RNA secondary structures and comparison of protein structures are important. As shown below, these two problems can be defined in similar ways as in protein threading with pair score functions. Indeed, the proofs of general hardness results and approximation algorithms for special cases in these three problems are similar to each other (Akutsu & Miyano, 1999; Goldman, Istrail, & Papadimitriou, 1999; Lin, Chen, Jiang, & Wen, 2002). These similarities suggest that a technique developed in one problem may be applied to the other problems. Here, we briefly explain relations between these two problems and protein threading.

First, we consider a comparison of RNA secondary structures. An RNA secondary structure can be considered as a combination of a sequence and a graph like a tree structure. Therefore, comparison of RNA secondary structures can be defined as graph-theoretic problems. Though various formulations have been proposed (Allali & Sagot, 2005; Hoechsmann, Voss, & Giegerich, 2004; Jiang, Lin, Ma, & Zhang, 2004; Zhang, 2004), we consider here the longest arc-preserving common subsequence problem (LAPCS) because it has been well-studied (Lin et al., 2002) and has a close relationship with protein threading (see Figure 6).

An RNA sequence is a string over $\Sigma=\{A,U,G,C\}$. In LAPCS, a set of arcs is associated with each sequence. Let $s = s_1 \ldots s_m$ be an RNA sequence. Then, an arc set A_s is defined as a set of pairs of positions (i,j) such that $1 \leq i < j \leq m$. LAPCS can be defined as a problem

Figure 6. Example of LAPCS *(longest arc-preserving common sequence); arcs shown by bold lines are preserved in this alignment*

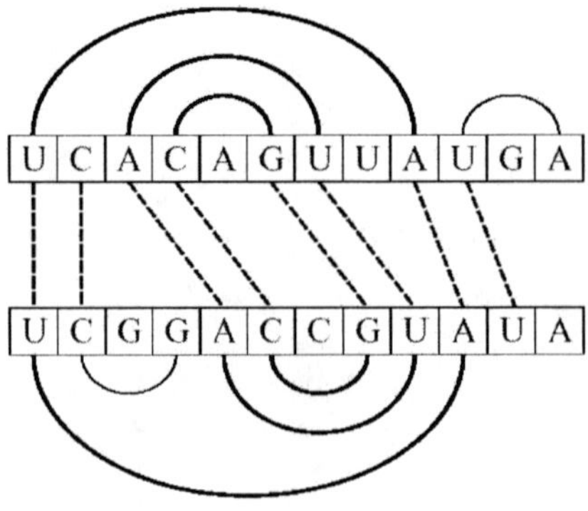

of finding an optimal (maximum) alignment between two RNA secondary structures (s, A_s) and (t, A_t) under the following conditions: (i) $f(x,y)=1$ if $x=y$, otherwise $f(x,y)=0$, (ii) any position pairs (i,j) in s and (i',j') in t that are aligned to the same positions must satisfy $(i,j) \in A_s$ iff. $(i',j') \in A_t$.

Although this definition looks different from the original definition given in Lin et al., these two are equivalent since the length of the longest common subsequence is equal to the score of the optimal alignment. Here, we re-formulate LAPCS as a threading-like problem. We define a function $g_1(i,j,i',j')$ by:

$$g_1(i,j,i',j') = \begin{cases} 0, & \text{if } ((i,j) \in A_s \text{ iff. } (i',j') \in A_t), \\ -\infty, & \text{otherwise.} \end{cases}$$

Then, we find an alignment with the maximum score where the score of an alignment is given by:

$$score(s',t') = \sum_{i=1}^{l} f(s'_i,t'_i) + \sum_{(i,i') \in AL, (j,j') \in AL, i<j} g_1(i,j,i',j'),$$

where $(i,i') \in AL$ means that the i-th nucleotide of s is aligned to the i'-th nucleotide of t in (s',t'). Readers can see that this score is similar to that in protein threading with pair score functions where the sign of the score should be inverted. If we allow general f and g_1, the above formulation can include protein threading with pair score functions. As in protein threading, LAPCS is proven to be NP-hard. Lin et al. considered the following restrictions: (1) no two arcs share an endpoint (i.e., each i appears at most once in A_s), (2) no two arcs cross each other (i.e., $\forall (i_1,i_2),(i_3,i_4) \in A_s\ (i_3 \le i_1 \le i_4 \text{ iff. } i_3 \le i_2 \le i_4)$), (3) no two arcs nest (i.e., $\forall (i_1,i_2),(i_3,i_4) \in A_s\ i_1 \le i_3 \text{ iff. } i_2 \le i_3)$), (4) there are no arcs at all. Based on these, they defined the following five levels: UNLIMITED: no restrictions; CROSSING: restriction 1; NESTED: restrictions 1 and 2; CHAIN: restrictions 1,2, and 3; and PLAIN: restriction 4. They used the notation LAPCS($L1$,$L2$) to represent the LAPCS problem where A_s is of level $L1$ and A_t is of level $L2$. It should be noted that LAPCS(PLAIN,PLAIN) corresponds to the longest common subsequence problem (without arcs). Various complexity results have been obtained depending on subclasses of LAPCS based on this notation. They and other people obtained theoretical results on various subclasses of LAPCS. For example, LAPCS(NESTED,NESTED) is NP-hard whereas LAPCS(CHAIN,NESTED) can be solved in polynomial time. For details, see Lin et al. and other references listed there. It should be noted that it is possible to obtain polynomial time algorithms for RNA secondary structure comparison using dynamic programming if we employ some other formulations (Allali & Sagot, 2005; Hoechsmann et al., 2004; Zhang, 2004). However, it is unclear which formulation is adequate for practice. Thus, rigorous comparison of existing algorithms should be done in the future.

Figure 7. Example of CMO *(contact map overlap); each arc means that the distance between two residues connected by the arc is less than some threshold, arcs shown by bold lines are common in two structures (i.e., the score of this alignment is 5)*

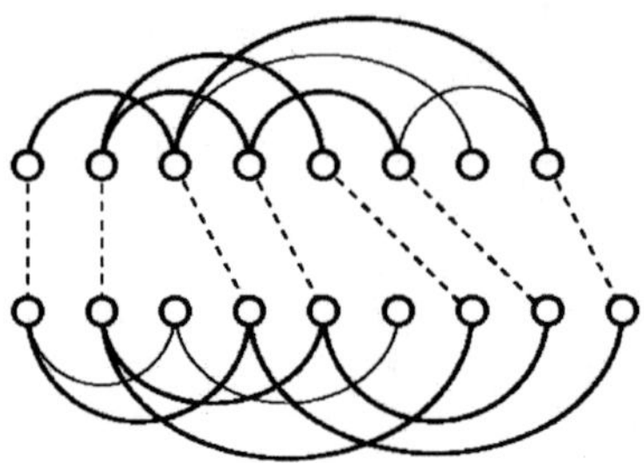

Next, we consider comparison of protein structures. Comparison of protein structures is important because protein structures are closely related to protein functions and it is known that there are many protein pairs with structural similarity but without sequence similarity. Therefore, various kinds of approaches have been proposed to compare protein structures. Protein structure alignment is one of the well-studied approaches. Though there exist various definitions on protein structure alignment, we consider here the contact map overlap (CMO) problem (Goldman et al., 1999; Caprara, Carr, Istrail, Lancia, & Walenz, 2004) because it has close relationships with protein threading and LAPCS (see Figure 7).

As in LAPCS, an arc set A_s is associated with each protein sequence s. Each arc (i,j) means that the distance between the i-th and j-th residues is small. Then, CMO is, given (s, A_s) and (t, A_t), to find an optimal alignment which maximizes the score given by:

$$score(s',t') = \sum_{(i,i')\in AL,(j,j')\in AL,i<j} g_2(i,j,i',j'),$$

where g_2 is defined by:

$$g_2(i,j,i',j') = \begin{cases} 1, & \text{if } (i,j)\in A_s \text{ and } (i',j')\in A_t, \\ 0, & \text{otherwise.} \end{cases}$$

From this definition, readers can see close relationships between CMO and other problems. As in other cases, CMO is proven to be NP-hard (Goldman et al., 1999). However, practical algorithms, in which optimal solutions are guaranteed, are developed using branch-and-bound and Lagrangian relaxation techniques (Caprara et al., 2004).

Summary

In this chapter, we overviewed algorithms for computing optimal solutions for protein threading and related problems. It is shown that various kinds of optimization techniques have been applied to these problems in non-trivial manners. Owing to these developments, it has become possible to compute optimal threadings in reasonable CPU time for most proteins. However, quick computation is required if we apply protein threading to annotation of genome sequences. Therefore, further improvements should be done. As seen in the latter part of this chapter, it is interesting that several important problems in bioinformatics have similar problem structures. Therefore, techniques developed for one of these problems may also be applied to other problems.

Computation of optimal solutions is important not only from an algorithmic viewpoint but also from a practical viewpoint. Indeed, PROSPECT and/or RAPTOR showed good performances in CASP competitions, which are the world wide competitions for protein structure prediction. These results suggest that computation of optimal solutions is important to get good prediction results in practice. Information about both PROSPECT and RAPTOR is available from http://www.bioinformaticssolutions.com/. It should also be noted that special issues (Lattman, 2001, 2003) in a protein journal and the Web pages (http://predictioncenter.llnl.gov/) for CASP are good sources to know state-of-the-art techniques, software, and benchmark data for protein structure prediction and protein threading.

In this chapter, we did not explain details of score functions. However, score functions greatly affect the quality of threading results. Thus, developing a good score function is important for improving the performance of protein threading. Indeed, a lot of studies have been done for deriving good score functions from training data based on statistical and/or heuristic techniques. RAPTOR also adopted SVM to tune score functions. However, a few studies have been done from an algorithmic viewpoint (Akutsu & Yagiura, 1998). Therefore, more algorithmic studies should be done for deriving good score functions.

References

Akutsu, T., Hayashida, M., Tomita, E., Suzuki, J., & Horimoto, K. (2004). Protein threading with profiles and constraints. *Proceedings of 4th IEEE International Symposium of BioInformatics and BioEngineering* (pp. 537-544). Los Alamitos: IEEE.

Akutsu, T., & Miyano, S. (1999). On the approximation of protein threading. *Theoretical Computer Science, 210*, 261-275.

Akutsu, T., & Yagiura, M. (1998). On the complexity of deriving score functions from examples for problems in molecular biology. *Lecture Notes in Computer Science, 1143*, 832-843. Berlin: Springer.

Albrecht, M., Hanisch, D., Zimmer, R., & Lengauer, T. (2002). Improving fold recognition of protein threading by experimental distance constraints. *In Silico Biology*, *2*, 0030.

Allali, J., & Sagot, M.-F. (2005). A new distance for high level RNA secondary structure comparison. *IEEE/ACM Trans. Computational Biology and Bioinformatics*, *2*, 3-14.

Altschul, S. F., Madden T. L., Schaffer A. A., Zhang, J., Zhang, Z., Miller, W., & Lipman, D. J. (1997). Gapped BLAST and PSI-BLAST: A new generation of protein database search programs. *Nucleic Acids Research*, *25*, 3389-3402.

Bowie, J. U., Luthy, R., & Eisenberg, D. (1991). A method to identify protein sequences that fold into a known three-dimensional structure. *Science*, *253*, 164-179.

Caprara, A., Carr, R., Istrail, S., Lancia, G., & Walenz, B. (2004). 1001 Optimal PDB structure alignments: Integer programming methods for finding the maximum contact map overlap. *Journal of Computational Biology*, *11*(1), 27-52.

Clote, P., & Backofen, R. (2000). *Computational molecular biology: An introduction*. New York: Wiley.

Durbin, R., Eddy, S. R., Krogh, A., & Mitchison, G. (1998). *Biological sequence analysis: Probabilistic models of proteins and nucleic acids*. Cambridge, UK: Cambridge University Press.

Edgar, R. C., & Sjolander, K. (2004). A comparison of scoring functions for protein sequence profile alignment. *Bioinformatics*, *20*, 1301-1308.

Garey, M. R., & Johnson, D. S. (1979). *Computers and intractability: A guide to the theory of NP-completeness*. New York: Freeman.

Goldman, D., Istrail, S., & Papadimitriou, C. H. (1999). Algorithmic aspects of protein structure similarity. *Proceedings of the 40th IEEE Symposium on Foundations of Computer Science* (pp. 512-522). Los Alamitos: IEEE.

Godzik, A., & Skolnick, J. (1992). Sequence-structure matching in globular proteins: Application to supersecondary and tertiary structure determination. *Proceedings of National Academy of Sciences USA, 89*, (pp. 12098-12102).

Hoeshmann, M., Voss, B., & Giegerich, R. (2004). Pure multiple RNA secondary structure alignments: A progressive profile approach. *IEEE/ACM Trans. Computational Biology and Bioinformatics*, *1*, 53-62.

Jiang, T., Lin, G-H., Ma, B., & Zhang, K. (2004). The longest common subsequence problem for arc-annotated sequences. *Journal of Discrete Algorithms*, *2*, 257-270.

Jones, D. T., Taylor, W. R., & Thornton, J. M. (1992). A new approach to protein fold recognition. *Nature*, *358*, 86-89.

Lathrop, R. H. (1994). The protein threading problem with sequence amino acid interaction preferences is NP-complete. *Protein Engineering*, *7*, 1059-1068.

Lathrop, R. H., & Smith, T. F. (1996). Global optimum protein threading with gapped alignment and empirical pair score functions. *Journal of Molecular Biology*, *255*, 641-665.

Lattman, E. E. (2001). CASP4 editorial. *Proteins: Structure, function, and genetics*, *45*(S5), 1-1.

Lattman, E. E. (2003). Fifth meeting on the critical assessment of techniques for protein structure prediction. *Proteins: Structure, Function, and Genetics*, *53*(S6), 333-333.

Li, W., Zhang, Y., & Skolnick, J. (2004). Application of sparse NMR restraints to large-scale protein structure prediction. *Biophysical Journal*, *87*, 1241-1248.

Lin, G-H., Chen, Z-Z, Jiang, T., & Wen, J. (2002). The longest common subsequence problem for sequences with nested arc annotations. *Journal of Computer and System Sciences*, *65*, 465-480.

Meiler, J., & Baker, D. (2003). Rapid protein fold determination using unassigned NMR data. *Proceedings of National Academy of Sciences USA, 100*, (pp. 15404-15409).

Xu, J., Li, M., Kim, D., & Xu, Y. (2003). RAPTOR: Optimal protein threading by linear programming. *Journal of Bioinformatics and Computational Biology*, *1*, 95-118.

Xu, Y., Xu, D., Crawford, O. H., & Einstein, J. R. (2000). A computational method for NMR-constrained protein threading. *Journal of Computational Biology*, *7*, 449-467.

Xu, Y., Xu, D., Uberbacher, E. C. (1998). An efficient computational method for globally optimal threading. *Journal of Computational Biology*, *5*, 597-614.

Young, M. M. et al. (2000). High throughput protein fold identification by using experimental constraints derived from intermolecular cross-links and mass spectrometry. *Proceedings of the National Academy of Sciences USA*, *97*, 5802-5806.

Zhang, K. (2004). RNA structure comparison and alignment. In Wang, J. T-L. et al. (Eds.), *Data mining in bioinformatics* (pp. 59-81). Berlin: Springer.

Chapter VIII

Pattern Differentiations and Formulations for Heterogeneous Genomic Data through Hybrid Approaches

Arpad Kelemen, The State University of New York at Buffalo, USA
& Niagara University, USA

Yulan Liang, The State University of New York at Buffalo, USA

Abstract

Pattern differentiations and formulations are two main research tracks for heterogeneous genomic data pattern analysis. In this chapter, we develop hybrid methods to tackle the major challenges of power and reproducibility of the dynamic differential gene temporal patterns. The significant differentially expressed genes are selected not only from significant statistical analysis of microarrays but also supergenes resulting from singular value decomposition for extracting the gene components which can maximize the total predictor variability. Furthermore, hybrid clustering methods are developed based on resulting profiles from several clustering methods. We demonstrate the developed hybrid analysis through an application to a time course gene expression data from interferon-β-1a treated multiple sclerosis patients. The resulting integrated-condensed clusters and overrepresented gene lists demonstrate that the hybrid methods can successfully be applied. The post analysis includes function analysis and pathway discovery to validate the findings of the hybrid methods.

Introduction

Progress in mapping the human genome and developments in microarray technologies have provided considerable amount of information for delineating the roles of genes in disease states. Since complex diseases typically involve multiple intercorrelated genetic and environmental factors that interact in a hierarchical fashion and the clinical characteristics of diseases are determined by a network of interrelated biological traits, microarrays hold tremendous latent information but their analysis is still a bottleneck. Pattern analysis can be useful for discovering the knowledge on gene array data related to certain diseases (Neal et al., 2000; Slonim, 2002). The associations between patterns and their causes are the bricks from which the wall of biological knowledge and medical decisions are built.

Pattern differentiations and pattern formulations are two major tracks of patterns analysis. Pattern differentiation of gene expressions is the first step to identify potential relevant genes in biological processes. The coordinated/temporal gene arrays are widely used for pattern formulation in order to study the common functionalities, co-regulations, and pathways that ultimately are responsible for the observed patterns. The identification of groups of genes with "similar" temporal patterns of expression is usually a critical step in the analysis of kinetic data because it provides insights into the gene-gene interactions and thereby facilitates the testing and development of mechanistic models for the regulation of the underlying biological processes. These temporal pattern analyses provide clues for genes that are related in their expression through linkage in a common developmental pathway.

There are several critical challenges in the pattern analyses. One is in the pattern differentiations, the notorious "large p small n" problem (West, 2000). The large number of irrelevant and redundant genes with high level noise measurements and uncertainty severely degrade both classification and prediction accuracy. The solution for the "large p" problem is through affine transformation and feature selection. Affine transformation such as principal component analysis (PCA) or singular value decomposition (SVD) has advantage of simplicity and it may remove non-discriminating and irrelevant features (i.e., genes) by extracting eigenfeatures corresponding to the large eigenvalues (Alter, Brown, & Botstein, 2000; Yeung and Ruzzo, 2001; Wall, Dyck, & Brettin, 2001). Yet it is very difficult to identify important genes with these methods and the inherent linear nature is their prominent disadvantage.

Feature selection consists of two strategies: screening and wrappers. In the screening approaches, all genes are analyzed and tested individually to see whether they have higher expression level in one class than in the other (Baldi & Long 2001; Tusher, Tibshirani, & Chu, 2001; Storey & Tibshirani 2003; Hastiel et al., 2000). The disadvantage of screening processes is that they are non-invertible and can cause multiple testing and model selection problems (Westfall & Young, 1993; Benjamini & Hochberg, 2002).

In wrapper methods, genes are tested not independently, but as ensembles, and according to their performance in the classification model (Golub et al., 1999; Khan et al., 2001; Wuju & Momiao, 2002). Since the number of feature subsets increases exponentially with the dimensions of the feature space, wrappers are computationally intractable for high-dimensional gene data.

The second critical challenge in the patterns' analysis is that there are some limitations of most of the existing pattern formulation methods such as clustering methods (Eisen, Spellman, Brown, & Botstein, 1998; Tamayo et al., 1999; Yeung et al., 2003). First, they may not detect genes with different functions showing similar profiles by chance or stochastic fluctuation from high level noise. Second, most of these approaches have not incorporated the temporal information, such as the features of significant nonlinear trend and the presence of sudden bursts of amplitude at irregular time intervals and the short-term dynamics of the data. Third, these methods may not detect indirectly correlated genes, which show weak correlations due to a time delay or nonlinear association. Last, most of them can not determine the number of patterns a priori and the final partition of the data requires evaluation.

In this chapter we develop hybrid methods and several interrelated integrative procedures for overcoming some limitations of existing approaches, increasing the power and the reproducibility of differentially expressed gene temporal pattern profiles from heterogeneous genomic data. These procedures include prior analysis with screening methods for mining the significant differential gene profiles, pattern formulations with different clustering methods, and post analysis for pathway and function study. The differentially expressed gene profiles were firstly selected not only from q-value computed from Significant Analysis of Microarray (SAM) but also the supergenes resulting from Singular Value Decomposition (SVD) methods (Tusher et al., 2001; Alter et al., 2001). Hybrid pattern formulation methods will be further developed based on the resulting profiles from different clustering methods, in our case including Hierarchical clustering, k-means, dynamic Bayesian clustering and singular value decomposition (SVDMAN) (Eisen et al., 1998; Tamayo et al., 1999, Ramoni, Sebastiani, & Kohane, 2002; Wall et al., 2001). Most existing clustering methods are based on resampling procedures and statistical validity testing (Kerr & Churchill, 2001; Dudoit & Fridlyand, 2002; Tibshirani et al., 2001; McShane et al., 2002). For example, Kerr and Churchill have proposed using bootstrap techniques and confidence interval to validate the reproducibility of the clustering results. McShane et al. developed hypothesis testing approaches and suggested using two-step approaches to evaluate the sample clustering. We propose hybrid clustering approaches (the clustering of the clusters themselves), which not only allow us visual examination of the degrees of heterogeneity between methods but also help to improve the clustering power and reproducibility.

Since different clustering algorithms optimize different objective functions and criteria, hybrid clustering will provide the agreement of various methods, help to detect true patterns and differentiations not by chance or stochastic fluctuations, and provide an Integrated-Condensed (I/C) overrepresented gene list with low false positive rate. The follow-up post analysis such as function analysis and pathway discovery for constructing the association of the informative genes is performed to validate the findings of the methods (Hosack, Hosack, Sherman, Lane, & Lempicki, 2003; Glynn et al., 2003). We also demonstrate the developed hybrid temporal analysis through an application to a time course of lymphocyte gene expression data from interferon-β-1a treated multiple sclerosis (MS) patients (Weinstock-Guttman et al., 2003).

Multiple Sclerosis Microarray Data Set

Multiple sclerosis (MS) is an autoimmune disease in which the body's immune system attacks the brain and spinal cord, resulting in neurological disabilities. Multiple sclerosis is an inflammatory-demyelinating disease of the central nervous system that affects over 1 million patients worldwide. It is a complex, variable disease that causes physical and cognitive disabilities: nearly 50% of the patients diagnosed with MS are unable to walk after 15 years. The etiology and pathogenesis of MS remains poorly understood (Jacobs et al., 1996). Recombinant human IFN-β is one of the most commonly prescribed forms of therapy for relapsing multiple sclerosis patients on the basis of several clinical trials (Steinman & Zamvil, 2003; Weinstock-Guttman et al., 2003). The effects of IFN-β treatment are complex, and its pharmaco-dynamics at the genomic level in humans are poorly understood. Patients with relapsing MS respond better to IFN-β treatment than patients with progressive disease. However, this better response is not homogeneous among relapsing patients. Studies were conducted to characterize the dynamics of the gene expression induced by IFN-β-1a treatment in MS patients and to examine the molecular mechanisms potentially capable of causing heterogeneity in response to therapy (Ramanathan et al., 2001).

Fourteen patients with active relapsing remitting MS were recruited for the study. These patients had not previously received IFN-β and were clinically stable for the preceding four weeks. Peripheral blood samples were obtained just before treatment and at 1, 2, 4, 8, 24, 48, 120, and 168 hrs. and three months after a 30 mg dose of i.m. IFN-β-1a was administered. The GeneFillers GF211 DNA arrays (Research Genetics, Huntsville, AL) were used in this study. Each filler contained multiple positive control spots and housekeeping genes. Each housekeeping gene was spotted in duplicate, and there were over 4000 known genes in the GF211 array (see ftp://ftp.resgen.com/pub/genefilters/). Fourteen patients with 4324 genes were measured at each time point, except the first patient who was not measured at three months. Because treatment can often cause changes in the expression of housekeeping genes, the global normalization option was selected, i.e., each of the filters were normalized using the intensity from all spots on the GF211 GeneFilter DNA array and for intensity ranges. In addition, it was discovered that 266 genes (henceforth referred to as unknowns) did not have titles but they all had clone numbers.

Methods and Results

Pattern Differentiations

In order to identify the significant differences in expression levels between two different conditions (time points) significant analysis of microarray (SAM) is performed first. SAMs have utilized the multi-class grouping option to facilitate comparison of gene expression levels across the entire time course of gene expression. There are several other major advantages using SAM algorithms to find a list of the most significant genes (i.e., the genes whose expression changed significantly in response to treatment). First, SAMs use q-value instead of p-value to control the false discovery rate where multiple testing problems and dependences among genes are taken into account. Second, they utilize the permutation methods in which no assumptions are made regarding the null distribution. The "tuning parameter", known as *delta* (Δ), is chosen based on an acceptable false positive rate. Since SAM does not provide any direct means of normalizing/centering the data, these steps are performed prior to inputting the data into SAM.

Due to the fact that complex diseases are contributed by many small effects of multiple genes, principal component analysis (PCA) is also used to characterize the data (including the variations and structures of the data) and to look for combination effects (supergenes) that have major contributions to the given disease. To deal with the computationally inhibitive "large p, small n" problem, we used singular value decomposition analysis of microarray (SVDMAN), which uses singular value decomposition as a computational shortcut of PCA. The general idea is that the genes by array space is transformed through a translation matrix into a reduced eigengene x eigenarray space to capture the inherent dependency and high variations of the data. One advantage of SVDMAN is that it can be further used for pattern formulations. Figure 1 provides the Venn diagram showing the relative proportions and intersection of the output of significant genes from the SAM and SVD outputs.

Pattern Formulations

The hierarchical clustering with mean centered, and the uncentered Pearson correlation coefficient (PCC) similarity metric, k-means, Bayesian cluster analysis of gene expression (CAGED), and SVDMAN were applied to differentially expressed genes selected by SAM and SVD methods. Hierarchical clustering and k-means methods have visualization

Figure 1. Venn Diagram showing the relative proportions and intersection of the output of significant genes from the SAM and SVD outputs

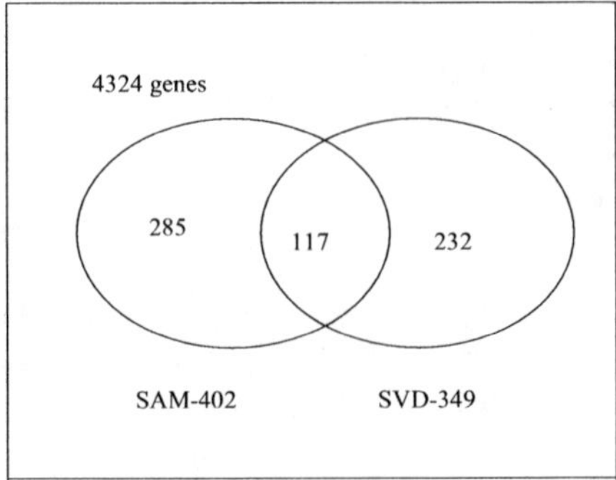

advantages and straightforward interpretations and are model free. The advantage of hierarchical clustering is that it places genes into non-distinct clusters and a biologist can look at the structure of the tree and judge which genes were clustered in a meaningful manner. For the hierarchical clustering, the data was hierarchically clustered using the usual uncentered PCC similarity metric and the centroid linkage clustering. At the risk of a sounding flippant, k-means clustering has the advantage of placing genes into (a predetermined number of) distinct clusters. That is, the biologists can force the genes into a predetermined number of clusters (10 clusters in this case). However, it is subjective for the choice of the number of clusters; both methods have no measures for the stochastic fluctuations and they disregard the dynamic features of the observations and temporal information. CAGED and SVDMAN methods are both model based and they utilize the significance of the similarity to identify the best number of clusters from the data (Ramoni et al., 2002). CAGED methods are also referred to as Bayesian model-based approaches that account for the dependences of the temporal information using autoregressive models and use agglomerative procedures to search for the most probable set of clusters. The CAGED method has two major advantages over other methods. First, it determines the number of clusters based on its internal modeling algorithm (see Figure 2). Second, it takes into account that the series is a time series and therefore the nature of a given point must take into consideration the point(s) before it. That is, if we would rearrange the order of the series similarly for all genes (e.g. switched

Figure 2. Part of the Dendogram from the model based clustering of the MS data using the CAGED 1.0 program; the "unknown" samples (numbers) can be seen (cluster 11)

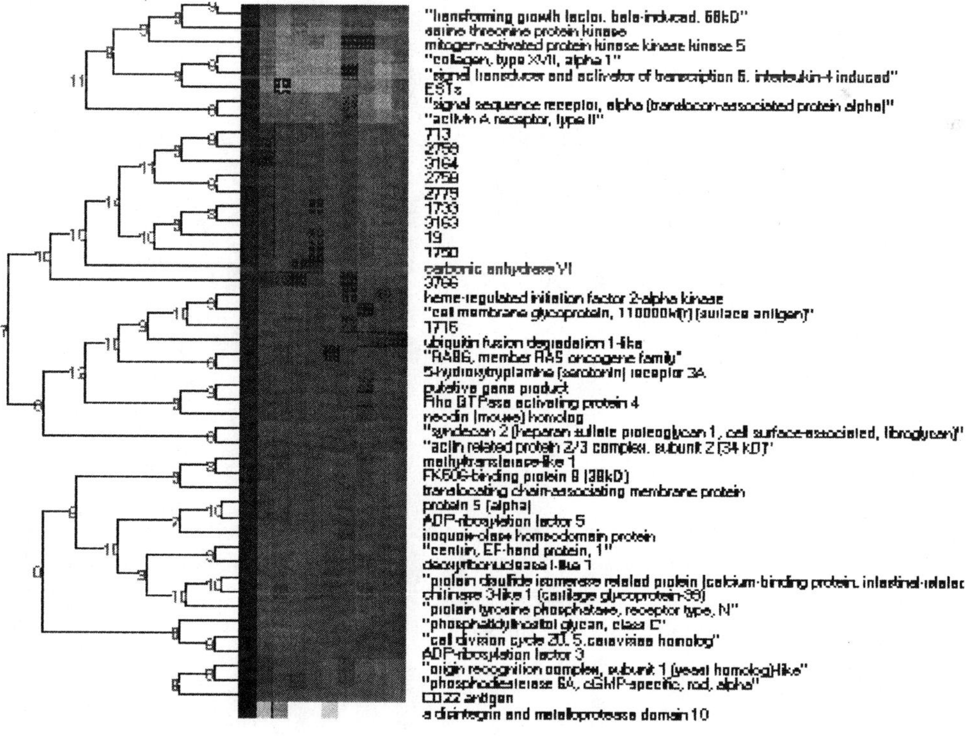

Figure 3(a). Graphs of the Right Singular Vectors from the SVDMAN program run on the MS data; the first four principal components zero through three are shown (note: Principal component zero appears flat because of the scale)

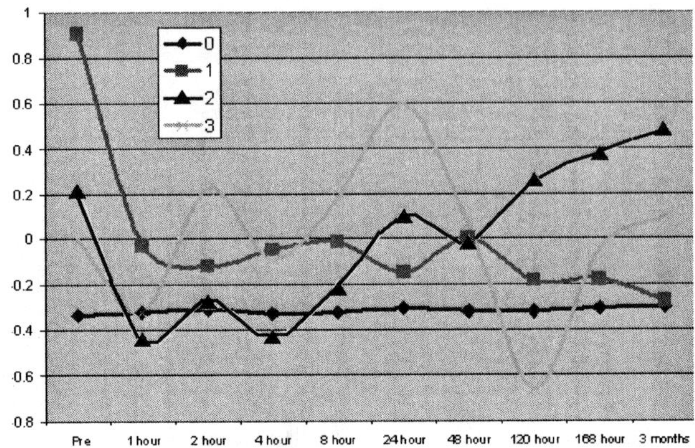

Figure 3(b). Eigenexpression Fractions of the ten principle components (in order), which capture the most variations of the data

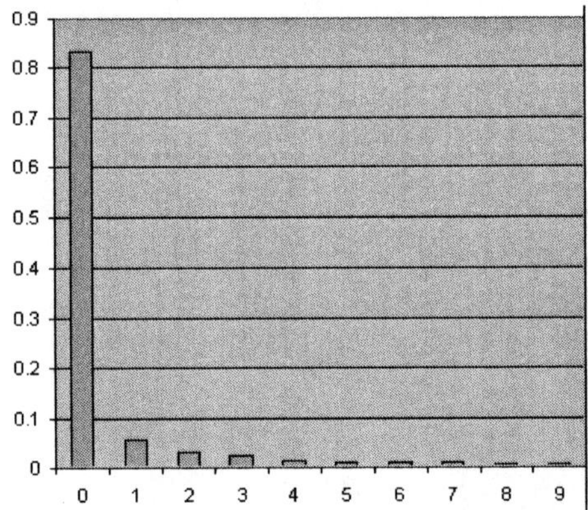

the one hour and seven day values), the hierarchical and k-means clustering results would be the same, but the results from the CAGED analysis would change. The advantage of SVD is that it can capture most of the variation and the distribution of the data and provide supergenes that are functionally not separable (see Figure 3(a) and Figure 3(b)).

Hybrid Methods for Pattern Differentiation and Formulations

Resampling techniques, such as permutation and bootstrap techniques, can improve statistical significance and the reproducibility of the results. However, the genes observed to be differentially expressed from a given method should be confirmed with other independent methods on independent samples. For instance, the wrapper methods can be utilized in the classification model and use the known classes' information and classification error to verify whether the best predictors have been obtained. In this chapter we develop another strategy, called hybrid methods to improve the significance and reproducibility of the results. Due to the fact that different methods apply different algorithms and criteria, the hybrid multiple criteria could help to improve the quality of the resulting informative gene profiles. For hybrid pattern differentiation, we applied SAM's and SVDMAN's results, in which controlling the false positive rate and capturing the majority of the variations of the data can be achieved. This also improved the identifications of the novel genes that are truly responsible for specific diseases and with low false-positive rate and good precision while maintaining sensitivity to relatively subtle changes.

Figure 4. Differential patterns from various existing clustering methods for MS disease: top left: Hierarchical clustering, profiles of HN1-5 (hierarchical negatively expressed clusters 1 to 5) and HP1-4 (hierarchical positively expressed clusters 1 to 4); top right: k-means (K0-K9); bottom left: Dynamic clustering from CAGED (C1-C14); bottom right: Singular value decomposition from SVDMAN, the nine principal components zero through eight are shown (S0-S8). x-axis: time, y-axis: gene expression level

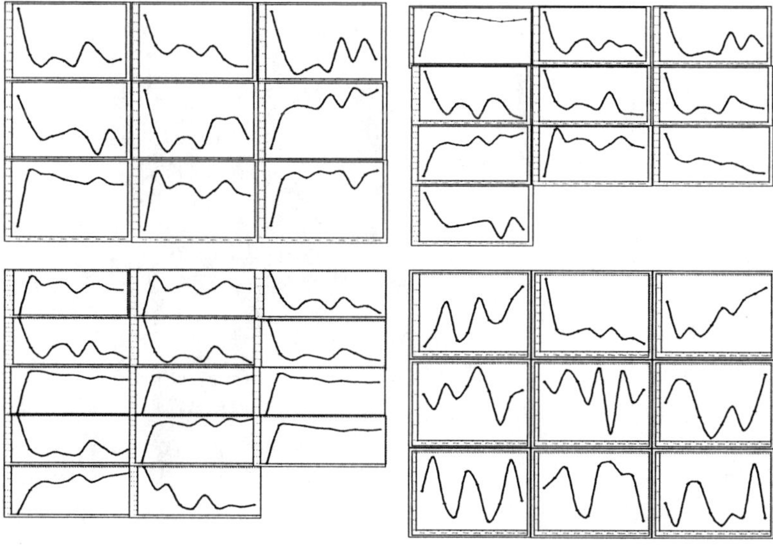

The determined numbers of clusters for each method and cluster profiles for our data were as follows: 9 for hierarchical: Profiles of HN1-5 (hierarchical negatively expressed clusters 1 to 5) and HP1-4 (hierarchical positively expressed clusters 1 to 4), 10 for k-means (K0-K9), 14 for CAGED (C1-C14), and 9 for SVDMAN (S0-S8) as shown in Figure 4. Accordingly, hybrid clustering (clustering of the clusters) from four clustering methods (hierarchical, k-means, CAGED, and SVDMAN) was developed and applied to our data sets to compare the results of different methods and to improve the power of reproducibility. That is, after the four different methods were performed, the individual profiles had to be matched up from one method to the next. Instead of doing this "by hand", it dawned on the authors of this chapter that a better way to do this is by clustering the clusters themselves. The profiles of the clusters from all four clustering methods (hierarchical, k-means, CAGED, and SVDMAN) were compiled into one spreadsheet and the spreadsheet was loaded into Cluster 3.0 (Eisen, 1998). The data was then hierarchically reclustered using the usual uncentered PCC similarity metric and the centroid linkage clustering. Tight clusters were handpicked and composite graphs from the mean centered data are illustrated in Figure 5. The resulting composite C/C's (Clustering of the Clusters) demonstrate that this was performed successfully (see Figure 5 and Figure 6).

Figure 5. Hybrid clustering (clustering of the resulting profiles from various methods); the global view window from JavaTreeView has been sized down to "zero" for this illustration

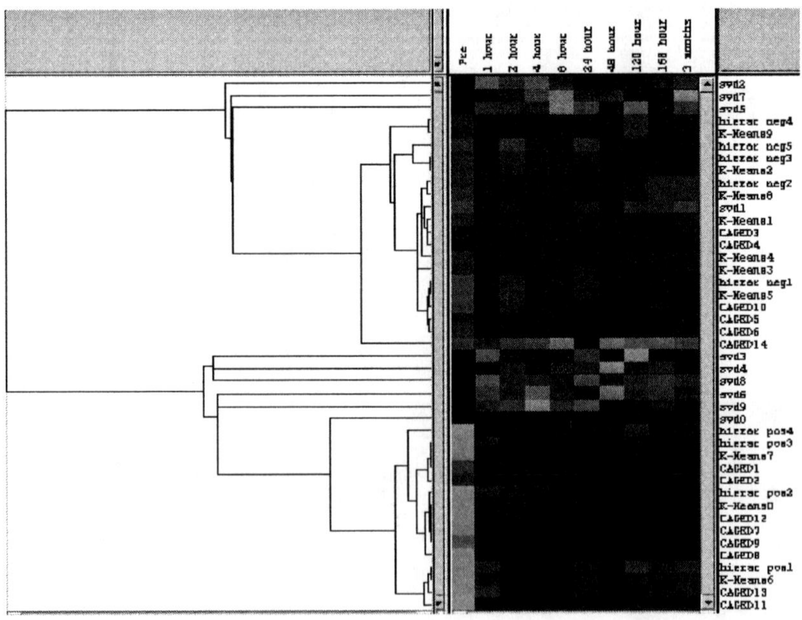

Figure 6. Hybrid clusters (C/C): clustering after clustering (hierarchical, k-means, Bayesian with CAGED, and SVD); overall general functional characteristics: left: down regulated (Top to bottom: negatively expressed C/C-1 proliferation; C/C-2 phosphorylation; C/C-3 signal receptors); right: positively expressed up regulated (top to bottom: C/C-4 stress; C/C-5 DNA replication, membrane; C/C-6 protein biosynthesis and transport)

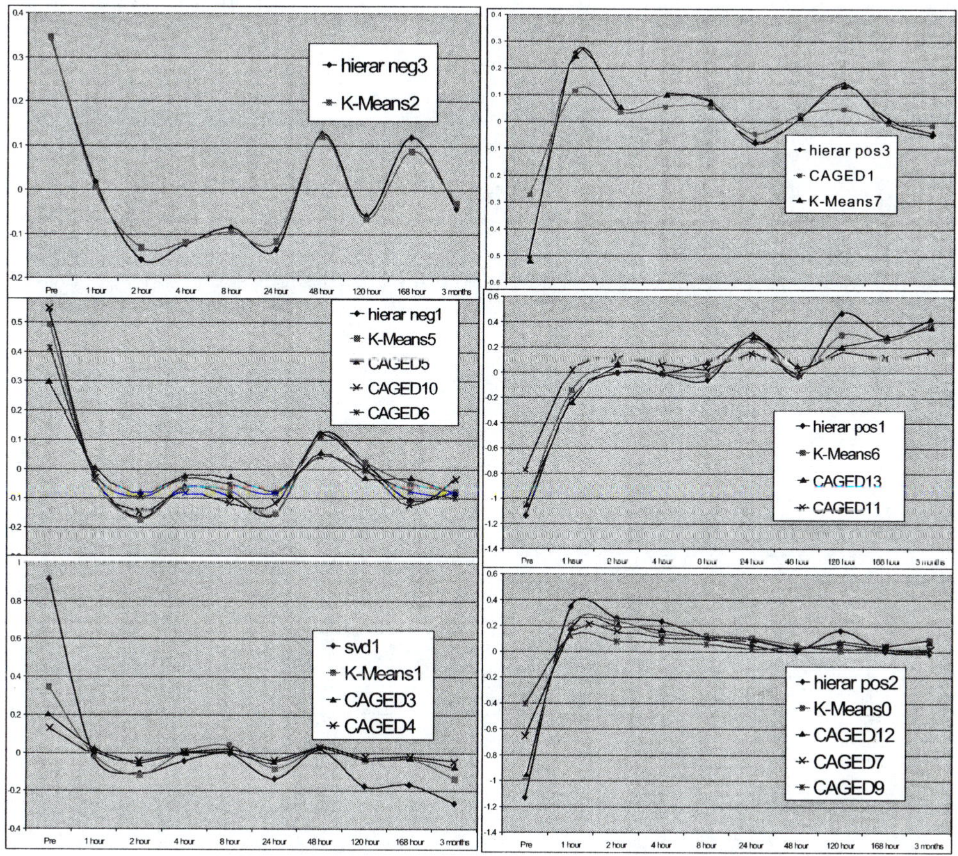

Comparison of the Four Clustering Methods via Hybrid Clustering (C/C)

The results from the hierarchical, k-means, and CAGED cluster analyses came up with similar profiles, as evidenced by the inclusion of all the three methods in four of the six C/Cs and the inclusion of two of the three methods in the other two C/Cs. SVD only found one profile through C/C (S1 in C/C-3). The CAGED program usually had multiple members in the handpicked clusters from the clustering of the clusters. That is, CAGED had a tendency to be divisive — breaking clusters excessively into subclusters. The SVD analysis resulted in a list of genes that did not coincide well with the list of significant genes from the SAM output. The SVD analysis resulted in a list of 351 genes distributed across 10 groups (the total length of the list was 929). Most genes appeared in multiple groups. The average gene appeared in 2.65 groups (929 total length of list/351 unique genes). The SVD analysis did, however, result in profiles that were markedly different from one another. That is, the trajectories of a given SVD profile were partial inversions in some way of the other SVD profiles. This can be seen visually by the fact that in the clustering of the profiles the SVDMAN clusters were all at the end of long branches — indicating that the SVDMAN analysis correctly extracted the principal components.

Post Analysis: Function and Pathway Analysis of Genes Resulting from Hybrid Methods

The functions of genes and the functional relationships between genes that were statistically significant from hybrid clustering were further investigated using the

Table 1(a). EASE Scores and overrepresented gene categories of negatively expressed hybrid clusters (C/C)

C/C	Cluster	Gene Category	EASE score
1	HN3	development	0.042
1	HN3	cell proliferation	0.071
1	K2	development	0.089
1	K2	cellular process	0.148
1	K2	coenzyme and prosthetic group biosynthesis	0.156
1	K2	coenzyme and prosthetic group metabolism	0.156
2	S1	organelle organization and biogenesis	0.047
2	HN1	protein modification	0.009
2	K5	protein modification	0.017
2	C5	signal transduction	0.020
2	HN1	phosphate metabolism	0.029
2	HN1	Phosphorus metabolism	0.029
2	C5	intracellular signaling cascade	0.032
2	HN1	protein amino acid phosphorylation	0.033
2	C5	cell surface receptor linked signal transduction	0.036
2	C5	cell communication	0.042
2	HN1	phosphorylation	0.047
2	C10	cell communication	0.049
3	S1	induction of apoptosis by extracellular signals	0.004
3	K1	G-protein coupled receptor protein signaling pathway	0.009
3	S1	EGF receptor signaling pathway	0.010
3	C3	vesicle-mediated transport	0.014
3	S1	cell organization and biogenesis	0.015
3	S1	cytoskeleton organization and biogenesis	0.022
3	S1	cytoplasm organization and biogenesis	0.033
3	K1	cell communication	0.039
3	K1	second-messenger-mediated signaling	0.043
3	C3	DNA metabolism	0.045
3	S1	nitric oxide mediated signal transduction	0.046

Table 1(b). EASE Scores and overrepresented gene categories of positively expressed hybrid clusters (C/C)

C/C	Cluster	Gene Category	EASE score
4	C1	Oxygen and reactive oxygen species metabolism	0.070
4	K7	response to oxidative stress	0.076
4	K7	physiological process	0.111
4	K7	Oxygen and reactive oxygen species metabolism	0.113
4	HP3	physiological process	0.157
4	K7	response to external stimulus	0.164
4	K7	response to stress	0.179
5	K6	DNA dependent DNA replication	0.104
5	K6	S phase of mitotic cell cycle	0.128
5	K6	DNA replication	0.128
5	K6	DNA replication and chromosome cycle	0.175
5-c	C11	Membrane	0.101
5-c	C11	Integral to membrane	0.110
5-c	C11	Unlocalized	0.158
5-c	K6	Integral to membrane	0.161
5-c	C11	plasma membrane	0.165
5-c	K6	Membrane	0.166
6	C12	protein metabolism	0.047
6	HP2	protein biosynthesis	0.054
6	C12	protein transport	0.085
6	C12	intracellular protein transport	0.085
6	HP2	macromolecule biosynthesis	0.131
6	C12	physiological process	0.132
6	C12	protein biosynthesis	0.144
6	C12	intracellular transport	0.157
6	C12	M phase of mitotic cell cycle	0.168
6	C12	Mitosis	0.168
6	K0	macromolecule biosynthesis	0.178
6	C12	Metabolism	0.194
6	HP2	Biosynthesis	0.199

Expression Analysis Systematic Explorer (EASE) software. A variant one-tailed Fisher exact probability — referred to as EASE score — is a statistical measure of the overrepresentation of a class of genes within the genes resulting from SAM and SVDMAN. The gene list was sorted using the accession numbers (kind of gene ID), and the accession numbers for each hybrid cluster were run through EASE to obtain the themes of the overrepresented genes (see Tables 1(a) and 1(b)). The "GenBank Human Supplemental" was selected in the "2. Input Genes Either Load" pulldown menu in the EASE program. The rows with the lowest EASE scores were copied and pasted to a new spreadsheet for each hybrid cluster (cluster of clusters) (C/C-1 through C/C-6). The cutoffs for the EASE scores for each cluster of clusters were given as follows: C/C-1<0.15, C/C-2<0.05, C/C-3<0.05, C/C-4<0.20, C/C-5<0.20, and C/C-6<0.20. Each Excel sheet of lowest EASE scores was sorted by the "EASE scores". The "GO cellular" and "GO molecular" rows in the EASE were deleted, leaving only the "GO biological" rows (except

Table 2. Overall general characteristics of hybrid clusters (C/C's)

C/C	Combined-Condensed Overrepresented Gene Categories	Overall EASE General Characteristic
1	development, cell proliferation, cellular process, coenzyme and prosthetic group biosynthesis/metabolism	PROLIFERATION
2	organelle organization and biogenesis, protein modification, signal transduction, phosphate/ phosphorous metabolism, intracellular signaling cascade, protein amino acid phosphorylation, cell surface receptor linked signal transduction, cell communication	PHOSPHORY-LATION
3	induction of apoptosis by extracellular signals, G-protein coupled receptor protein / EGF receptor signaling pathways, vesicle mediated transport, cell/cytoskeleton/cytoplasm organization and biogenesis, cell communication, second-messenger mediated signaling, DNA metabolism, nitric oxide mediated signal transduction	SIGNAL RECEPTORS
4	oxygen and reactive oxygen species metabolism, response to oxidative stress/external stimulus/stress, physiological process	STRESS
5	DNA dependent DNA replication, S phase of mitotic cell cycle, DNA replication and chromosome cycle	DNA REPLICATION
5-c	integral to membrane, unlocalized	MEMBRANE
6	protein metabolism/biosynthesis/transport, intracellular protein transport, macromolecule/protein biosynthesis, physiological process, M phase of mitotic cell cycle	PROTEIN BIOSYNTHESIS AND TRANSPORT

for C/C-5 — only the GO molecular rows were deleted, leaving both the biological and cellular rows (this was done because the cellular rows seemed to have a general theme also)).

Binding of IFN-β to the cell receptor eventually results in various cross-phosphorylations and other sequential phosphorylations. The peculiar result of the EASE analysis is that the genes that tended to be associated with phosphorylation and signal receptors (C/C-2 and C/C-3 resp.) were in the down regulated clusters — the opposite of what would be expected. Perhaps somewhere in the reduction of the information from EASE, the meaning was lost. This data and the summary of the data for each cluster and cluster of clusters (C/C) are given in Tables 1(a) and 1(b). The results in Tables 1(a) and 1(b) were combined and condensed for each C/C (two times over). The results of these two "waves" of combination and condensation are given in Table 2 as "Combined-Condensed Overrepresented Gene Categories" and "Overall General Characteristic", respectively.

DAVID-KEGG Analysis

DAVID (Database for Annotation, Visualization and Integrated Discovery) is an on-line tool that allows rapid biological interpretation of gene lists and performs theme discovery and annotation. All the pathways of all the clusters contained in each hybrid cluster were compiled and submitted to EASE; then the obtained output was sent trough DAVID via the accession numbers of genes. The resulting list of pathways was then reduced by eliminating the less promising pathways (see Tables 2, 3). For example, if a given cluster produced a list of pathways, which contained 3, 2, and 1 gene(s), the pathways containing only 1 gene were deleted. If the list contained only pathways that contained one gene,

Table 3(a). DAVID-KEGG Pathways for negative hybrid clusters (C/C's)

C/C	1	2	3
EASE General Characteristic	Proliferation	Phosphorylation	Signal Receptors
DAVID-KEGG Pathways	Folatebiosynthesis (1) MAPKsignaling pathway (1) Porphyrin and chlorophyll metabolism (1)	Purine metabolism (3) MAPKsignaling pathway (2) Oxidative phosphorylation (2) N-Glycansbiosynthesis (2) Tryptophan metabolism (2) Tyrosine metabolism (2) beta-Alanine metabolism (2) Arginine and proline metabolism (1) Integrin-mediated celladhesion (1)	Purine metabolism (3) Ribosome (3) Proteasome (3) Phosphatidylinositol signaling system (3) Glycerolipid metabolism (2) Glutathione metabolism (2) MAPKsignaling pathway (2) Oxidative phosphorylation (2) Starch and sucrosemetabolism (2)

Table 3(b). DAVID-KEGG Pathways for positive hybrid clusters (C/C's)

C/C	**4**	**5**	**6**
EASE General Characteristic	**Stress**	**DNA replication, Membrane**	**Protein Biosynthesis and Transport**
DAVID-KEGG Pathways	Glutathione metabolism (2)	Cell cycle (1) Nitrogen metabolism (1) Purine metabolism (1)	Ribosome (1) Glycosylphosphatidyl inositol (GPI)-anchor biosynthesis (1)

the list was left alone. Next, the pathways that were found in only one cluster for a given C/C were deleted.

When the gene lists from our hybrid clusters were run in DAVID, only three or fewer genes from a given cluster were found in a given pathway. This is not an unusually low number, and not surprisingly some of the pathways returned in the DAVID-KEGG analysis did coincide with the EASE results. That is, C/C-2 (phosphorylation) returned the pathway "oxidative phosphorylation (2)", C/C-3 (signal receptors) returned the pathways "phosphatidylinositol signaling system (3)" and "MAPK signaling pathway (2)", C/C-5 (DNA replication, membrane) returned the pathway "cell cycle (1)", and C/C-6 (protein biosynthesis and transport) returned "ribosome (1)". One definite future direction is determining which genes from the gene lists (accession numbers) are the ones matching up in the pathways returned by DAVID-KEGG.

Interferon Related Genes

Figure 7 illustrates the interferon pathway of Multiple Sclerosis disease, which is part of the generalized JAK-STAT pathway [http://stke.sciencemag.org/cgi/cm/CMP_8390]. Any gene title from the list of overrepresented genes that contained the word "interferon" was investigated as to what C/C and what cluster(s) it was a member of, and what the general direction of change of expression of that gene was. Results are given in Table 4. It is surprising that four of the five interferon-related genes were actually down regulated.

Figure 7. The interferon pathway for MS disease

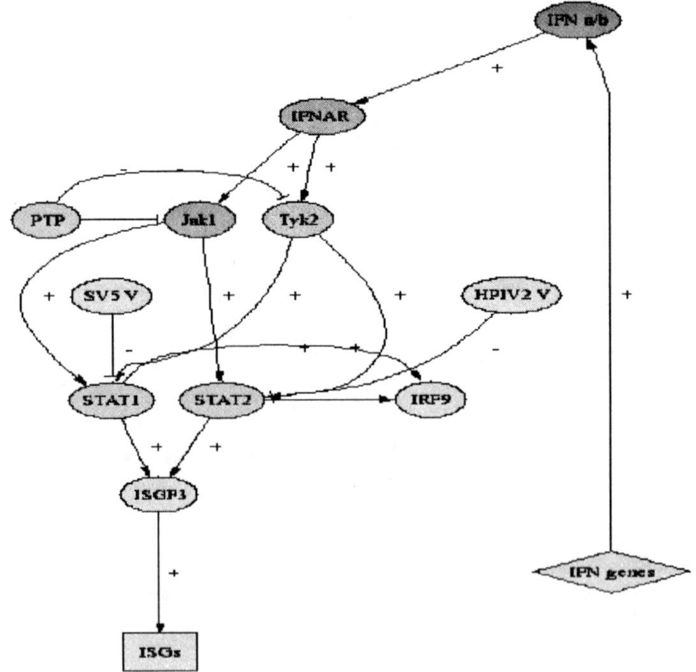

Table 4. Interferon induced or related genes and their corresponding hybrid clusters (C/C's)

ACCESSION NUMBER	TITLE	CLUSTER	C/C	Gene Expression Response
AA464417	interferon induced transmembrane protein 3 (1-8U)	K1, C3	3,3	Negative
N98563	retinoic acid- and interferon-inducible protein (58kD)	C5	2	Negative
AA827287	interferon-induced protein 35	C5	2	Negative
AA157813	interferon, alpha-inducible protein 27	K6	5	Positive
AA676598	interferon-related developmental regulator 1	HN1, K5, C5	2,2,2	Negative

Table 5. Various software and publicly available resources used in this chapter

Name	Type	Reference	Web-Based Resources
SAM	Significance Analysis of Microarrays	Tusher et al., (2001)	http://www-stat-class.stanford.edu/SAM/SAMServlet
Cluster Treeview	Hierarchical and k-means, etc.	Eisen, et al. (1998) *Cluster and Treeview Manual.*	http://bonsai.ims.u-tokyo.ac.jp/~mdehoon/software/cluster/software.htm http://genetics.stanford.edu/~alok/TreeView/
CAGED	Model	*CAGED v 1.0 user manual* (2002)	http://genomethods.org/caged/about.htm
SVDMAN	Modified PCA	Alter, et al. (2000) and Wall et al. (2001)	http://public.lanl.gov/svdman/
EASE	Function analysis	Hosack, et al., (2003)	http://david.niaid.nih.gov/david/ease.htm
DAVID-KEGG	Pathway analysis	Glynn et al. (2003)	http://david.niaid.nih.gov

Web-Based Resources

A summary of the Web-based software support used in this chapter is provided in Table 5. In the table we list the names of the methods, a brief description, the corresponding references, and the freely available Web-based software support.

Summary

Our goals include the investigation and assessment of the reproducibility among different pattern differential and formulation methods, the optimization of the performance of a single model, and most importantly the provision of the procedures that integrate various methods and the derivations of their results. Through the developed hybrid methods we can increase the power and highlight the quality of the systematic analysis. Our results demonstrate that this idea is feasible and the resulting I/C clusters demonstrate that this hybrid method can be successfully achieved. The most novel innovation in this chapter is the successful clustering of the clusters themselves (hybrid clustering) in order to compare the results of different methods and the improvement of the power of reproducibility.

The biologist is faced with a "dizzying" array of choices when it comes to choosing the proper algorithm for pattern discovery. One approach is to utilize various algorithms and combine the results by hybrid methods such as clustering the clusters as was done in this chapter. Afterwards, Web-based resources such as EASE and DAVID-KEGG can be used to determine the tendency of the biological function of the genes in a discovered hybrid pattern as was also performed in this chapter. This may aid the determination of the function of unknown genes based on their inclusion in a given cluster.

This chapter discusses a promising way to find consistent, reproducible, nearly optimal solutions for genomic data analysis. It is clearly the way future scientists will have to investigate by putting together most available methods and all of the research results in order to make progress in a diverse, massively data intensive field, such as genomics or bioinformatics. Undoubtedly, such analysis is better equipped to discover new, biologically meaningful associations than separate semi-heuristic selection and application from a pool of specialized algorithms, each offering unique features, abilities. Also, questions can be addressed in a more informative way both during and after the analysis and could easily lead to insights beyond what is provided by the calculation of a single method and can significantly contribute to the generalization of study results and lead to the potential identification of the most promising or urgent research question. It may permit a more accurate calculation of some parameters such as the minimum sample size required for future studies. It can provide a more precise estimate of a treatment effect, and may explain heterogeneity between the results of separate studies. Results show that our analysis show lower false positive rate, higher power, and more stable results than most existing approaches.

References

Alter, O., Brown, P. O., & Botstein, D. (2000). Singular value decomposition for genome-wide expression data processing and modeling. *Proceedings of the National Academy of Science, 97*, (pp. 10101-10106).

Baldi, P., & Long, A. D. (2001). A bayesian framework for the analysis of microarray expression data: Regularized t-test and statistical inferences of gene changes. *Bioinformatics*, *17*(6), 509-519.

Benjamini, Y., & Hochberg, Y. (1995). Controlling false discovery rate: A practical and powerful approach to multiple testing. *Journal of Royal Statistical Society*, *B 57*, 289-300.

Dudoit, S., & Fridlyand, J. (2002). A prediction-based resampling method to estimate the number of clusters in a dataset. *Genome Biology*, *3*, RESEARCH0036.

Eisen, M. B., Spellman, P. T., Brown, P. O., & Botstein, D. (1998). Cluster analysis and display of genome-wide expression patterns. *Proceedings of the National Academy of Science,* 95, (pp. 14863-14868).

Glynn, D., Sherman, B. T., Hosack, D. A., Yang, J., Gao, W., Lane, H. C., et al. (2003). DAVID: Database for annotation, visualization, and integrated discovery. *Genome Biology*, *4*, R60.

Golub, T. R., Slonim, D. K., Tamayo, P., Huard, C., Gaasenbeek, M., Mesirov, J. P., et al. (1999). Molecular classification of cancer: Class discovery and class prediction by gene expression monitoring. *Science*, *286*, 531-537.

Hastiel, T., Tibshirani, R., Eisen, M. B., Alizadeh, A., Levy, R, Staudt, L., et al. (2000). Gene shaving as a method for identifying distinct sets of genes with similar expression patterns. *Genome Biology, 1*(2), research0003.1-0003.21.

Hosack, D., Hosack, A., Jr., G. D., Sherman, B. T., Lane, H. C., & Lempicki, R. A. (2003). Identifying biological themes within lists of genes with EASE. *Genome Biology*, *4*, R70.

IFNB Multiple Sclerosis Study Group. (1993). Interferon b-1b is effective in relapsing-remitting multiple sclerosis. I. Clinical results of a multicenter, randomized double-blind, placebo-controlled trail. *Neurology*, *43*, 655.

Jacobs, L. D., Cookfair, D. L., et. al. (1996). Intramuscular interferon b-1a for disease progression in relapsing multiple sclerosis: The Multiple Sclerosis Collaborative Research Group (MSCRG). *Annual Neurology*, *39*, 285.

Kerr, M., & Churchill, G. (2001). Bootstrapping cluster analysis: Assessing the reliability of conclusions from microarray experiments. *Proceedings of the National Academy of Science*, *97*, (pp. 8961-8965).

Khan, J., Wei, J. S., Ringner, M., Saal, L. H., Ladanyi, M., Westermann, F., et al. (2001). Classification and diagnostic prediction of cancers using gene expression profiling and artificial neural networks. *Nature Medicine*, *7*(6), 673-679.

McShane, L. M., Radmacher, M. D., Friedlin, B., Yu, R., Li, M. C., & Simon, R. (2002). Methods for assessing reproducibility of clustering patterns observed in analyses of microarray data. *Bioinformatics*, *18*, 1462-1469.

Neal, S. H., Madhusmita, M., Holter, N. S., Mitra, M., Maritan, A., Cieplak, M., et al. (2000). Fundamental patterns underlying gene expression profiles: Simplicity from complexicity. *Proceedings of the National Academy of Science, 97*, 8409-8414.

Ramanathan, M., Weinstock-Guttman, B., Nguyen, L. T., Badgett, D., Miller, C., Patrick, K., et al. (2001). Gene expression revealed using DNA arrays: The pattern in multiple sclerosis patients compared with normal subjects. *Journal of Neuroimmunology*, *116*(2), 213-219.

Ramoni, M., Sebastiani, P., & Kohane, I. (2002). Cluster analysis of gene expression dynamics. *Proceedings of the National Academy of Science, 99*, 9121- 9126.

Slonim, D. K. (2002). From patterns to pathways: Gene expression data analysis comes of age. *Nature genetics supplement*, *32*, 502.

Steinman, L., & Zamvil, S. (2003). Transcriptional analysis of targets in multiple sclerosis, *Nature Review Immunology, June 3*(6), 483-92.

Storey, J. D., & Tibshirani, R. (2003). Statistical significance for genomewide studies. *Proceedings of the National Academy of Science, 100*(16), 9440-9445.

Tamayo, T., Slonim, D., Mesirov, J., Zhu, Q., Kitareewan, S., Dmitrovsky, E., Lander, E. S., & Golub, T. R. (1999). Interpreting patterns of gene expression with self-organizing maps: Methods and application to hematopoietic differentiation. *Proceedings of the National Academy of Science, 96*, (pp. 2907-2912).

Tibshirani, R., Walter, G., & Hastie, T. (2001). Estimating the number of clusters in a data set via the gap statistic. *Journal of Royal Statistical Society, B63*, (pp. 411-423).

Tusher, V. G., Tibshirani, R., & Chu, G. (2001). Significance analysis of microarrays applied to the ioniaing radiation response. *Proceedings of the National Academy of Science, 9*, 5116-5121.

Wall, M. E., Dyck, P. A., & Brettin, T. S. (2001). SVDMAN - Singular value decomposition analysis of microarray data. *Bioinformatics, 17*, 566-568.

Weinstock-Guttman, et al. (2003). Genomic effects of IFN-beta in multiple sclerosis patients. *Journal of Immunology, 171*, 2694-702.

West, M. (2000). *Bayesian regression analysis in the "Large p, Small n" paradigm* (Technical Report 00-22). Institute of Statistics and Decision Sciences, Duke University, CSE-2000-08-01.

Westfall, P., & Young, S. (1993). *Resampling-based multiple testing*. Wiley-Interscience.

Wuju, L., & Momiao, X. T. (2002). Class: Tumor classification system based on gene expression profile. *Bioinformatics, 18*, 325-326.

Yeung, K. Y., & Ruzzo, W. L. (2001). Principal component analysis for clustering gene expression data. *Bioinformatics, 17*, 763-774.

Chapter IX

Parameterless Clustering Techniques for Gene Expression Analysis

Vincent S. Tseng, National Cheng Kung University, Taiwan
Ching Pin Kao, National Cheng Kung University, Taiwan

Abstract

In recent years, clustering analysis has even become a valuable and useful tool for in-silico analysis of microarray or gene expression data. Although a number of clustering methods have been proposed, they are confronted with difficulties in meeting the requirements of automation, high quality, and high efficiency at the same time. In this chapter, we discuss the issue of parameterless clustering technique for gene expression analysis. We introduce two novel, parameterless and efficient clustering methods that fit for analysis of gene expression data. The unique feature of our methods is they incorporate the validation techniques into the clustering process so that high quality results can be obtained. Through experimental evaluation, these methods are shown to outperform other clustering methods greatly in terms of clustering quality, efficiency, and automation on both of synthetic and real data sets.

Introduction

Clustering analysis has been applied in a wide variety of fields such as biology, medicine, psychology, economics, sociology, and astrophysics. The main goal of clustering analysis is to partition a given set of objects into homogeneous groups based on their features such that objects within a group are more similar to each other and more different from those in other groups (Chen, Han, & Yu, 1996). Clustering groups the genes into biologically relevant clusters with similar expression patterns. The genes clustered together tend to be functionally related, hence clustering can reveal the co-expression of genes which were previously uncharacterized. In recent years, clustering analysis has even become a valuable and useful tool for in-silico analysis of microarray or gene expression data (Eisen et al., 1998; Alon et al., 1999; Ben-Dor & Yakhini, 1999; Tamayo et al., 1999). For example, Eisen et al. (1998) applied a variant of hierarchical clustering to identify groups of co-expressed yeast genes. Alon et al. (1999) used a two-way clustering technique to detect groups of correlated genes and tissues. Self-organizing maps were used by Tamayo et al. (1999) to identify clusters in the yeast cell cycle data set and human hematopoietic differentiation data set.

Thoroughly and extensive overviews of clustering algorithms are given by Aldenderfer and Blashfield (1984) and Jain and Dubes (1998). Although a number of clustering methods have been studied in the literature (Carpenter & Grossberg, 1987; Kohonen, 1990; Zhang, Ramakrishnan, & Livny, 1996; Chen, Han, & Yu, 1996; Guha, Rastogi, & Shim, 1998; Guha, Rastogi, & Shim, 1999), they are not satisfactory in terms of: (1) automation, (2) quality, and (3) efficiency. First, most clustering algorithms request users to specify some parameters. In real applications, however, it is hard for biologists to determine the suitable parameters manually. Thus an automated clustering method is required. Second, most clustering algorithms aim to produce the clustering results based on the input parameters and their own criterions. Hence, they are incapable of producing optimal clustering result. Third, the existing clustering algorithms may not perform well when the optimal or near-optimal clustering result is enforced from the universal criterions.

On the other hand, a variety of clustering validation measures are applied to evaluate the validity of the clustering results, the suitability of parameters, or the reliability of clustering algorithms. A good overview of clustering validation can be found in the book "Cluster Analysis" (Aldenderfer & Blashfield, 1984), in which numerous validation index are considered, like DB-index (Davies & Bouldin, 1979; Aldenderfer & Blashfield, 1984), Simple matching coefficient (Aldenderfer & Blashfield, 1984; Jain & Dubes, 1998), Jaccard coefficient (Aldenderfer & Blashfield, 1984; Jain & Dubes, 1998), Hubert's Γ statistic (Aldenderfer & Blashfield, 1984; Jain & Dubes, 1998), etc. There also exist several other measures, like ANOVA (Kerr & Churchill, 2001), FOM (Yeung, Haynor, & Ruzzo, 2001), VCV (Hathaway & Bezdek, 2003). Nevertheless, the roles of them are placed only on the phase of "post-validation". The study on how to integrate validation techniques with clustering methods tightly for improving the clustering quality has been absent.

In this chapter, we will describe new clustering methods that are integrated with validation techniques. We introduce two novel, parameterless and efficient clustering methods that are fit for analysis of gene expression data. The methods determine the best grouping of genes on-the-fly without any user-input parameter.

Background

The related work of clustering analysis can be divided into three types (Aldenderfer & Blashfield, 1984). The first type is similarity measurements that will affect the final clustering results directly. Euclidean distance and correlation coefficient are most popular similarity measures. Surely this is an important task for clustering analysis, but it is not the focus of this chapter. The second type is clustering methods that are the core of clustering analysis and have received extensive attentions. The third type is validation techniques that are applied to evaluate the validity of the clustering results, the suitability of parameters, or the reliability of clustering algorithms. In addition, all validation techniques may be roughly divided into two main types: scalar measurements and intensity image methods (Hathaway & Bezdek, 2003). Clustering methods and validation techniques are introduced briefly in the following section.

Clustering Methods

The most well-known clustering method is probably the k-means algorithm (Aldenderfer & Blashfield, 1984; Jain & Dubes, 1998), which is a partitioning method. The k-means algorithm partitions the dataset into k groups, based primarily on the distance between data items, where k is a parameter specified by the user. Hierarchical methods, for example, UPGMA (Rohlf, 1970), BIRCH (Zhang, Ramakrishnan, & Livny, 1996), CURE (Guha, Rastogi, & Shim, 1998), and ROCK (Guha, Rastogi, & Shim, 1999), are another popular kind of clustering methods. Agglomerative or divisive are used to represent distance matrices as ultrametric trees. In recent years, artificial neural networks, such as self-organizing maps (SOM) (Kohonen, 1990), competitive learning networks (Rumelhart & Zipser, 1985), and adaptive resonance theory (ART) networks (Carpenter & Grossberg, 1987), have also been used for clustering.

Cluster affinity search technique (CAST) (Ben-Dor & Yakhini, 1999) takes as input a parameter called the affinity threshold t, where $0 < t < 1$, and tries to guarantee that the average similarity in each generated cluster is higher than the threshold t. CAST generates one cluster at a time and selects the object with the most neighbors as a seed for the current cluster. It adds un-clustered objects with high affinity (the average similarity between the object and the cluster is greater than t) to the current cluster and removes objects with low affinity from the current cluster iteratively. The main advantage of CAST is that it can detect outliers more effectively.

Scalar Measurements

Jain and Dubes (1998) divided the cluster validation procedure into two main parts: external and internal criterion analysis. External indices are used to measure the extent to which cluster labels (a clustering result) match externally supplied class labels (a partition that is known a priori). There are many statistical measures that assess the agreement between an external criterion and a clustering result. For example, Milligan, Soon, and Sokol (1983) and Milligan and Cooper (1986) evaluated the performance of different clustering algorithms and different statistical measures of agreement for both synthetic and real data. The problem of external criterion analysis is that reliable external criteria are rarely available when gene expression data are analyzed. In contrast, internal indices are used to measure the goodness of a clustering structure without respect to external information. For example, compactness and isolation of clusters are possible measures of goodness of fit. A measure called the Figure of Merit (FOM) was used by Yeung, Haynor, and Ruzzo (2001) to evaluate the quality of clustering performed on a number of real datasets.

We introduce an internal index, namely *Hubert's Γ (gamma) statistic* (Aldenderfer & Blashfield, 1984; Jain & Dubes, 1998) and two external indices, namely *Simple matching coefficient* and *Jaccard coefficient* (Aldenderfer & Blashfield, 1984; Jain & Dubes, 1998). The definition of Hubert's Γ statistic is as follows:

Let $X=[X(i,j)]$ and $Y=[Y(i,j)]$ be two $n \times n$ proximity matrices on the same n genes. From the viewpoint of correlation coefficient, $X(i, j)$ indicates the observed correlation coefficient of genes i and j and $Y(i,j)$ is defined as follows:

$$Y(i,j) = \begin{cases} 1 & \text{if genes } i \text{ and } j \text{ are clustered in the same cluster,} \\ 0 & \text{otherwise.} \end{cases} \quad (1)$$

The Hubert's Γ statistic represents the point serial correlation between the matrices X and Y, and is defined as follows when the two matrices are symmetric:

$$\Gamma = \frac{1}{M}\sum_{i=1}^{n-1}\sum_{j=i+1}^{n}\left(\frac{X(i,j)-\overline{X}}{\sigma_X}\right)\left(\frac{Y(i,j)-\overline{Y}}{\sigma_Y}\right), \quad (2)$$

where $M = n\,(n-1)/2$ is the number of entries in the double sum, and σ_X and σ_Y denote the sample standard deviations while $\overline{X}$ and $\overline{Y}$ denote the sample means of the entries of matrices X and Y.

From the viewpoint of distance, $X(i,j)$ indicates the observed distance of genes i and j and $Y(i,j)$ is defined as (1) by exchange the "1" and "0". The value of Γ is between [-1, 1] and a higher value of Γ represents the better clustering quality.

Figure 1. The concept of Hubert's Γ statistic

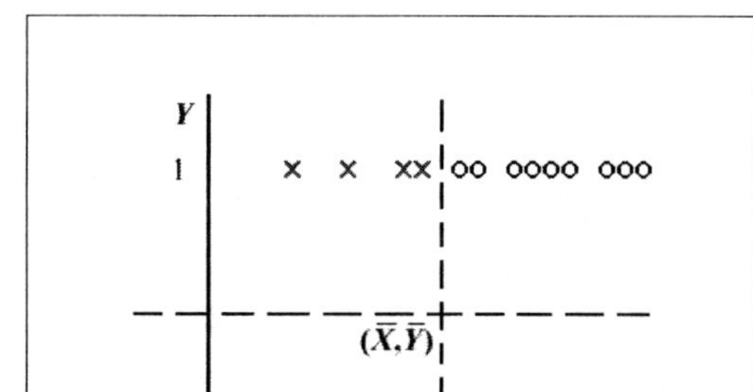

Figure 1 illustrates the concept of Hubert's Γ statistic. For matrices *X* and *Y* as described above, the solid lines represent the original coordinate and the dotted lines represent the coordinate centered by $(\bar{X},\bar{Y})$. A point $A_{i,j}$ that falls in quadrant I of dotted coordinate indicate that genes *i* and *j* have higher similarity and are clustered in the same cluster. A point $B_{i,j}$ that falls in quadrant II indicate that genes *i* and j have lower similarity but are clustered in the same cluster. In contrast, a point $C_{i,j}$ that falls in quadrant III of dotted coordinate indicate that genes *i* and *j* have lower similarity and are clustered in different clusters. A point $D_{i,j}$ that falls in quadrant IV indicate that genes *i* and *j* have higher similarity but are clustered in different clusters. In brief, the more points that fall in quadrants I and III, the better the quality of clustering results are. Therefore, the point serial correlation between the matrices *X* and *Y* can be used to measure the quality of clustering results.

The definitions of Simple matching coefficient and Jaccard coefficient (Aldenderfer & Blashfield, 1984; Jain & Dubes, 1998) are as follows:

Let $U=[U(i,j)]$ and $V=[V(i,j)]$ be two $n \times n$ binary matrices on the same *n* genes. The matrices indicate two distinct clustering results and are defined as follows:

Table 1. Association table

		Matrix ***U***	
		1	0
Matrix ***V***	1	*a*	*b*
	0	*c*	*d*

$$G(i, j) = \begin{cases} 1 & \text{if genes } i \text{ and } j \text{ are clustered in the same cluster,} \\ 0 & \text{otherwise} \end{cases}, G{=}U, V. \quad (3)$$

Let us consider the contingency table of the matrices U and V as shown in Table 1. Where a indicates the number of entries on which U and V both have values "1" and b indicates the number of entries on which U have values "1" and V have values "0", and so on.

The Simple matching coefficient represents the point serial correlation between the matrices $\boldsymbol{X}$ and $\boldsymbol{Y}$, and is defined as follows when the two matrices are symmetric:

The Simple matching coefficient is defined by:

$$S = \frac{a + d}{a + b + c + d}. \quad (4)$$

that is total number of matching entries divided by total number of entries. The Jaccard coefficient is defined by:

$$S = \frac{a + d}{a + b + c}. \quad (5)$$

which is similar to the Simple matching coefficient, only with "negative" matches (d) are ignored. The Simple matching coefficient usually varies over a smaller range than the Jaccard coefficient because "negative" matches are often a dominant factor.

Intensity Image Methods

One main problem of scalar validity measures is that representing the accuracy of different clustering methods by a single real number may lose much information. Accordingly, intensity image methods use all of the information generated by the clustering methods to produce an $n{\times}n$ gray scale intensity image. It converts the similarity (or dissimilarity) into gray scale and reorders the data of similarity matrix to display the clustering result. For instance, VCV (Hathaway & Bezdek, 2003) approach retains and organizes the information that is lost through the massive aggregation of information by scalar measures.

Smart-Cast and Correlation Search Technique (CST)

In this section, we will present the ideas and algorithms of our novel and parameterless clustering methods, namely Smart-CAST (Tseng & Kao, 2005) and Correlation Search Technique (CST) (Tseng & Kao, 2005). Before the methods can be applied, we have to produce a similarity matrix *S* based on the original dataset. The matrix *S* stores the degree of similarity between each pair of genes in the dataset, and the range of the degree is [0, 1]. This much reduces the computation overhead incurred by some clustering algorithms that calculate the similarities dynamically. You may obtain the similarity by using any similarity measurements (e.g., Euclidean distance, Pearson's correlation coefficient, etc.) for various purposes. Then, Smart-CAST and CST can automatically cluster the genes according to the similarity matrix *S* without any user input parameters.

Tseng & Chen (2002) evaluated a number of validity indexes for measuring the quality of clustering results. They concluded from the experimental results that *Hubert's Γ (gamma) statistic* (Aldenderfer & Blashfield, 1984; Jain & Dubes, 1998) might be the best index for both partition-based clustering methods and density-based methods for both low-similarity and high-similarity data. So we use the validation index, namely Hubert's Γ statistic, as the key measure of clustering quality during the clustering process.

Smart-CAST

Preview of Smart-CAST

The main ideas of the Smart-CAST method are as follows. In the first step, a density-and-affinity based algorithm is applied as the basic clustering algorithm. Along with a specified input parameter, the basic clustering algorithm utilizes the similarity matrix *S*

Figure 2. Hubert's Γ statistic vs. values of t

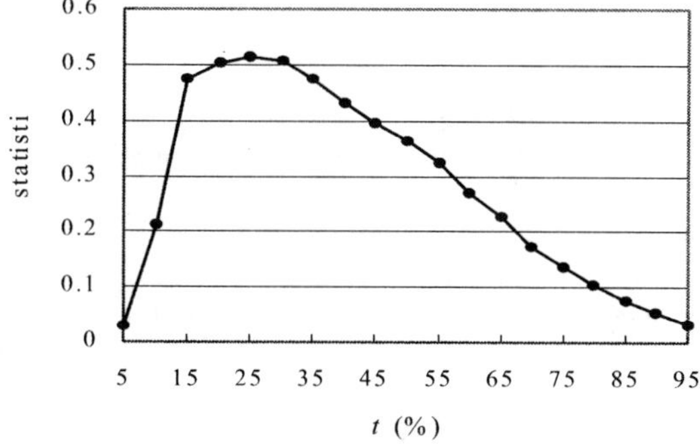

to conduct the clustering task. Thus, a clustering result will be produced by the basic clustering algorithm based on the given input parameter. A good candidate for the base clustering algorithm is CAST (Ben-Dor & Yakhini, 1999), which needs only one input parameter, called the affinity threshold t, where $0 < t < 1$. The reasons why we prefer to use CAST as the base clustering method are as follows:

1. CAST has the capability of isolating outliers.
2. Unlike other algorithms that are unstable (e.g., k-means), CAST is a stable algorithm.
3. The execution time of CAST is shorter than most of the other clustering algorithms.

In the second step, a validation test is performed to evaluate the quality of the clustering result produced in step two. We adopt Hubert's Γ statistic (Aldenderfer & Blashfield, 1984; Jain & Dubes, 1998) to measure the quality of clustering, because Tseng & Chen (2002) concluded from the experimental results that Hubert's Γ might be the best index.

With the above steps, it is clear that high quality clustering can be achieved by applying a number of different values of the affinity threshold t as input parameters to the CAST algorithm, by calculating the Hubert's Γ statistic of each clustering result, and by choosing the one with the highest value of the Hubert's Γ statistic as the output. In this way, a local-optimal clustering result can be obtained by users automatically. An example is shown in Figure 2, where the X axis represents the values of the affinity threshold t input to CAST and the Y axis shows the obtained Hubert's Γ statistic for each of the clustering results. The peak in the curve corresponds to the best clustering result, which has a Γ statistic value of around 0.52 when t is set to be 0.25.

This approach is feasible in that firstly, CAST executes very quickly since the similarity matrix of gene expressions is obtained in advance; secondly, the Hubert's Γ statistic for each clustering result can be calculated easily. However, one problem encountered is how to determine suitable values of the affinity threshold t. The easiest way is to increment the value of the affinity threshold t with a fixed interval. For example, we may increase the value of t from 0.05 to 0.95 in increments of 0.05. We call this approach CAST-FI (Fixed Increment) in the following. The main disadvantage of CAST-FI is that many iterations of computations are required. Therefore, a new method is introduced in the next section to reduce the computation overhead.

Computation Reduction of Smart-CST

The idea behind the method is to reduce the amount of computation by eliminating unnecessary executions of clustering so as to obtain a "nearly-optimal" clustering result instead of the optimal one. That is, we try to execute CAST as few times as possible.

Therefore, we need to narrow down the range of the parameter affinity threshold t effectively. The method works as follows:

1. Initially, a testing range R for setting the affinity threshold t is set to be [0, 1]. We divide R equally into m parts with the points $P_1, P_2, \ldots, P_{m-1}$, where $P_1 < P_2 < \ldots < P_{m-1}$, $m \geq 3$. Then, the value of each P_i is taken as the affinity threshold t for executing CAST, and the G statistic of the clustering result for each of P_i is calculated. We call this process a "run."
2. When a clustering run is completed, the clustering at point P_b that produces the highest G statistic is considered to be the best clustering. The testing range R is then replaced by the range $[P_{b-1}, P_{b+1}]$ that contains the point P_b.
3. The above process is repeated until the testing range R is smaller than a threshold δ or the difference between the maximal and minimal quality values is smaller than another threshold σ.
4. The best quality clustering result obtained through the tested process is output as the answer.

In this way, we can obtain a clustering result that has "nearly-optimal" clustering quality with much less computation. In the next section, through empirical evaluation, we shall evaluate how good the generated clustering result is and to what extent the amount of computation could be reduced by our approach.

Correlation Search Technique (CST)

Preview of CST

The main idea behind CST is to integrate clustering method with validation technique so that it can cluster the genes quickly and automatically. Since validation technique is embedded, CST can obtain a "near-optimal" clustering result.

The validation index we used is Hubert's Γ statistic (Aldenderfer & Blashfield, 1984; Jain & Dubes, 1998). Since the computing of Γ statistic may spend considerable time, we simplify it to enhance the performance of our approach. We expand Hubert's Γ statistic as follows:

$$\Gamma = \frac{\left(M \sum_{i=1}^{n-1} \sum_{j=i+1}^{n} X(i,j)Y(i,j) - \sum_{i=1}^{n-1} \sum_{j=i+1}^{n} X(i,j) \sum_{i=1}^{n-1} \sum_{j=i+1}^{n} Y(i,j) \right)}{\sqrt{M \sum_{i=1}^{n-1} \sum_{j=i+1}^{n} X(i,j)^2 - \left(\sum_{i=1}^{n-1} \sum_{j=i+1}^{n} X(i,j) \right)^2} \sqrt{M \sum_{i=1}^{n-1} \sum_{j=i+1}^{n} Y(i,j)^2 - \left(\sum_{i=1}^{n-1} \sum_{j=i+1}^{n} Y(i,j) \right)^2}}. \quad (6)$$

While we decide which clustering result is better, the measurement Γ' of quality of clustering result can be simplified as follows:

$$\Gamma' = \frac{M\sum_{i=1}^{n-1}\sum_{j=i+1}^{n} X(i,j)Y(i,j) - \sum_{i=1}^{n-1}\sum_{j=i+1}^{n} X(i,j)\sum_{i=1}^{n-1}\sum_{j=i+1}^{n} Y(i,j)}{\sqrt{M\sum_{i=1}^{n-1}\sum_{j=i+1}^{n} Y(i,j) - \left(\sum_{i=1}^{n-1}\sum_{j=i+1}^{n} Y(i,j)\right)^2}}. \tag{7}$$

We call (7) as simplistic Hubert's Γ statistic. For more details, readers are referred to the literature by Tseng and Kao (2005).

CST Algorithm

The input of CST is a symmetric similarity matrix $\boldsymbol{X}$, where $X(i,j) \in [0, 1]$. CST is a greedy algorithm that constructs clusters one at a time, and the currently constructed cluster is denoted by C_{open}. Each cluster is started by a seed and is constructed incrementally by adding (or removing) elements to (or from) C_{open} one at a time. The temporary clustering result of each addition (or removal) of x is computed by simplistic Hubert's Γ statistic, i.e. (7), and is denoted by $\Gamma_{add}(x)$ (or $\Gamma_{remove}(x)$). In addition, the currently maximum of simplistic Hubert's Γ statistic is denoted by Γ_{max}. We say that an element x has high positive correlation if $\Gamma_{add}(x) \geq \Gamma_{max}$, and x has high negative correlation if $\Gamma_{remove}(x) \geq \Gamma_{max}$. CST takes turns between adding high positive correlation elements to C_{open}, and removing high negative correlation elements from it. When C_{open} is stabilized by addition and removal procedure, this cluster is complete and next one is started.

The addition and removal procedure strengthen the quality of C_{open} gradually. Moreover, the removal procedure exterminates the cluster members that were inaccurately added at early clustering stages. In addition, a heuristics is added to CST for selecting an element with the maximum number of neighbors to start a new cluster.

To save computation time, we simplify the simplistic Hubert's Γ statistic, i.e. (7), further. While we decide which element will be added to (or removed from) C_{open}, the measurement Γ'' of effect of each added (or removed) element is:

$$\Gamma'' = \sum_{i=1}^{n-1}\sum_{j=i+1}^{n} X(i,j)Y(i,j) \tag{8}$$

Figure 3. A pseudo-code of CST

```
Input: An n-by-n similarity matrix X

0. Initialization:
    M = n (n - 1) / 2
    S_X = Σ_{i=1}^{n-1} Σ_{j=i+1}^{n} X(i, j)
    S_Y = 0
    S_XY = 0
    C = Ø                          /* The collection of closed clusters */
    U = {1, 2, ..., n}             /* Elements not yet assigned to any cluster */
    Γ_max = 0
1. while (U ≠ Ø) do
    C_open = Ø
    a (•) = 0
    1.1. SEED: Pick an element u ∈ U with most neighbors
        U = U - {u}                                    /* Remove u from U */
        For all i ∈ U set a(i) = X(u, i)               /* Update the affinity*/
        C_open = {u}                                   /* Insert u into C_open */
    1.2. ADD: while MaxValidaty( ) ≥ Γ_max do
        Pick an element u ∈ U with maximum a(•)
        U = U - {u}                                    /* Remove u from U */
        S_Y = S_Y + |C_open|
        S_XY = S_XY + a(u)
        For all i ∈ U ∪ C_open set a(i) = a(i) + X(u, i)   /* Update the affinity*/
        C_open = C_open ∪ {u}                          /* Insert u into C_open */
        Γ_max = MaxValidaty( )
    1.3. REMOVE: while MaxValidaty( ) > Γ_max do
        Pick an element v ∈ C_open with minimum a(•)
        C_open = C_open - {v}                          /* Remove v from C_open */
        S_Y = S_Y - |C_open|
        S_XY = S_XY - a(u)
        For all i ∈ U ∪ C_open set a(i) = a(i) - X(u, i)   /* Update the affinity*/
        U = U ∪ {v}                                    /* Insert v into U */
        Γ_max = MaxValidaty()
    1.4. Repeat steps ADD and REMOVE as long as there are no elements been removed.
    1.5. C = C ∪ {C_open}
  end
2. Done, return the collection of cluster, C.
```

More details on the CST algorithm can be found in the literature by Tseng & Kao (2005). A pseudo-code of CST is shown in Figure 3. The subroutine MaxValidaty(•) computes the possible maximum of measurement Γ yet, i.e. (7), while a certain element is added (or removed). In the addition stage, it is equal to:

$$\frac{\left(M*(S_{XY}+\max\{a(u)\mid u\in U\})-S_X*(S_Y+|C_{open}|)\right)}{\sqrt{M*(S_Y+|C_{open}|)-(S_Y+|C_{open}|)^2}} \tag{9}$$

For the removal stage, it becomes:

$$\frac{\left(M*(S_{XY} - \min\{a(v) \mid v \in C_{open}\}) - S_X*(S_Y - |C_{open}| + 1)\right)}{\sqrt{M*(S_Y - |C_{open}| + 1) - (S_Y - |C_{open}| + 1)^2}} \qquad (10)$$

Figure 4. Profiles of the four seed sets: (a) Dataset I, (b) Dataset II, (c) Dataset III, (d) Dataset IV

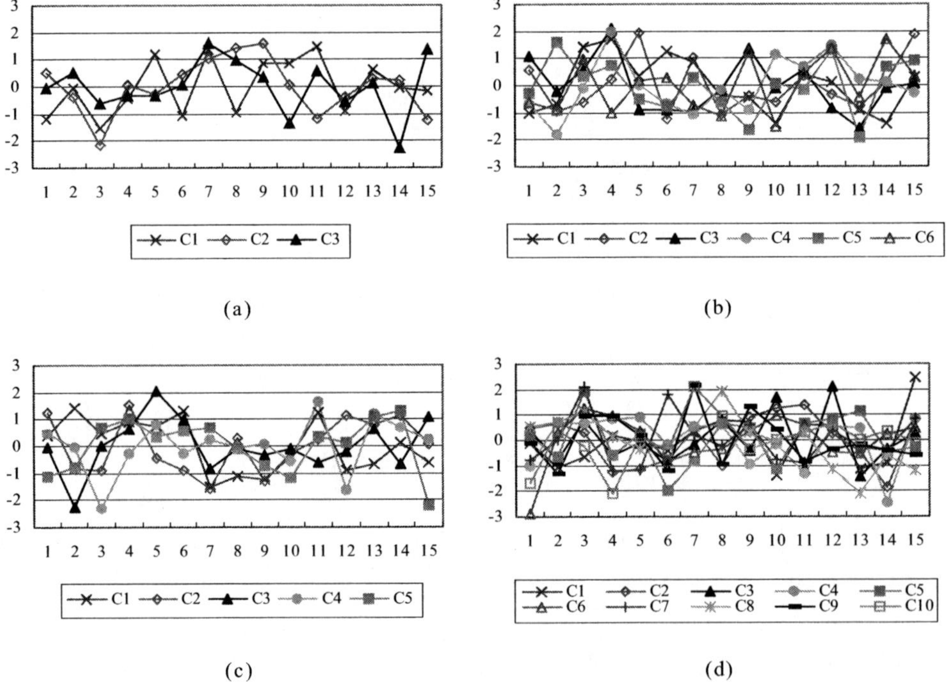

Empirical Evaluation

We process a series of experiments to evaluate the performance and accuracy of the Smart-CAST and CST methods. Accordingly, we tested the methods on four synthetic datasets generated by our correlation-inclined cluster dataset generator. Meanwhile, a real dataset, namely the yeast cell-cycle dataset by Spellman et al. (1998), was also used for testing the methods. In the synthetic data generator, the users can set up some parameters for generating various kinds of gene expression datasets with variations in terms of the number of clusters, the number of genes in each cluster, the number of outliers (or noise), etc. Firstly, a number of seed genes are produced by the generator or input by the user. All the seed genes must have same dimensions (number of conditions) and very low similarity mutually so as to ensure that the generated clusters will not overlap. Second, the generator produces clusters one by one. It randomly produces genes in each cluster, and all produced genes are similar to the seed gene of this cluster in the viewpoint of correlation. Third, the outliers are produced randomly and added into the dataset. It should be noted that the outliers may fall in some clusters. In this way, all genes in the same cluster will have very high similarity and they will have high dissimilarity with genes in other clusters.

We first generate four seed sets with size three, six, five, and ten, respectively. The seeds in each set have 15 dimensions and their correlation coefficient is less than 0.1. The profiles of these four seed sets are shown in Figure 1. Then, we load these four seed sets into our dataset generator to generate four synthetic gene expression datasets for testing, respectively, namely Dataset I, Dataset II, Dataset III and Dataset IV. The Dataset I contains three gene clusters of sizes 900, 700, and 500, with additional 400 outliers. The Dataset II contains six gene clusters of sizes 500, 450, 400, 300, 250, and 200, with additional 400 outliers. The Dataset III contains five gene clusters of sizes 1200, 1000, 800, 700, and 500, with 800 additional outliers. The Dataset IV contains ten gene clusters of sizes 650, 550, 550, 500, 450, 400, 350, 300, 250, and 200, with 800 additional outliers. Table 2 shows the cluster structures of these four datasets.

For the real dataset, the yeast cell-cycle dataset contains 72 expressions (includes alpha factor, cdc15, cdc28, and elutriation). Spellman et al. (1998) used hierarchical clustering and analysis of the 5' regions of the genes to identify 8 groups, namely CLN2 (76 genes),

Table 2. Cluster structure of datasets

	Dataset I	Dataset II	Dataset III	Dataset IV
# Clusters	3	6	5	10
Cluster Size	900, 700, 500.	500, 450, 400, 300, 250, 200.	1200, 1000, 800, 700, 500.	650, 550, 550, 500, 450, 400, 350, 300, 250, 200.
# Outliers	400	400	800	800
# Total Patterns	2500	2500	5000	5000

Y' (31 genes), Histone (9 genes), MET (20 genes), CLB2 (35 genes), MCM (34 genes), SIC1 (27 genes), and MAT (13 genes). Since several groups are similar in terms of the expression profiles, the 8 groups are further grouped into 4 main clusters according to the hierarchical structure of Spellman's analysis, namely G1 (CLN2 and Y' groups), S/G2 (Histone and MET groups), M (CLB2 and MCM groups), and M/G1 (SCI1 and MAT groups). Since we would like to cluster the genes with main considerations on the expression profiles, we take the G1, S/G2, M, and M/G1 groups as the standard for the clustering results.

We compare the Smart-CAST and CST methods with the well-known clustering method, namely k-means (Aldenderfer & Blashfield, 1984; Jain & Dubes, 1998), and the CAST-FI methods we introduced before. For k-means, the value of k is varied from 2 to 21 in step of 1, and from 2 to 41 in step of 1, respectively. For CAST-FI, the value of affinity threshold t is varied from 0.05 to 0.95 in fixed increment of 0.05. Furthermore, the parameters m, δ and σ were set to default values of 4, 0.01 and 0.01, respectively The quality of clustering results was measured by using Hubert's Γ statistic, Simple matching coefficient (Aldenderfer & Blashfield, 1984; Jain & Dubes, 1998), and Jaccard coefficient (Aldenderfer & Blashfield, 1984; Jain & Dubes, 1998). Simple matching coefficient and Jaccard coefficient can evaluate the quality of clustering results in pre-clustering dataset. We also use intensity image to exhibit the visualized results of the clustering methods. Since correlation coefficient is widely adopted in most studies on gene expression analysis, we evaluate only correlation coefficient for the similarity measure.

Simulated Datasets

Table 3 presents the total execution time and the best clustering quality of the tested methods on Dataset I. The notation "M" indicates the number of main clusters, and we consider clusters whose sizes are smaller than 50 and 5 as outliers for simulated datasets and the yeast cell-cycle dataset, respectively.

It is observed that CST, Smart-CAST and CAST-FI outperform k-means substantially in both of execution time and clustering quality, i.e., Hubert's Γ statistic, Simple matching coefficient, and Jaccard coefficient. In particular, CST performs 396 times to 1638 times faster on Dataset I than k-means with k ranged from 2 to 21 and 2 to 41, respectively. In

Table 3. Experimental results (Dataset I)

Methods	Time (s)	# Clusters	Γ Statistic	Matching Coefficient	Jaccard Coefficient
CST	< 1	65 (M=3)	0.800	0.981	0.926
Smart-CAST	12	85 (M=3)	0.800	0.986	0.944
CAST-FI	54	91 (M=3)	0.799	0.986	0.945
k-means (k=2-21)	396	6 (M=6)	0.456	0.825	0.427
k-means (k=2-41)	1638	6 (M=6)	0.456	0.825	0.427

Figure 5. Intensity images on Dataset I: (a) The prior cluster structure, (b) the actual input to the algorithms, (c) clustering results of CST, (d) clustering results of CAST

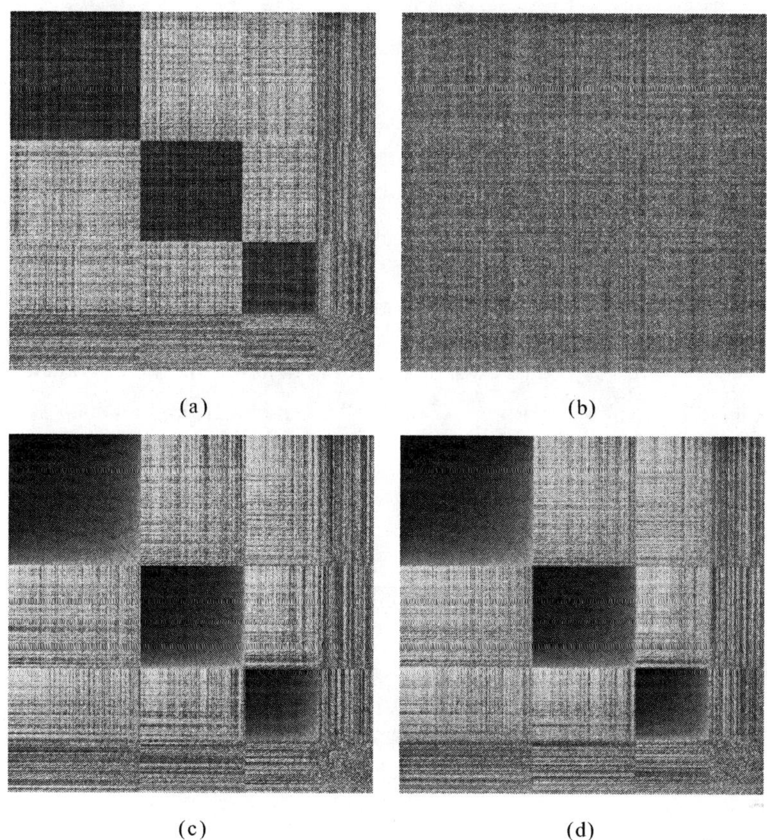

addition, the results also show that the clustering quality generated by CST is very close to that of Smart-CAST and CAST-FI in all of Hubert's Γ statistic, Simple matching coefficient, and Jaccard coefficient. It means that the clustering quality of our approach is as good as Smart-CAST and CAST-FI even though the computation time of our approach is reduced substantially.

Table 3 also shows that k-means produced six main clusters as the best clustering result for Dataset I, while the sizes of clusters are 580, 547, 457, 450, 352, and 114. This is true no matter k is varied from 2 to 21 and from 2 to 41 and does not match the original cluster structure. In contrast, CST produces 65 clusters as the best clustering result. In particular, it is clear that three main clusters are generated, with sizes 912, 717, and 503. This tightly matches the original cluster structure.

The intensity images of the original cluster structure and the clustering results for Dataset I are shown in Figure 5. From Figure 5(a), we easily observe that there are three

Table 4. Experimental results (Yeast dataset)

Methods	Time (s)	# Clusters	Γ Statistic	Matching Coefficient	Jaccard Coefficient
CST	< 1	5 (M=4)	0.703	0.983	0.941
Smart-CAST	< 1	5 (M=5)	0.706	0.978	0.922
CAST-FI	3	5 (M=5)	0.705	0.973	0.908
k-means (k=2-21)	75	5 (M=5)	0.683	0.914	0.724
k-means (k=2-41)	325	5 (M=5)	0.683	0.914	0.724

main clusters and quite a number of outliers in Dataset I. It is observed that Figure 5(c) and Figure 5(d) are very similar to Figure 5(a), meaning that the clustering results of CST and Smart-CAST are very similar to the original cluster structure in Dataset I.

It is obvious that the observations on Dataset II, Dataset III and Dataset IV are similar to that on Dataset I. For more detail, readers are referred to the literature by Tseng & Kao (2005).

The Yeast Cell-Cycle Dataset

Table 4 presents the total execution time and the best clustering quality of the tested methods on the real dataset, i.e., the yeast cell-cycle dataset as described previously. The CST method produced four main clusters with one outlier, while other methods generated five main clusters. It is observed that CST, Smart-CAST and CAST-FI outperform k-

Figure 6. Intensity images of the yeast dataset: (a) The original cluster structure, (b) Clustering results of CST

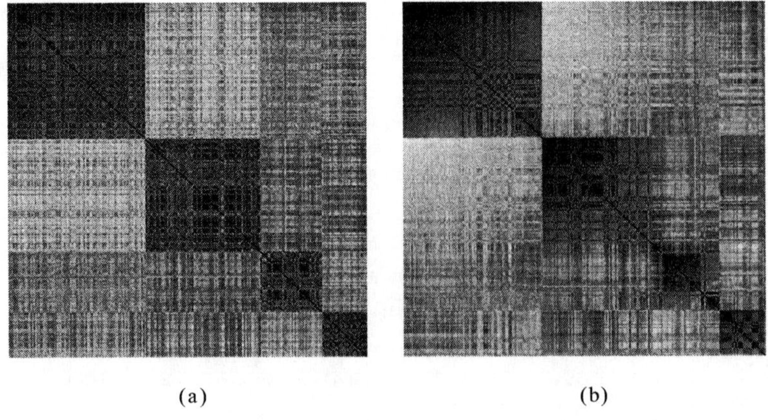

(a) (b)

means substantially in execution time and clustering quality. In particular, CST delivers the best execution time and clustering quality. Specifically, CST performs 75 to 325 times faster than k-means with *k* ranged as [2, 21] and [2, 41], respectively.

To investigate the clustering quality, Figure 6 shows the intensity images for the original cluster structure of the dataset and the clustering results. From Figure 6(a), we can easily observe that there are four main clusters in the yeast dataset. It is observed that Figure 6(b) is very similar to Figure 6(a), meaning that the clustering result generated by CST is very close to the original cluster structure. In fact, CST produced four main clusters, namely G1, S/G2, M and M/G1 clusters, with one outlier named SRD1. This matches the actual structure of this yeast dataset, as described previously. In contrast, Smart-CAST, CAST-FI, and k-means produced five main clusters as the best clustering results, which is kind of deviated from the actual structure of the dataset. Hence, it is shown that CST outperforms other methods substantially in terms of execution time and clustering quality on the real yeast cell-cycle dataset.

Summary

Clustering analysis is a valuable and useful technique for in-silico analysis of microarray or gene expression data, but most clustering methods incur problems in the aspects of automation, quality, and efficiency. In this chapter, we explore the effective integration between clustering methods and validation techniques. We introduce two novel, parameterless, and efficient clustering methods, namely Smart-CAST and CST, that are fit for analysis of gene expression data.

The primary characteristics of the introduced methods are in as follows: First, the Smart-CAST and CST methods are parameterless clustering algorithms. Secondly, the quality of clustering results generated by the introduced methods is "nearly optimal". Through performance evaluation on synthetic and real gene expression datasets, Smart-CAST and CST methods are shown to achieve higher efficiency and clustering quality than other methods without requesting the users to input any parameter. Therefore, the methods can provide high degree of automation, efficiency and clustering quality, which are lacked in other clustering methods for gene expression mining.

In the future, the following issues may be further explored:

1. **Reduction of memory requirements:** Smart-CAST and CST need memory to store the similarity matrix. Hence, a memory-efficient and memory reduction scheme is needed, especially when the number of genes in the microarry is large.
2. **Applications of CST on more real microarray datasets:** Everyone can apply Smart-CAST and CST on real microarray datasets to evaluate the validity of the clustering results, with the aim to promote related bioinformatics research like disease marker discovery.

References

Aldenderfer, M. S., & Blashfield, R. K. (1984). *Cluster analysis*. Newbury Park, CA: Sage Publications.

Alon, U., Barkai, N., Notterman, D. A., Gish, K., Ybarra, S., Mack, D., & Levine, A. J. (1999). Broad patterns of gene expression revealed by clustering analysis of tumor and normal colon tissues probed by oligonucleotide arrays. *Proceedings of National Academy of Science, 96*(12), 6745-6750.

Ben-Dor, A., & Yakhini, Z. (1999). Clustering gene expression patterns. *Journal of Computational Biology, 6*(3-4), 281-297.

Carpenter, G. A., & Grossberg, S. (1987). A massive parallel architecture for a self-organizing neural pattern recognition machine. *Computer Vision, Graphics, and Image Processing, 37*, 54-115.

Chen, M. S., Han, J., & Yu, P. S. (1996). Data mining: An overview from a database perspective. *IEEE Transactions on Knowledge and Data Engineering*, 8(6), 866-883.

Davies, D. L., & Bouldin, D. W. (1979). A cluster separation measure. *IEEE Transactions on Pattern Analysis and Machine Intelligence, 1*(2), 224-227.

Eisen, M. B., Spellman, P. T., Brown, P. O., & Botstein, D. (1998). Clustering analysis and display of genome wide expression patterns. *Proceedings of the National Academy of Sciences,* (Vol. 95, pp. 14863-14868).

Guha, S., Rastogi, R., & Shim, K. (1998). CURE: An efficient clustering algorithm for large databases. *Proceedings of the 1998 ACM SIGMOD International Conference on Management of Data,* (pp. 73-84).

Guha, S., Rastogi, R., & Shim, K. (1999). ROCK: A robust clustering algorithm for categorical attributes. *Proceedings of the 15th International Conference on Data Engineering,* Australia, (pp. 512-521).

Hathaway, R. J., & Bezdek, J. C. (2003). Visual cluster validity for prototype generator clustering models. *Pattern Recognition Letters,* 24(9-10), 1563-1569.

Jain, A. K., & Dubes, R. C. (1998). *Algorithms for clustering data.* Englewood Cliffs, NJ: Prentice Hall.

Kerr, M. K., & Churchill, G. A. (2001). Churchill. Bootstrapping cluster analysis: Assessing the reliability of conclusions from microarray experiments. *Proceedings of National Academy of Science, 98*(16), 8961-8965.

Kohonen, T. (1990). The self-organizing map. *Proceedings of the IEEE, 78*(9), 1464-1479.

Milligan, G. W., & Cooper, M. C. (1986). A study of the comparability of external criteria for hierarchical cluster analysis. *Multivariate Behavioral Research, 21*, 441-458.

Milligan, G. W., Soon, S. C., & Sokol, L. M. (1983). The effect of cluster size, dimensionality and the number of clusters on recovery of true cluster structure. *IEEE Transactions on Pattern Analysis and Machine Intelligence, 5*, 40-47.

Rohlf, F. J. (1970). Adaptive hierarchical clustering schemes. *Systematic Zoology, 19*, 58-82.

Rumelhart, D. E., & Zipser, D. (1985). Feature discovery by competitive learning. *Cognitive Science, 9*, 75-112.

Spellman, P. T., Sherlock, G., Zhang, M. Q., Iyer, V. R., Anders, K., Eisen, M. B., et al. (1998). Comprehensive identification of cell cycle-regulated genes of the yeast Saccharomyces Cerevisiae by Microarray Hybridization. *Molecular Biology of the Cell, 9*(12), 3273-3297.

Tamayo, P., Slonim, D., Mesirov, J., Zhu, Q., Kitareewan, S., Dmitrovsky, E., et al. (1999). Interpreting patterns of gene expression with self-organizing maps: Methods and application to hematopoietic differentiation. *Proceedings of National Academy of Science, 96*(6), 2907-2912.

Tseng, S. M., & Chen, L. J. (2002). An empirical study of the validity of gene expression clustering. *Proceedings of the 2002 International Conference on Mathematics and Engineering Techniques in Medicine and Biological Sciences (METMBS'02)*, USA.

Tseng, S. M., & Kao, C. P. (2005). Mining and validating gene expression patterns: An integrated approach and applications. *Informatica,* 27, 21-27.

Tseng, V. S. (n.d.). Parameter-less clustering method. *IEEE/ACM Transactions on Computational Biology and Bioinformatics (TCBB).*

Yeung, K. Y., Haynor, D. R., & Ruzzo, W. L. (2001). Validating clustering for gene expression data. *Bioinformatics, 17*(4), 309-318.

Zhang, T., Ramakrishnan, R., & Livny, M. (1996). BIRCH: An efficient data clustering method for very large databases. *Proceedings of the 1998 ACM SIGMOD International Conference on Management of Data,* (pp. 103-114).

Chapter X

Joint Discriminatory Gene Selection for Molecular Classification of Cancer

Junying Zhang, Xidian University, China

Abstract

This chapter introduces gene selection approaches in microarray data analysis for two purposes: cancer classification and tissue heterogeneity correction and hence is divided into two respective parts. In the first part, we search for jointly discriminatory genes which are most responsible to classification of tissue samples for diagnosis. In the second part, we study tissue heterogeneity correction techniques, in which independent component analysis is applied to tissue samples with the expression levels of only selected genes, the genes which are functionally independent and/or jointly discriminatory; we also employ non-negative matrix factorization (NMF) to computationally decompose molecular signatures based on the fact that the expression values in microarray profiling are non-negative. Throughout the chapter, a real world gene expression profile data was used for experiments, which consists of 88 tissue samples of 2308 effective gene expressions obtained from 88 patients of 4 different neuroblastoma and non-hodgkin lymphoma cell tumors.

Introduction

Spotted cDNA microarray, along with recent advances in machine learning and pattern recognition methods, promise cost effective and powerful new tools for the large-scale analysis of gene expressions. Still, the biological and medical research advances present more challenging topics to science and engineering. The most interesting ones among them include discriminatory gene selection, tissue heterogeneity correction, tissue and or gene clustering and tissue classification for cancer diagnosis.

As a common feature in microarray profiling, gene expression profiles represent a composite of more than one distinct source (i.e., the observed signal intensity will consist of the weighted sum of activities of the various sources). For example, molecular analysis of cells in their native tissue environment provides the most accurate picture of the *in vivo* disease state. However, tissues are complicated three-dimensional structures and the cell subpopulation of interest (e.g., malignant cells in tumor) may constitute only a small fraction of the total tissue volume. More specifically, in the case of solid tumors, the related issue is called partial volume effect, in other words, the heterogeneity within the tumor samples caused by stromal contamination. Blind application of the microarray profiling could result in extracting signatures reflecting the proportion of stromal contamination in the sample, rather than underlying tumor biology. Such "artifacts" would be real, reproducible, and potentially misleading, but would not be of biological or clinical interest. As a result, the overlap of source signals can severely decrease the sensitivity and specificity for the measurement of molecular signatures associated with different disease processes. Despite its critical importance to almost all the follow-up analysis steps, this issue (i.e., heterogeneity correction) is often less emphasized or at least has not been rigorously addressed as compared to the overwhelming interest and effort in pheno/geno-clustering and class prediction.

As a counterpart to heterogeneity correction, feature selection, the task of selecting (hopefully) the best subset of the input feature set, is becoming more and more important, especially since the improvement of biology and the advent of spotted cDNA microarray, which permit scientists to screen thousands of genes simultaneously to determine whether those genes are active, hyperactive or silent in normal or cancerous tissue.

Gene selection is one of the most important inseparable step from the other steps in microarray data analysis, and is featured with multiple purposes. At first, the genes with their expression levels not stably expressed, in other words, the genes we refer to as the unstable expressed genes (UEG), should be removed; also, in carrying out comparisons of expression data using measurements from multiple samples or arrays, the question of normalizing data arises, where the control gene set, i.e., the designated subset expected to be unchanged over most circumstances, what we call the constantly expressed genes (CEG), should be obtained for normalization; differentially expressed genes are those which result in differences in expression levels due to tissue samples which come from tissues of different diseases or tumors/tumor levels, which may functionally tell the biological/clinical cause of the diseases; finally, jointly discriminatory genes (JDG) are those we should select from gene set via some discriminatory/separability measurement such that the best classification/prediction performance can be reached when cancer classification and/or cancer diagnosis is expected.

On the other hand, gene selection is strongly encouraged from tissue clustering/classification point of view. Since the number of genes is huge compared with the number of tissue samples, tissue clustering and tissue classification are such tasks which will undergo severely the curse of dimensionality if it were fulfilled directly in the huge dimensional gene space in which much less number of samples exist (in most cases, this is unrealistic due to the super-high dimensionality of the space). Gene selection, as a tool to reduce dimensionality of the space, is widely applicable. Among the most widely cited reasons to reduce data dimensionality are (1) to eliminate redundancy, (2) to reduce the computational time and cost of performing the analysis and (3) to reduce misclassifications.

Perhaps the most important reason is the ability of a reduced data dimensionality to reduce the misclassification rate. This rate is reduced because as the number of measurements in a data set increases, the misclassification rate first decreases and then increases. With high data dimensionality, the classifier will have to allocate computational resources to deal with the noise dimensions, leaving fewer available to encode the knowledge embedded in the informative dimensions. Limiting dimensionality also reduces the impact of the curse of dimensionality. Thus, by excluding genes that clearly show little likelihood of correlation with phenotype/outcome, the precision of analyses can be improved and the unnecessary variation associated with clusters that are not relevant to the goal of the analysis is reduced.

This chapter is devoted to some work done in microarray data analysis, relating to jointly discriminatory gene selection, and tissue heterogeneity correction, as well as cancer classification/prediction. The chapter is divided into two parts, focused on gene selection for cancer diagnosis, and heterogeneity correction respectively. The first part looks for jointly discriminatory genes which are most responsible to classification of tissue samples for diagnosis. In the second part, we study heterogeneity correction techniques, in which independent component analysis (ICA) is applied to tissue samples with the expression levels of only selected genes, the genes which are functionally independent, and/or jointly discriminatory. Also, we employ non-negative matrix factorization (NMF) to computationally decompose molecular signatures based on the fact that the expression values in microarray profiling are non-negative. The real-world data used for experiment in this chapter is NCI data (Khan et al., 2001). The data set consists of 88 tissue samples of 2308 effective gene expressions obtained from 88 patients of four different neuroblastoma and non-hodgkin lymphoma cell tumors.

Problem of Finding Joint Discriminatory Genes for Cancer Diagnosis

The advent of DNA microarray technology has brought to data analysis broad patterns of gene expression simultaneously recorded in a single experiment. In the past few months, many data sets have become publicly available on the Internet. These data sets present multiple challenges (Anil, Robert, & Mar, 2000), including a huge number of gene expression values per experiment (several thousands to tens of thousands), and a relatively small number of experiments (a few dozen) belonging to several disease classes. The gene expression data set is generally expressed in a matrix, where each row indicates an experiment (a tissue sample), with its elements being the expression levels of different genes. Generally speaking, the number of genes is over 2000, while the number of samples is comparatively small, say, less than 100.

Gene selection becomes more and more important especially at present for DNA microarray data set since the number of dimensions (genes) becomes higher and higher compared with a very limited number of samples (experiments) in gene space (G-space) where the curse of dimensionality problem emerges (Trunk, 1979). In fact, the data set itself has its intrinsic dimensionality. Gene selection is an approach of selecting a subgroup of genes based on some criterion, such as discriminative power of the samples belonging to different classes. There are three main reasons to keep the dimensionality of the pattern representation (i.e., the number of features/genes) as small as possible: biologists' interest, measurement cost, and classification accuracy. A limited yet salient feature set simplifies both the pattern representation and the classifiers that are built on the selected representation. Consequently, the resulting classifier will be faster and will use less memory. Also, a small number of features/genes can alleviate the curse of dimensionality when the number of training samples is limited. Most importantly, on the other hand, seeking a subset of genes, the genes that have the greatest discriminatory power of different diseases, is one of the main interests of biologists.

A serious curse of dimensionality emerges in the case of DNA microarray profiles (Toronen, 2004). Let m_0 be the dimensions of the space, N be the number of tissue samples in the space, and the tissue samples belong to K tumor classes. If the distribution of the data set is known in this space, the larger the ratio of m_0 to is, N the less the Bayesian decision error would be, which will approach zero when the number of dimensions of the space is infinite; however, when the distribution of the data set in this space is unknown, which is the case for real DNA microarray data set, the Bayesian decision error would converge to its possible maximum value as m0 approaches infinity, the phenomenon referred to as the curse of dimensionality (Haykin, 1999; Choi, 2002). Notice that an acceptable number of samples in the space should be more than 10 times the number of dimensions of the space, a good practice to follow in classifier design, for avoiding the curse of dimensionality if the distribution of the data is unknown (Haykin, 1999). For the case when each dimension corresponds to a gene and each sample corresponds to an experiment for the spotted cDNA microarray dataset, the space is called gene space, or simply G-space. From this viewpoint, a serious curse of dimensionality occurs in the spotted cDNA microarray dataset, where in general the number of dimensions (genes) is larger than 2000, and ever increasing as the development of advanced biochip techniques, while the number of samples obtained for clinical experiments is very limited and comparatively small, say, less than 100. In fact, the data has its intrinsic dimensions: when the number of training samples is limited and fixed, too many dimensions will not benefit classification but bring more misclassification. On the other hand, what the biologists are interested in is to select dimensions (genes) which are responsible for the classification of cancer diseases, then a biochip of much smaller scale could be produced, which includes only the expression profile of the cancer-related genes. This will induce much less disgnostic cost for cancer disgnosis.

Recently many researchers do research work on gene selection and gene expression pattern clustering. Hierarchical growing neural network was presented for clustering gene expression patterns (Herrero, Valencia, & Dopazo, 2001), support vector machine (SVM) method was used in selecting genes (Guyon, Weston, Barnhill, & Vapnik, 2002), and statistical PCA method was presented for biomarker identification of the diseases (Xiong, Fang, & Zhao, 2001).

The next two sections will be focused on two gene selection methods. One, referred to as class-space method (C-space method), transforms the high-dimensional sample space into a low-dimensional class space to tackle the curse of dimensionality, and the discriminatory genes are selected via principal component analysis (PCA) in the class space. Another, referred to as contribution space method, is based on the fact that a small number of samples, under some trivial condition, is generally linearly separable in a high-dimensional space especially when the number of samples is much less than that of the dimensions of the sample space, which is just the situation of the microarray profiles. This characteristic encourages us to simply design a linear support vector machine (SVM) to classify tissue samples which belong to pair-wise classes, since a simple linear SVM is sufficient as a classifier with the greatest generalization potential, while the margin vector of the SVM is directly applicable to select genes which are responsible to the classification of these two classes. The method is then extended to gene selection for the case of more than two cancer classes. Diagnostic genes for real DNA microarray data set were selected and compared with the ones selected with SNR method, TSM method and SVM/MLP-CV method.

Finding Discriminatory Genes for Molecular Classification of Cancer in Class Space

In this section, an effective gene selection method is presented based on the transformation of gene space (G-space) into its dual space, referred to as class space (C-space) for tackling the curse of dimensionality. The genes are selected so that the largest discriminative power is attained in separating samples from belonging to different classes with the help of principal component analysis (PCA). Our experimental results on real DNA microarray data set are evaluated with Fisher criterion, weighted pairwise Fisher criterion and leave-one-out cross validation, showing the effectiveness and efficiency of the proposed method.

Space Transformation: From Gene Space to Class Space

Assume that the gene expression data set at hand is expressed in the matrix of:

$$\begin{array}{c} \\ s_1 \\ s_2 \\ \vdots \\ s_{N_s} \end{array} \begin{array}{c} \begin{array}{ccccc} g_1 & g_2 & \cdots & g_{N_g} \end{array} \\ \begin{bmatrix} g_{11} & g_{21} & \cdots & g_{N_g 1} \\ g_{12} & g_{22} & \cdots & g_{N_g 2} \\ \vdots & \vdots & \vdots & \vdots \\ g_{1N_s} & g_{2N_s} & \cdots & g_{N_g N_s} \end{bmatrix} \end{array},$$

where each row corresponds to the gene expression levels over all genes in a specific experiment (sample), and each column corresponds to gene expression levels of a specific gene over all the experiments; the number of dimensions N_g of the G-space is much larger than the number of samples N_g in the space; also, each sample belongs to one of K classes. For avoiding the curse of dimensionality of the G-space, we make a space transformation, from the gene space (G-space) to class space (C-space). Construct a class space (C-space) each dimension of which corresponds to a class, and each sample of which is a gene in G-space. Since each gene is expressed as:

$$g_i = g_{i,1}s_1 + g_{i,2}s_2 + \cdots + g_{i,N_s}s_{N_s}, \tag{1}$$

it can also be expressed in C-space as,

$$g_i = g'_{i,1}S_1 + g'_{i,2}S_2 + \cdots + g'_{i,K}S_K, \tag{2}$$

Figure 1. Demonstration for absolute/relative contribution of genes to separation of two different deceases

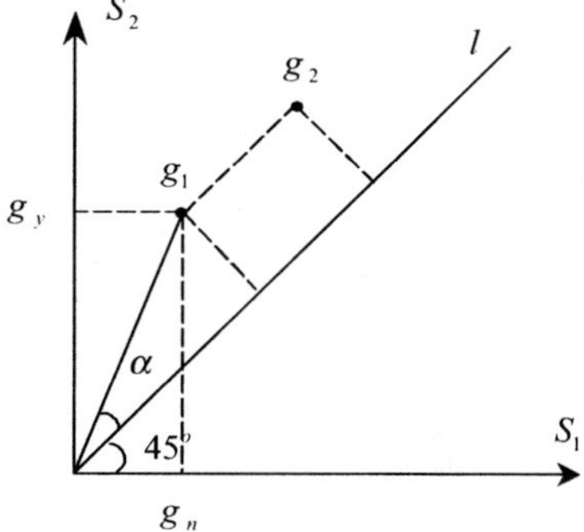

where Sj is the jth class axis, and,

$$g'_{i,j} = (g^2_{i,k_1} + g^2_{i,k_2} + \cdots + g^2_{i,k_m})^{1/2} \tag{3}$$

if the samples belonging to the ith class are $s_{i,k_1}, s_{i,k_2}, \cdots, s_{i,k_m}$. Evidently, $g'_{i,j}$ indicates virtually the expression level of gene gi over all samples belonging to the *j*th class (expression level is always positive). Since the number of dimensions is and the number of samples is N_g in C-space, curse of dimensionality does not occur in C-space.

Now let us consider discriminatory gene selection problem. A simple example is given first where there are only two genes ga and gb, and only two tissue classes, normal and abnormal, for introducing the main principle of selecting genes. Assume that the gene expression level is $(g_a, g_b) = (3, 3)$ for normal tissue, and is $(g_a, g_b) = (30, 3)$ for abnormal tissue, i.e.,

	g_a	g_b
normal	3	3
abnormal	30	3

Accordingly, the expression level of g_a changing from 3 to 30 is the reason of change from normal tissue to abnormal tissue which can be used for the diagnosis of the disease, while that of g_b without any change in its expression level indicates that the expression level of g_b is not responsible to the classification and diagnosis of normal tissue and abnormal tissue. From this viewpoint, gene g_a rather than g_b should be selected to distinguish normal tissue and abnormal tissue, i.e., gene ga has its contribution to the separation of normal tissue and abnormal tissue, while the contribution of gene g_b to this separation is 0. Figure 1 shows the two genes in C-space for two class situation, where none of the genes on the 45-degree line *l* have contribution to separation of these two classes: their contribution values are all 0s.

It is obvious that the genes which are closer to one have smaller contribution values, while those which are closer to the normal class axis S1/abnormal class axis S2 have larger contribution values, and the two genes which are symmetric against *l* should have same contribution value. Now we divide contribution value of gene *g* to the separation of these two classes into absolute contribution and relative contribution, denoted by $c_a(g)$ and $c_r(g)$ respectively, and use their multiplication to indicate the contribution of gene to this separation, i.e.,

$$c(g) = c_r(g) \times c_a(g). \tag{4}$$

For a gene $g=(g_y, g_n)$ in C-space, where g_y and g_n are the projections of in normal and abnormal class axis respectively, the absolute contribution value of *g* to separation of normal and abnormal classes is defined as the distance from *g* to the 45-degree line *l*, i.e.,

Figure 2. Possible quasi-hyperplanes in two-class situation

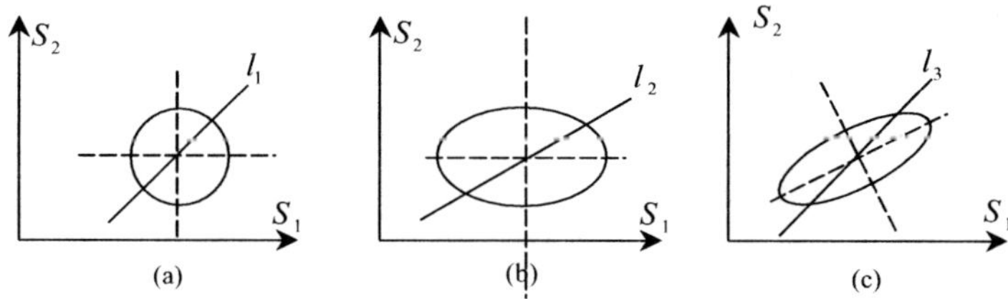

$c_a(g)=d(g,l)$. In Figure 1, g_l and g_2 have the same absolute contribution value since they have the same distance to the line l. However, their closeness to the normal class axis/abnormal class axis is different. Thus they should have different relative contribution values. The relative contribution value of gene g, $C_r(g)$, is defined as the angle α between vector g and the line l, since it reflects the ratio of expression levels of normal to abnormal situation.

The line l separates the C-space into two areas, which normal/abnormal class axis is inside, referred to as normal/abnormal class area. The genes that are in normal class area have contributions to separating normal class to abnormal class, while those that are in abnormal class area have contributions to separating abnormal class to normal class.

Finding Joint Discriminatory Genes in Class Space

Figure 2 demonstrates three different gene distribution situations in 2 dimensional C-space. For Figure 2a, the genes which have no contributions should be on 45-degree line l_l, while for Figure 2b and c, such genes should be on lines l_2 and l_3 respectively according to the distribution of the majority of the data set. Here we call and quasi 45-degree lines. In higher dimensional C-space, they would be quasi 45-degree hyperplanes, or simply qusi-hyperplanes.

Figure 3. Derivation of quasi 45-degree hyperplane equations

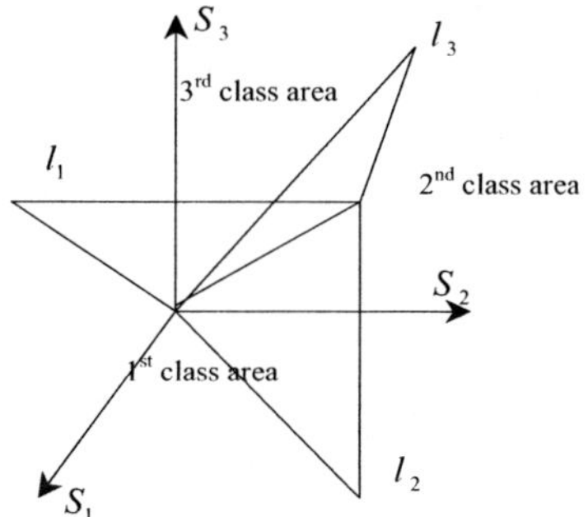

We perform PCA on gene data set in C-space, find the principal component direction, and then find the quasi-hyperplane for separation of the data in one class and the data in all other classes. For doing this, the following four steps are necessary:

Step 1. Centering by,

$$g^{(1)} = g - \bar{g}, \tag{5}$$

where $\bar{g}$ is the mean of all of the genes in C-space.

Step 2. Performing PCA by,

$$g^{(2)} = Q^T g^{(1)}, \tag{6}$$

where $Q = (q_1, q_2, \ldots, q_K)$ is a matrix whose columns are the eigenvectors of the covariance matrix of $g^{(1)}$.

Step 3. Whitening by,

$$g^{(3)} = D^{-1/2} g^{(2)}, \tag{7}$$

where D is the diagonal matrix with its diagonal elements being the eigenvalues of the covariance matrix of $g^{(1)}$.

Step 4. Finding the 45-degree hyperplanes in space.

In fact, the 45-degree hyperplanes in space are just the quasi-hyperplanes in C-space. For the case of K = 3 shown in Figure 3, the equations of the 45-degree hyperplanes in $g^{(3)}$ space, i.e., the quasi-hyperplanes in C-space, are:

$$l_1 : g_1^{(3)} \quad g_2^{(3)},\ g_1^{(3)} \quad g_3^{(3)},$$
$$l_2 : g_2^{(3)} \quad g_1^{(3)},\ g_2^{(3)} \quad g_3^{(3)},$$
$$l_3 : g_3^{(3)} \quad g_1^{(3)},\ g_3^{(3)} \quad g_2^{(3)}.$$

These equations can also be written in the matrix form of:

$$l_1: \begin{matrix} 1 & 1 & 0 \\ 1 & 0 & 1 \end{matrix} \; g^{(3)} \quad 0,$$

$$l_2: \begin{matrix} 1 & 1 & 0 \\ 0 & 1 & 1 \end{matrix} \; g^{(3)} \quad 0,$$

$$l_3: \begin{matrix} 1 & 0 & 1 \\ 0 & 1 & 1 \end{matrix} \; g^{(3)} \quad 0.$$

Similarly, the equations of K - 1 quasi-hyperplanes in K dimensional C-space, i.e., the 45-degree hyperplanes in $g^{(3)}$ space for separating the ith class and all other classes, can be expressed in the matrix form of:

$$G(i)g^{(3)} = [0, 0, \cdots, 0]^T, \tag{8}$$

where

$$G(i) = i \begin{bmatrix} -1 & & & 1 & & & \\ & \ddots & & \vdots & & & \\ & & -1 & 1 & & & \\ & & & 1 & & & \\ & & & \vdots & -1 & & \\ & & & & & \ddots & \\ & & & 1 & & & -1 \end{bmatrix}$$

(with the column of 1s at column i)

and all of the other elements in G(i) are 0s. Combining equations (5)-(8), we get all the quasi-hyperplanes in C-space as:

$$G(i)D^{-1/2}Q^T(g - \bar{g}) = 0, \; i = 1, 2, \cdots, K. \tag{9}$$

For each i, i = 1, 2, ..., K, the ith class area is referred to the area surrounded by the quasi-hyperplanes shown in equation (8) (i.e., the area which satisfies $G(i)D^{-1/2}Q^T(g - \bar{g}) \quad [0, 0, \cdots, 0]^T$).

Figure 4. Gene selection performance vs. the number of genes selected by c-space ranking

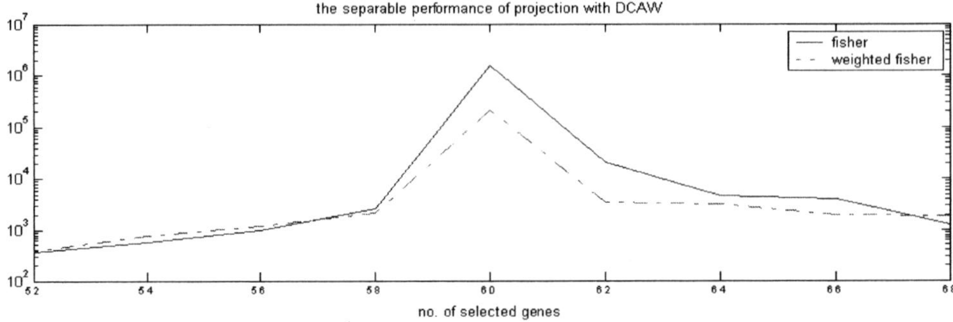

In order to calculate the contribution of each gene to separation, it is necessary to know which class area the gene belongs to, and then this gene has a contribution to separate this class from all other classes. Otherwise, it does not have this contribution. The class area that gene g belongs to can be obtained by finding i, i = 1, 2, …, K, such:

$$G(i)D^{-1/2}Q^T(g-\bar{g}) \geq [0,0,\cdots,0]^T. \tag{10}$$

Then gene g has contribution to separate ith class from all other classes. The key point K - 1 for calculating this contribution is to calculate the closeness between vector g and the quasi-hyperplanes presented by equation (8) in C-space. The closer they are, the less contribution it has, while no genes on quasi hyperplanes have contribution to separation. Here relative and absolute contribution of a gene g are defined, but both are dimensional vectors. The ith element of the relative contribution and the absolute contribution of gene g means the relative contribution and the absolute contribution of gene g to separating the ith class from all other classes respectively. The absolute contribution vector is defined as a vector the ith element of which is the distance between $g-\bar{g}$ and l_i, and the relative contribution vector of g is defined as a vector the ith element of which is the angle between the g(i) and $g(i)$, where is the projection point of g on the ith quasi-hyperplane.

The contribution of gene g to separation is the sum of contributions of gene g to separating the ith class to all other classes for i = 1, 2, …, K, that is:

$$c(g) = c_r(g)^T c_a(g). \tag{11}$$

Gene selection is performed by ranking all the genes according to their contributions to separation, and selecting the top M genes, where M is the number of genes one wants to select. Another way for gene selection, referred to as within-class ranking, is to rank genes according to their contributions to separation in each class area and select M_1, M_2,

..., M_K genes in the K class areas respectively, where $M_1, M_2, \ldots, M_K$ are predetermined numbers of genes one wants to select in the corresponding class areas.

Experiments and Results

We performed gene selection on the real NCI microarray data set (Khan et al., 2001). For this dataset, the dimensionality of the C-space is K = 4, and there are altogether 2308 sample points (genes) in the C-space. Our experiments include three parts: (1) calculate the linear separability criterion of the data on the selected gene subspace according to different gene contribution ranking method; (2) study the linear separability of the data on selected gene subspace with different gene selection methods (SNR method, three-step method (TSM) and C-space methods); (3) study the nonlinear separability of the data on the selected gene subspace by understanding the relation between the number of genes selected with C-space method and the nonlinear separability of the data in the selected gene subspace.

Experimental Results on Different Contribution Ranking Methods

For NCI data, gene selection is performed in a K = 4 dimensional C-space, where there are altogether 2308 genes (samples). Figure 4 presents the number of selected genes versus the linear separability criterion of data in the selected gene subspace, where $M_1 = M_2 = M_3 = M_4 = 1$—50 for within-class ranking, and M = 1—200 for all-gene ranking and select top M. In our experiments, the linear separability is measured with Fisher criterion and weighted Fisher criterion. Shown in Figure 4 are two very consistent curves, which get to the largest when M = 60 for both, indicating that for NCI data, both the ranking methods get approximately the same linear separability criterion for the same number of selected genes.

However, notice that the selected subgroup of 60 genes with Fisher criterion and that with weighted Fisher criterion are not the same. In our following experiments, only all-gene ranking is used. In order not to lose genes which may play an important role in biology, the predetermined number of genes selected in our following experiments is 150, which is much larger than 60, and also larger than the number of samples.

Comparison with Other Gene Selection Methods

Finding 150 genes from 2308 genes which can guarantee the best separability of the data in this selected gene subspace, is an NP complete problem (Pudil & Novovicova, 1998). Thus, some methods have been developed for finding the suboptimal gene subgroup. A simple method biologists often used for gene selection, SNR method (Dudoit, Fridlyand, & Speed, 2002), performs gene selection directly in gene space, and consider the separability of the data in each gene expression level independently, i.e., to select genes with the top largest SNR. That is, for gene i, the SNR of the data set in this gene is calculated by:

Table 1. Separability criterion of the projected data from selected gene subspace into top three principal axes

Projection method	DCAF projection		DCAW projection		PCA projection		Number of genes selected	Computation time
Separability measure	Fisher	Weighted fisher	Fisher	Weighted fisher	Fisher	Weighted fisher		
C-space method	266.8972	463.8205	269.2144	463.3149	1.6933	0.032578	150	10 s
TSM (2138,10,10)	285.6568	475.5676	286.1407	480.8895	2.6211	0.028189	150	20 hr
TSM (2138,15,5)	186.0106	370.8758	186.5364	384.5828	0.83254	0.013029	150	9 hr
TSM (2138,20,0)	115.0260	287.3228	115.3815	291.6633	1.8997	0.009975 1	150	30 min
SNR method	147.2021	327.4194	150.3970	342.1500	5.6257	0.048028	150	15 min

$$SNR_i = \frac{\sum_k p_k (\mu_{ki} - \mu_i)^2}{\sum_k p_k \sigma_{ki}^2},$$

where μ_i is the mean of the *i*th gene expression levels of the data, μ_{ki} ,σ_{ki}^2 , and σ_{ki}^2 are respectively the mean and variance of the ith gene expression levels of the data which belong to the kth class, and pk is the proportion of the kth class data to all the data.

The three-step method (TSM) (Wang, Zhang, Wang, Clarke, & Khan, 2001) divides gene selection process into three steps with three parameters (K_1, K_2, K_3). K_1, K_2, K_3 indicate the number of the genes removed in the first, the second and the third step respectively. The first step is just SNR method. The second step is to find a gene subspace from the remained K - 1 dimensional gene subspace by calculating the within-scatter matrix and between-scatter matrix:

$$S_w = \sum_k p_k C_k,$$

$$S_b = \sum_k p_k (\mu_k - \mu)(\mu_k - \mu)^T$$

as well as the eigenvalue λ_m and eigenvector W_m of matrix $S_w^{-1} S_b$, where μ_k and C_k is the means and the covariance matrix of the samples belonging to the *k*th class, and μ the means of overall training samples. Rank the remained K - 1 genes according to:

$$c_j = \sum_{m=1}^{n-K_1} \lambda_m W_{mj},$$

and remove the last K_2 genes. The third step is to remove K_3 genes, one at a time, by removing the one which induces the greatest decreasing of Fisher criterion/weighted Fisher criterion value of the data on the unremoved gene subset.

For TSM, its first step, just SNR, is very simple and time saving, but the separability of the data on the selected gene subspace may not be satisfied. Hence, a small K_1 should be selected, such that important genes could not be lost. The second step of TSM is to consider the scatter of data on all genes, while the third step to consider the separability of the whole data. Therefore, to get better gene selection result, K_2, K_3 should be larger, but the computation for step 2 and step 3 will increase very seriously. In our three experiments with (K_1, K_2, K_3) being (2138, 10, 10), (2138, 15, 5), (2138, 20, 0) respectively, it took 20 hours to find 150 genes with TSM of (2138,10,10) while it took only 10 seconds to find 150 genes with our C-space method.

In order to visualize the separability of the data in the selected gene subspace, PCA/DCAF/DCAW are used respectively (PCA — principal component analysis (Haykin, 1999), DCAF/DCAW — discriminate component analysis based on Fisher (Wang et al., 2001)/weighted Fisher criterion (Loog, Duin, & R.P.W., 2001). That is, first, calculate,

$$\text{PCA}: \quad J_p - C,$$

$$\text{DCAF}: \quad J_F = S_w^{-1} S_b,$$

$$\text{DCAW}: J_w = S_w^{-1} \sum_{i=1}^{K_0-1} \sum_{j=i+1}^{K_0} p_i p_j w(\Delta_{ij})(\mu_i - \mu_j)(\mu_i - \mu_j)^T$$

Figure 5. Overlap situation of gene subsets selected with different methods. (a) Overlap for three TSM selected gene subsets; (b) overlap for three different gene selection methods

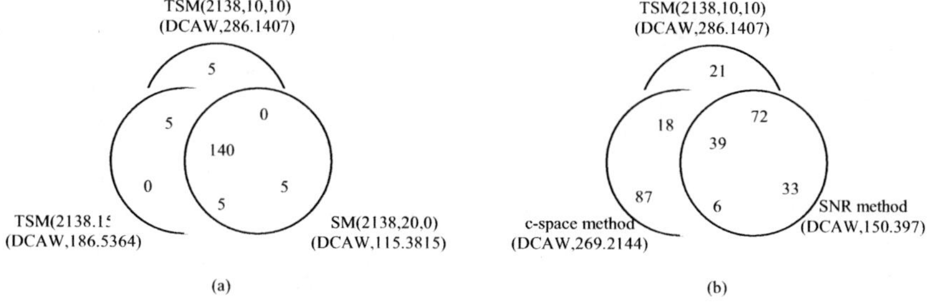

Figure 6. The number of genes selected vs. misclassification rate with a L-3-4 MLP, L=2~20

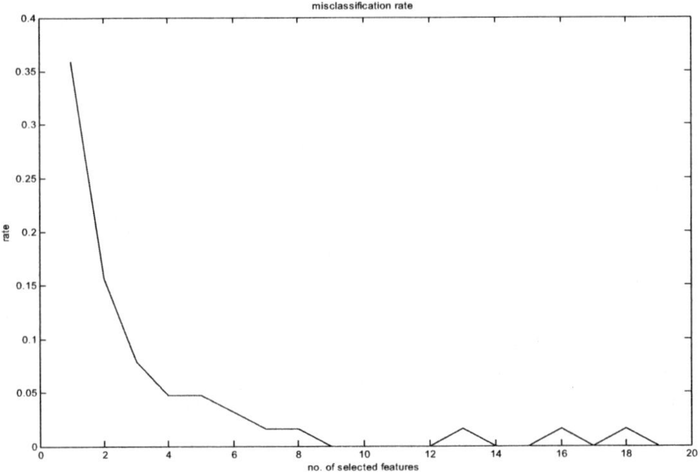

where C is the covariance matrix of the data in gene space, $w(\Delta_{ij}) = erf\left(\frac{\Delta_{ij}}{2\sqrt{2}}\right) \Big/ (2\Delta_{ij}^2)$ is the weight added to the class i and class j between-class scatter matrix, Δ_{ij} is the Mahalanobis distance between centers of class i and class j. Then, get the eigenvalue and eigenvector of J, and project the data into the top three principal eigenvector axes of the matrix J. Table 1 shows the linear separability Fisher criterion/weighted Fisher criterion of the projected data from the selected 150 gene subspaces with the above three gene selection methods. Notice that the subsets of the genes selected with the above methods are not the same and have some overlaps.

It is seen from Table 1 that TSM is decreasing in its Fisher criterion value in the order of(2138, 10, 10), (2138, 15, 5), (2138, 20, 0), which is in consistency with our analysis. The Fisher criterion value does not decrease in the same order for PCA method since PCA is an unsupervised learning which can guarantee that the reconstruction structure error is minimum while DCAF/DCAW is a supervised learning. It is also seen from Table 1 that the Fisher criterion value of C-space method is only a little bit less than that of (2138, 10, 10) TSM, while the computation time decreased greatly from 20 hours for (2138, 10, 10) TSM to 10 seconds for C-space method, indicating that the C-space method is greatly effective and efficient.

It is also seen from Table 1 that DCAW is superior in finding the projection axes which preserve separability of the data better compared with DCAF.

The overlap situation of the selected gene subsets with different methods was studied and is shown in Figure 5, where each number indicates the number of genes in the corresponding subset. It is shown from Figure 5(a) that among 150 genes selected with (2138, 10, 10) TSM, 140 genes are also selected with both (2138, 15, 5) TSM and (2138, 20, 0) TSM, but the separability performances decrease greatly from 286.1407 to 186.5364 to 115.3815. This seems to indicate that the remaining 10 genes are significant for the separability of the data. However, all of these 10 genes are not included in the 150 genes selected by C-space method. This indicates that the separability of the data is obtained by the combination of genes, rather than obtained by genes independently. This can also be seen from the overlaps of the gene subsets obtained by (2138, 10, 10) TSM, C-space method and SNR method shown in Figure 5(b) where the overlap of genes selected with TSM and with C-space method is only 39+18 in size but the separability of data by selecting genes with C-space method is already 269.2144, much better than SNR method where the separability is only 150.397.

The Number of Selected Genes vs. Nonlinear Classification

The number of selected genes is studied for nonlinear separability performance. L genes were selected with the C-space method, and a multiplayer perceptron (MLP) network (L-3-4) with L inputs, three hidden neurons, and four output neurons was trained for classification of samples belonging to four classes. The hidden and output neurons are activated with sigmoidal function and linear activation function respectively. Figure 6 shows the number of genes selected vs. misclassification rate of the trained network, where each point in Figure 6 is the one with the lowest misclassification rate among 30 trials with random initiation of the network parameters for avoiding getting stuck into local minimum of the training process. It is shown from Figure 6 that only nine genes rather than 150 genes are enough if a nonlinear classifier is used.

The generalization performance of the MLP is studied with leave-one-out cross validation approach for the 60 genes selected with the C-space method. The process was: (1) 60 genes were selected with C-space method; (2) DCAW was used to project the data with the selected 60 dimensions (genes) into top three principal axes (a linear projection); (3) a 3-3-4 MLP was trained (a nonlinear classifier). Each sample among 64 samples was used as a testing sample, while all other samples were used as training samples, to train a MLP. Hence, there were altogether 64 MLPs trained and tested. Our experimental result shows that the misclassification rate of all of these 64 MLPs for testing samples is zero. This indicates that these 64 genes selected with C-space method is very effective and efficient.

Notice that this method makes a quasi 45 degree hyperplane found with PCA method be a reference for overall genes, and calculates contribution of each gene for classification independently. This makes the method effective in both the separation performance of data on the selected genes and in computation complexity.

Finding Discriminatory Genes for Molecular Classification of Cancer in Contribution Space

Noticing the fact that a small number of samples, under some trivial condition, is generally linearly separable in a high-dimensional space especially when the number of samples is much less than that of the dimensions of the sample space, which is just the situation of the microarray profiles, and also the fact that support vector machine (SVM) is such a classifier designed for binary classification that brings the least empirical risk and best

Figure 7. Absolute contribution and relative contribution of a gene for overall classification

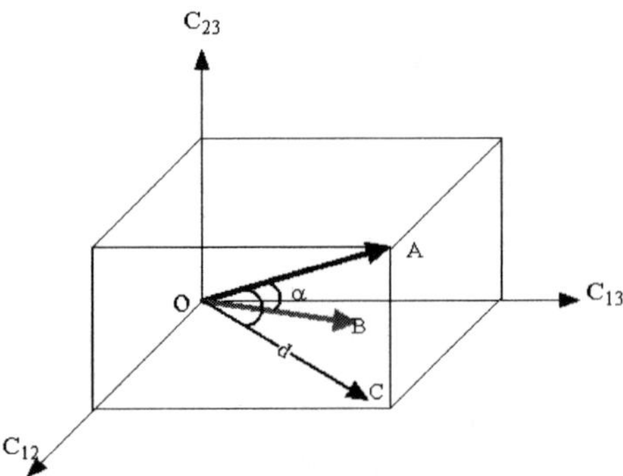

generalization performance when only a limited number of training samples are used for designing it, we simply design a linear support vector machine (SVM) to classify tissue samples which belong to pair-wise classes, with the margin vector of the SVM directly used to select genes which are responsible to the classification of these two classes. The method is then extended to gene selection for the case of more than two cancer classes.

Contribution Space and Gene Selection Method

As is known, in gene space (G-space), the number of dimensions (genes) m_0 is much larger than the number of samples n which belong to k disease classes. This makes it practical that the samples in G-space be linearly separable. Thus, for the case of only two classes (cancer/normal), a linear SVM can be trained as a linear classifier for classification of two class samples in G-space. The ith element g^i of the margin vector $M=[g^1, g^2, \text{L}, g^{m0}]$ tells the margin value defined by the SVM classifier in the ith gene axis, where m_0 is the total number of genes in microarray profiles. Notice that the larger the g^i is, the ith gene is more possible to be selected since the SVM results in larger margin for separation of samples in two classes compared with the other genes. For multiclass situation, say, k class situation, the binary class-pair includes ,(1,2), (1,3), (1,4),…,(k - 1,k), and k(k - 1)/2 SVMs can be trained as classifiers for classification of each pair of samples belonging to two different classes respectively. The resultant margin vectors can be obtained and expressed as follows:

$$M = \begin{bmatrix} M_{12} \\ \vdots \\ M_{1K} \\ \vdots \\ M_{(K-1)K} \end{bmatrix} = \begin{bmatrix} g_{12}^1 & g_{12}^2 & \cdots & g_{12}^{m_0} \\ \vdots & \vdots & \vdots & \vdots \\ g_{1K}^1 & g_{1K}^2 & \cdots & g_{1K}^{m_0} \\ \vdots & \vdots & \vdots & \vdots \\ g_{(K-1)K}^1 & g_{(K-1)K}^2 & \cdots & g_{(K-1)K}^{m_0} \end{bmatrix}. \tag{12}$$

We define a contribution space where each axis C_{ij} $(i, j = 1, 2, \quad , k, i \neq j)$ corresponds to the margin component derived from the SVM which classifies the ith class and the *j*th class. Evidently, the contribution space has $k(k - 1)/2$ dimensions, and each gene is a sample or a vector in the space. The *i*th column vector in matrix *M* defines a contribution vector of gene *i*, for $i=1,2,...m_0$. Firstly, take a look at a simple gene selection problem: select only one gene *i* from three genes through the following contribution matrix:

$$\begin{array}{cccc} & g^1 & g^2 & g^3 \\ C_{12} & 10 & 8 & 20 \\ C_{13} & 9 & 8 & 3 \\ C_{23} & 10 & 8 & 2 \end{array}. \quad (13)$$

Even though gene 3 contributes the classification of class 1 and class 2 greatly (the contribution value is 20) compared with all of other genes for this classification (the contribution value is only 10 and 8 respectively), its contributions to classification of class 1 and 3, and of class 2 and 3 are very small (the values are only 3 and 2 respectively). In contrast, gene 1 has moderately large contributions to classification of all the class-pairs. Hence, gene 1 would be selected as a gene selection result. Shown in Figure 7 is a simple 3-dimensional contribution space for illustration. *A* is a vector with same components, say, 1s, in each axis of the space, corresponding to the largest normalized contribution to classification of overall class-pairs. *O* is a zero vector with zero components in each axis of the space, corresponding to no contribution to overall class-pairs. A novel idea for gene selection is to calculate the overall classification contribution of each gene by comparing the corresponding column of contribution matrix *M* and the reference gene *A*, rank the overall classification contribution of all the genes in an descendent order, and select the first *r* genes as gene selection result.

We measure the overall classification contribution (OCC) of the *i*th gene g^i by two contribution issues: absolute contribution $c_a(g^i)$ and relative contribution. The absolute contribution $c_s(g)$ is defined as the length of the projected vector which is obtained by projection of the *i*th column vector in *M* contribution matrix to the reference vector *A*, and $c_r(g^i)$ is defined as the cosine function value of the angle between the ith column vector in and the reference vector *A*. The final OCC of the *i*th gene g^i is defined as a combination of absolute contribution and the relative contribution as follows:

$$OCC(g^i) = c_a(g^i) \times c_r(g^i). \quad (14)$$

Notice that the reference vector *A* does not affect the gene selection result as long as the elements in *A* are all the same. Thus, we have our gene selection algorithm as follows.

Step 1. Train a linear SVM for each class pair, class *i* and class *j*, and obtain the margin vector M_{ij} of the SVM, for *i,j=1,2, ...,k, i≠j*. Normalize the margin vectors M_{ij} such that each vector has a standard norm of 1. Then the contribution matrix *M* defined in equation 12 is obtained.

Step 2. Let the reference vector in contribution space to be ones in its elements. Calculate the OCC of each gene in contribution space by computing the absolute contribution and the relative contribution, and combining them with equation 14.

Step 3. Sequence the genes according to their OCCs in descendent order, and let the number of selected genes $r=1$;

Step 4. Select the first r genes from the gene sequence as the selected gene subset $G(r)$. Both Leave-one-out and leave-4-out cross validations (for leave-four-out, we leave a sample out from each class as training data to avoiding too large computation) are applied and the largest misclassification rate $\delta(r)$, for testing samples is used for the evaluation of the generalization performance of the linear classifier structure. Notice that we still use SVM based method for cross validation, i.e., train a linear SVM for classification of samples in each class-pair but expressed with only expression levels of the selected genes. The largest misclassification rate $\delta(r)$ for testing samples is calculated.

Step 5. If $\delta(r)\neq 0$ and $r+1\rightarrow r$ go to step 4; otherwise, $G(r)$ is the final selected gene subset.

Figure 8a. the scatter plot of samples expressed with gene expression levels of 20 selected genes obtained by (a) SVM/MLP-CV method and

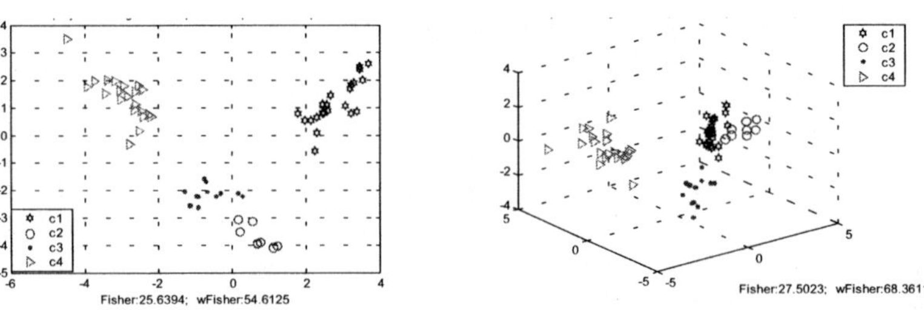

Figure 8b. CS method and projected onto top two(left) and top three(right) DCA space for visualization

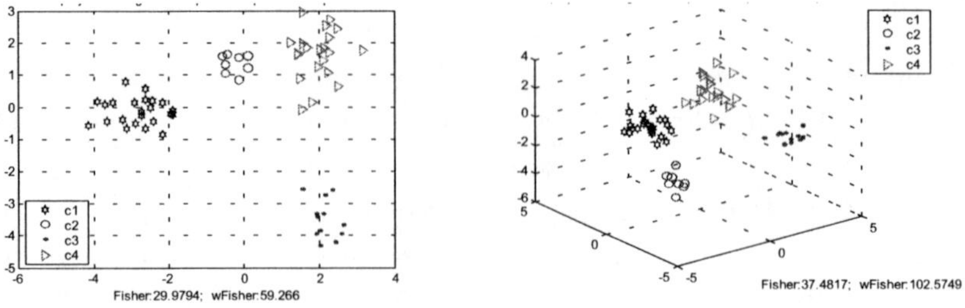

Experiment on Real Microarray Data

For NCI data, we constructed contribution space, which is of six dimensions, and there are altogether 2308 genes in the space for NCI data. We used the proposed contribution space based gene selection method (CS method) for gene selection. In order to compare the separability performance of the samples in the selected gene subgroup with the proposed method, we introduce another gene selection method briefly, called SVM/MLP cross validation method (SVM/MLP-CV). By noticing that the dimension of the gene space is too high while there are only a comparatively small number of training samples, for the situation of the samples belonging to only two classes, class i and class j, support vector machine (SVM) is used respectively for gene selection directly in gene space. The absolute value of the elements in the margin vector obtained from SVM is ordered decreasingly, and the first genes (determined by leave-one-out and leave-four-out cross validation with zero misclassification) are selected according to the first r_{ij} ordered absolute margin elements since it will bring the largest generalization power for classification of samples belonging to class i and class j in the selected gene subspace. After the genes are selected for samples belonging to each pair of classes, the selected genes are merged into a larger group of gene subset with, say, r genes. Then, sequential backward selection is used for gene subgroup selection until the largest misclassification rate for leave-one-out and leave-four-out cross validation with an MLP of r inputs (notice that by sequential backward selection, the number r is decreasing during gene selection process), two hidden neurons and four output neurons (each corresponds to a type of disease) is larger than zero. In this way, the final gene selection result is obtained.

In our experiment, Fisher criterion and weighted pairwise Fisher criterion were employed for evaluation of the separability performance of the samples in the selected gene subspace. Table 2 shows the selected gene indices, and the Fisher and weighted Fisher criterion values obtained from our experiment on this data set, when the number of selected genes is 20. It is seen from the Fisher and weighted Fisher criterion values in Table 1 that the separability of the samples in the selected gene subspace is much larger

Table 2. Comparison of SVM/MLP-CV method and the CS method for selecting 20 genes and the separability of samples in the selected gene subspace and their projection to the top 2/top 3 DCA/wDCA directions

Method	Selected gene indices	Separability evaluation				
		20-D selected gene space	Project to Top 2 DCA	Project to Top 3 DCA	Project to Top 2 wDCA	Project to Top 3 wDCA
SVM/MLP-CV	135,151,187,246,276,469,509,523,545,1093,1105,1389,1557,1645,1750,1915,1954,1955,2046,2050	31.8404 78.8581	25.7403 55.9319	27.5091 68.4060	25.6394 54.6125	27.5023 68.3611
CS	1,107,151,187,246,276,469,509,523,545,572,742,842,1084,1093,1389,1601,1915,1954,1955	40.9699 109.1131	29.6184 60.1432	36.6256 99.3981	29.9794 59.2660	37.4817 102.5749

when the genes are selected with the proposed CS method, compared with the previously proposed SVM/MLP-CV method. Notice that it was verified that in Table 2, the misclassification rates for leave-one-out and leave-4-out cross validation are all zeros when the group of genes was selected with both methods.

We also projected the samples with gene expression levels of only selected genes onto top 2/3 DCA subspace (i.e., top 2/3 discriminate component directions with discriminate component analysis (DCA)), and top 2/3 wDCA subspace (i.e., top 2/3 weighted discriminate component directions with weighted discriminate component analysis (wDCA)) for visualization of the separability of the samples in the selected gene subset. Figure 15(a) shows the scatter plot of the samples in the selected gene subspace obtained by SVM/MLP-CV method projected onto top 2/3 wDCA subspace, and Figure 15 (b) shows that of the samples in the gene subspace obtained by CS method projected onto top 2/3 wDCA subspace. By comparison of Figure 15(a) and Figure 15 (b), it is seen that the data belonging to different classes are more separable for the gene subset selected by CS method than that for the gene subset selected by SVM/MLP-CV method. This is because the CS method takes into account the contribution of a gene to classification ability on separating all class-pairs, rather than on separating only some specific pair of classes. This makes it possible not to loss the genes which have relatively small contribution values for separating some pairs of classes, but have great contribution values for overall classification.

Problem of Correcting Tissue Heterogenity Effect in Molecular Microarray Profiles

As a common feature in microarray profiling, gene expression profiles represent a composite of more than one distinct sources (i.e., the observed signal intensity will consist of the weighted sum of activities of the various sources). Such a heterogeneity effect would be real, reproducible, and potentially misleading, but would not be of biological or clinical interest. This is why heterogeneity correction should be performed to obtain microarray profiling which reflects the real biological molecular signature of the underlying tissues.

Various methods have been proposed to correct tissue heterogeneity in microarray studies. One of the main approaches is laser capture microdissection (LCM). The method uses specialized technology to separate and isolate the cancer cells (or other tissue cell subpopulations of interest) directly from the tissue. The molecular signatures in the isolated subpopulations can then be analyzed individually. The disadvantage of LCM is at the level of resources and expertise. Although LCM obtains pure tissue cell subpopulations, the amount of material can be quite small and has to be analyzed with appropriately sensitive analysis systems.

Independent component analysis (ICA) (Hyvarinen et al., 2000 , 2001; Chung et al., 2005; Choudrey & Roberts, 2003; Deniz et al., 2003; Lu et al., 2000; Lotlikar & Kothari, 2000;

Ristaniemi & Joutsensalo, 2000) is a newly-developed method that has been applied to blind source separation (BSS) for decomposition of composite signals. It aims at recovering the unobservable independent sources (or signals) from multiple-observed data masked by linear or nonlinear mixing. Most existing algorithms for linear mixing models stem from the theory of ICA. One of the basic assumptions for ICA model is the statistical independence between components. However, the dependent components are the often situation in the real world. Relaxing the very assumption of independence, thus explicitly formulating new data models, is an approach in this point. Three recently developed methods in this category are multidimensional ICA (MICA) (Hyvarinen et al., 2001; Cardoso, 1998), independent subspace analysis (ISA) (Hyvarinen et al., 2000, 2001) and topographic ICA (TICA) (Hyvarinen et al., 2001). In multidimensional ICA, it is assumed that only certain sets (subspaces) of the components are mutually independent. ISA is a closely related method where a particular distribution structure inside such subspaces is defined. TICA, on the other hand, attempts to utilize the dependence of the estimated "independent" components to define a topographic order, where the closer components in topography are more dependent to each other.

Notice that ICA can only work well with a strict restriction that each component should be statistically independent and non-gaussian distributed. This restriction does not hold in many applications, especially in real microarray profiles. Of course, if independence does not hold for some of component pairs, any connection to ICA would be lost, and the model would not be useful in those applications where ICA has proved useful. Notice that ICA can decompose the statistically independent signals, while the most genes in microarray profiles function dependently. If the independent genes are selected at first and ICA is applied to the microarray data with the expression values of only these selected genes (this is what is referred to as partially independent component analysis, PICA), the microarray profiling which reflects the real biological molecular signature of the underlying tissues can still be obtained. However, how to select such genes still remains an open problem due to its relation to up to infinitely-high-order statistical properties. Notice that computational decomposition of molecular signatures for heterogeneity correction and molecular classification of cancer tissue for cancer diagnosis are similar in their first step, or, gene selection. For computational decomposition of molecular signatures to correct heterogeneity with partially independent component analysis, we need to select genes which are of most functional independence; for molecular classification of cancer tissues, we need to select genes that the samples with expression levels of these genes are most separable. A novel computational heterogeneity correction method is developed in which informative genes are firstly selected with the largest separability criterion, and then ICA is conducted on tissue samples with the expression levels of only the selected genes for blind decomposition of the molecular signatures in the microarray profiles.

By noticing that the expression values in gene microarray profiling are only non-negative, a novel idea for heterogeneity correction is: it is a non-negative matrix factorization (NMF) problem without enforcement of statistical characteristics on sources and NMF technique is applied to heterogeneity correction of microarray profiles. Simulations and experiments on real microarray data show effectiveness of the proposed method.

Heterogenity Correction with Independent Component Analysis

Basic ICA Model

In basic ICA model, it is assumed that each observation is a linear mixture of independent components s_i, i.e.:

$$x_i = a_{i1}s_1 + a_{i2}s_2 + \cdots + a_{in}s_n \quad i = 1, 2, \cdots, n,$$

$$\text{or} \begin{bmatrix} x_1 \\ x_2 \\ \vdots \\ x_n \end{bmatrix} = \mathbf{A} \begin{bmatrix} s_1 \\ s_2 \\ \vdots \\ s_n \end{bmatrix}, \text{or} \;, \; \mathbf{x} = \mathbf{As}\,, \tag{15}$$

where s is an n-dimensional random component vector, **x** is an n-dimensional random observation vector, and **A** is an n by n mixing matrix. The statistically independence of the components is one of the most critical assumptions in ICA model (the nongaussianity of the components is another one). That means that the joint probability density function (pdf) of the components and the marginal pdf's satisfy:

$$p(s_1, s_2, \cdots, s_n) = \prod_{i=1}^{n} p(s_i). \tag{16}$$

However, this restriction does not hold in many cases in the real world. Several people talking to each other with a same topic is an example; The mixtures of several images with a same or similar background is another example; Of course, if independence would not hold for most component pairs, any connection to ICA would be lost, and the model would not be very useful in those applications where ICA has proved useful.

Methodology of PICA

In contrast to basic ICA model, PICA model assumes that the components expressed by **s**, i.e., s_1, s_2, L, s_n may be statistically dependent, the dependence may be linear, or nonlinear.

At the beginning, let us review the mechanism of independent component analysis on basic ICA model. From the very starting point, the components are statistically independent, while based on the Central Limit Theorem, the distribution of a sum (observations) of independent random variables (components) tends toward a Gaussian distribution, under certain conditions. Therefore, the procedure that ICA searches for estimates of the components is to find directions, such that the projections of observations on each direction are distributed with most nongaussian distribution.

It is clear that ICA algorithm can help find the precise estimate (statistically precise) of the components if the components are completely independent (except two ambiguities of ICA, i.e., scale and order of the components). However, it will mislead direction finding if the components are dependent but ICA algorithm is superimposed on the observations which are mixtures from the dependent components. This results in estimates of virtual components (the estimates of components), the components which are as independent as possible. Then, what is the relation between the virtual components and the real components (dependent)?

Let $\mathbf{s}'$ and $\mathbf{A}'$ be the virtual component vector and virtual mixing matrix obtained by applying ICA directly on observations. We have:

$$\mathbf{s}'=(\mathbf{A}')^{-1}\mathbf{A}\mathbf{s} \tag{17}$$

meaning that the virtual components are still linear mixture of real components. Compared with observations, $\mathbf{x}$, which are still linear mixtures of the real components, ordinarily speaking, the virtual components are rough estimates of real components since they are separation results by virtue of ICA.

Denote the component vector $\mathbf{s}$ as

$$\mathbf{s}=[S_{1,}S_{2}, \mathrm{L}, S_{m}] \tag{18}$$

where each column is a realization of random vector . We assume there are m realizations in all. In order to find the precise estimates of the real components, a set of realization indices I, $I \subset \{1,2,\ldots \boldsymbol{m}\}$, should be found, such that

(a) $\mathbf{s}_I=\{S_i, i\in I\}$ are independent, or say, the joint pdf and the marginal pdf's of random variables over the realizations of is

$$p(s_{1I}, s_{2I}, \cdots, s_{nI}) = \prod_{i=1}^{n} p(s_{iI}) \tag{19}$$

(b) $\{\mathbf{s}_1, S_j\}$are dependent for $\forall j \notin I$.

In other words, if we find the largest independent part of the real dependent components, ICA model will be precisely satisfied on this part of the real components. Then ICA could be performed on this part of the observations and this part of the real components together with the real mixing matrix could be estimated precisely. Finally, precise estimates of the real components can be obtained by $\mathbf{s}=\mathbf{A}^{-1}\mathbf{x}$.

The great difficulty for the resolution of this problem is that we want to find the largest independent part of the real components, but the real components are unknown and are the ones which remain to be estimated.

ICA is one of the approaches for blind separation of signals, while classification is an approach to separate patterns. If each signal or random variable is viewed as a pattern which represents that signal or random variable, it seems that there is some relation between separation of signals and the pattern expression of the signals. This will be studied in our next section.

Feature Analysis on Component Patterns

Now we denote the real components as:

$$\mathbf{s} = \begin{bmatrix} s_1 \\ s_2 \\ \vdots \\ s_n \end{bmatrix} = \begin{bmatrix} s_{11} & s_{12} & \cdots & s_{1m} \\ s_{21} & s_{22} & \cdots & s_{2m} \\ \cdots & \cdots & \cdots & \cdots \\ s_{n1} & s_{n2} & \cdots & s_{nm} \end{bmatrix} = [S_1, S_2, \cdots, S_m], \tag{20}$$

where s_i is the ith row of $\mathbf{s}$ indicating the ith component, S_j(j=1,2,L,...$\boldsymbol{m}$) is the jth column of $\boldsymbol{s}$ indicating the jth realization of random vector $\mathbf{s}$, and s_{ij} is the jth realization of $i$the th components. Thus, each component s_i is expressed by

$$s_i = (s_{i1}, s_{i2}, \cdots, s_{im}). \tag{21}$$

This means that it is s_i rather than sjsince its pattern corresponds to $(s_{i1}, s_{i2}, \cdots, s_{im})$. Therefore, each component s_i can be thought of as a pattern defined by a vector in m-dimensional feature space and its jth element s_{ij} is the jth feature of this component. The feature which can be induced from another or other features should be removed for more compact representation of the components and for dimension reduction (Tenenbaum, Silva, & Langford, 2000; Pudil & Novovicova, 1998). Of course, the only features which are independent to each other are not enough for representation of the components except they are the largest independent features to wholly express the features of the components.

Feature selection becomes more and more important since the number of dimensions (the number of features) of the data set met in the real world becomes greater and greater compared with the number of samples in the feature space (Haykin, 1999). Otherwise, the curse of dimensionality problem emerges. The curse is that as the dimensionality of data increases, they become dramatically more and more sparse in a multidimensional space (Cardoso , 1998). One example is microarray data set, where there are thousands of genes (dimensions) and only a few samples (experiments) which belong to several classes (disease categories). Only a few components but each have so many realizations (each realization corresponds to a dimension of the feature space that component falls in) shows us another example.

Pattern recognition is basically a classification problem combined with dimensionality reduction of pattern features that serve as the input to the pattern classifier. This reduction is a preprocessing of the data for obtaining a smaller set of representative features and retaining the optimal salient characteristics of the data. This will result in not only decreases of the processing time but also leads to more compactness of the model for classification task and better generalization of the designed classifier. The principle of feature selection for classification is mainly based on maximizing the separability of patterns in the reduced dimensional feature subspace.

For statistical independence of feature *i* and feature *j*, we can have the following two understandings:

(a) They are independent over projections of all data samples onto the subspace spanned by feature *i* and feature *j* (class unrelated).

(b) They are independent over projections of all class centers onto the subspace spanned by feature *i* and feature *j* (class related).

The feature selection for pattern classification should be the latter (class related). In our analysis, each real component is viewed as a class center which is the center of several virtual samples in the class, and the representation of the component, $s_i=(s_{i1},s_{i2},\ldots,s_{im})$, is the feature pattern of that class.

In order to understand the statistical relation of feature and feature , to know if they are independent or not, a large number of class centers is needed since only a large enough class centers can generate a distribution. However, in our ICA situation, the number of components is much less than the number of features which represent those components. The situation here is exactly the same with microarray data situation where the number of samples in feature space is too small to generate a distribution in the feature space for understanding the statistical relation of each pair of features. To tackle this problem, a component space is defined, where each axis of the space corresponds to a component, and each sample in this space corresponds to a feature.

The dependence of components is defined by the distribution of the features (feature scatter plot) in component space and the dependence of the features is defined by the distribution of component vectors in feature space (even though the number of components is small compared with the number of dimensions of the feature space). ICA is an approach to tackle the blind separation of components, which has an assumption that the components should be independent to each other, but we assume the components are not independent to each other. Feature selection aims at finding a feature subset to maximize the separable power of the classes. We have made a connection of component and its feature representation in our above study. Thus, we can study the independence of the components in its feature subspace and the separability of the components in this subspace.

Assume that I $(\subset \{1,2,\cdots,m\})$ be features which are selected by some feature selection method, and span an m_1-dimensional feature subspace. It is reasonable that we use Euclidian distance between s_1, s_2, i.e., $[d(s_1,s_2)]^2 = \sum_j (s_{1j} - s_{2j})^2$, to measure the separability of components in feature space. We have

$$\begin{aligned}[d(s_1,s_2)]^2 &= \sum_j (s_{1j} - s_{2j})^2 \\ &\approx m^2 E(s_1 - s_2)^2 \\ &= m^2 \iint (s_1 - s_2)^2 p(s_1,s_2) ds_1 ds_2\end{aligned} \tag{22}$$

and

$$E(s_{1I} - s_{2I})^2 = E(s_{1I}^2) + E(s_{2I}^2) - 2E(s_{1I} s_{2I}) \tag{23}$$

where the first two terms are the separability contributed by features I with an assumption that the projections of the components s_1, s_2, i.e., s_{1I}, s_{2I}, are independent in component space, and the last term is the separability contributed by the correlation of components s_{1I}, s_{2I} in the same space. Obviously, due to the possibility of $E(s_{1I}, s_{2I})$ either greater or less

Figure 9. Blind separation with ICA and PICA of the two dependent sources, where one source is a transpose of the other source: (a) Observations, (b) dexmixed with ICA, (c) observations (s.i.), (d) demixed with PICA (s.i.), (e) demixed with PICA, (f) sources

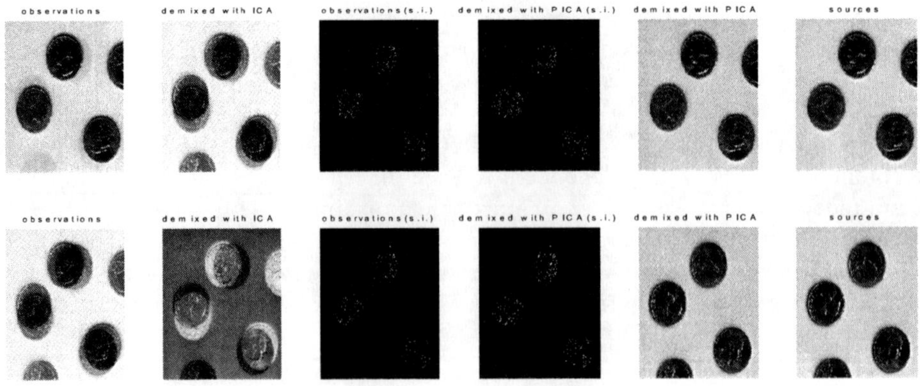

Figure 10. Blind separation of tire and coin images with ICA and PICA: (a) Observations, (b) results, (c) observations (s.i.),(d) ICA results (s.i.), (e) PICA results, (f) sources

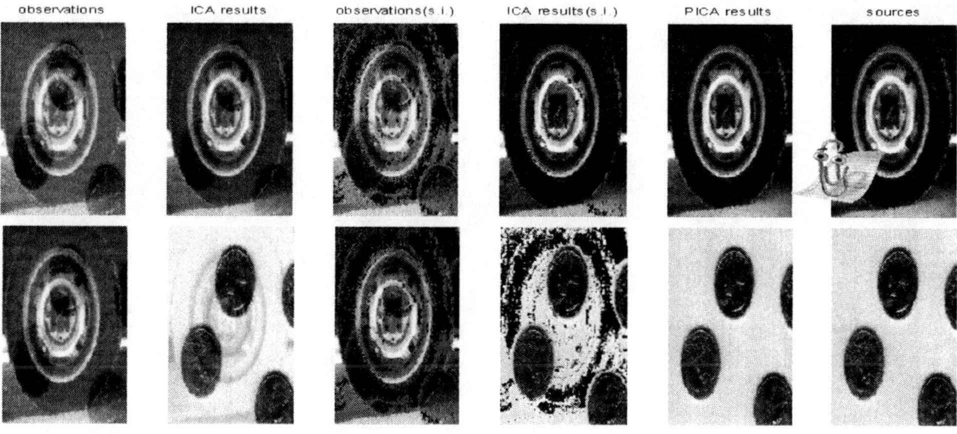

than 0, the separability of s_{11},s_{21} can either be better or worse than that when no correlation exists between s_{11},s_{21}.

Assume that I_1/I_2 are two separate feature subsets such that $s_{1I_1}, s_{2I_1} / s_{1I_2}, s_{2I_{12}}$ are independent in component space, and assume $I_2 \supset I_1$. We have

$$E[(s_{1I_2} - s_{2I_2})^2] > E[(s_{1I_1} - s_{2I_1})^2]. \tag{24}$$

Figure 11. Scatter plots of the sources(a), observations(b), ICA estimated sources(c) and PICA estimated sources(d) as well as the scatter plots of the observations(e)/ICA estimated sources(f) with only the selected indices

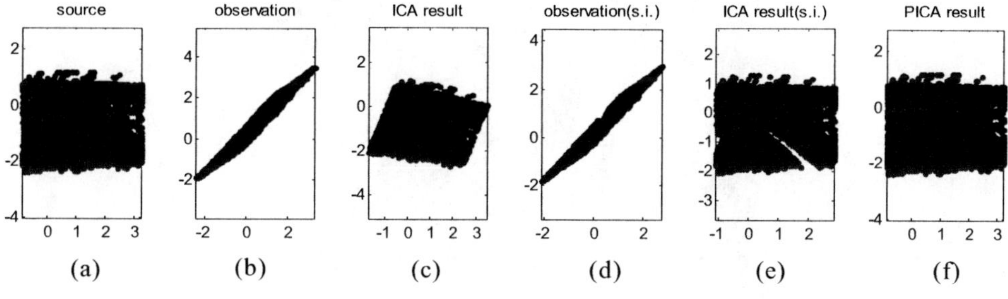

Figure 12. Comparison of the mixing matrix between the one estimated via ICA, PICA as well as the real one

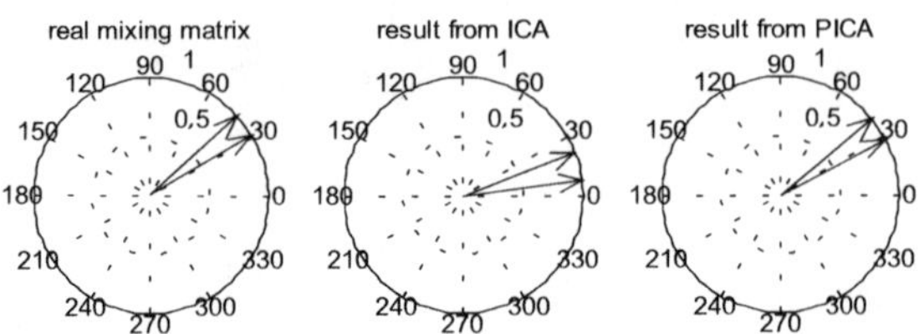

Therefore, the largest independent part of the components (in component space) results in the greatest separability of the component patterns (in feature space).

All in one, under the assumption that we have a compact representation of component features, it is coincident to find the largest independent part of the components (in component space) and to select a feature subset (in feature space) which can maximize the separability of the component patterns in feature subspace defined by the selected features.

Figure 13. Heterogeneity correction result (c) from the microarray profiles (b) and its comparison with the ground truth (a)

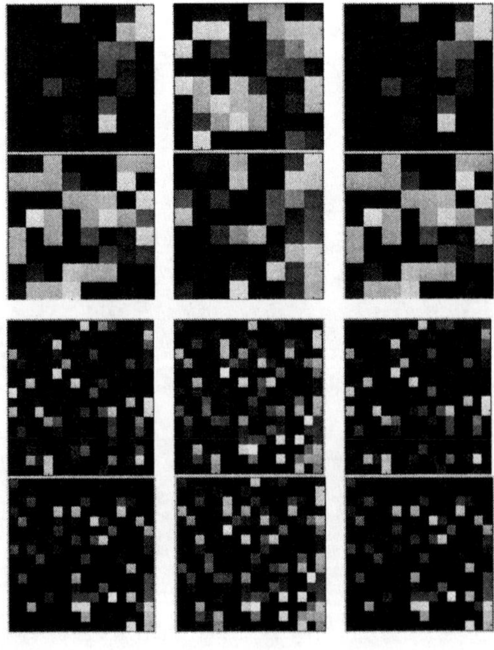

PICA Algorithm

C-space method is a feature selection method we developed for gene selection, which has been studied in the third section. It converts the feature selection problem in feature space into contribution calculation of each feature to classification task in class space and the top M features in their contributions are selected. M is an integer parameter for our C-space feature selection method. Since selecting the features that make the component patterns to be best separable in the feature subspace defined by these features is coincident with finding the greatest independent part of the components in feature space, we apply our C-space method for PICA. The PICA algorithm is as follows:

Step1. perform ICA on observations. A very rough estimate of the real component vector, i.e., , and a very rough estimate of the real mixing matrix , i.e., $\mathbf{s}'$ are obtained, due to the dependence of the real components $\mathbf{A}$; $\mathbf{A}'$ each row of $\mathbf{s}'$ is the feature pattern of the corresponding component.

Step 2. calculate the contribution of each feature for separation of all the component patterns by applying C-space method, and rank the features with descending order according to their contributions; Select top M features as the feature subset I.

Step 3. perform ICA on rough estimates of the components $\mathbf{s}'$, i.e., $\mathbf{s}'_I$. Then the estimate of mixing matrix $\mathbf{A}_2$ is obtained.

Step 4. the final estimate of the real mixing matrix is $\mathbf{A}'\mathbf{A}_2$, and the final estimate of the real component vector is $(\mathbf{A}'\mathbf{A}_2)^{-1}x$.

In the above procedure, ICA is performed twice, first to blindly separate observations into rough estimates of the real components, and then to blindly separate observations which are represented with only compact features into precise estimates of the real

Figure 14. Comparison of the scatter plots between the ground truth and the heterogeneity correction result obtained by PICA for the number of genes selected equals to64, 12(a), 128, 12 (b), and 2308, 12(c)

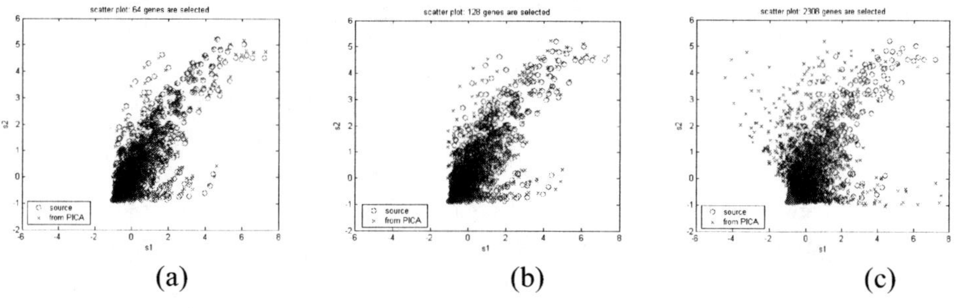

(a) (b) (c)

components for those compact features. With this method, the mixing matrix and the components are finally estimated precisely.

The features which will be fed to the second ICA calculation are selected with the principle of selecting features with the greatest separability of the component patterns in feature subspace defined by the selected features. This is compatible to finding the greatest independent part of the components. Therefore, we can think of the features selected to be the independent part of the components (we do not declare any proof on this point). Then we can make an alternative for step 3 and step 4 with the following steps:

Step 3'. perform ICA on the projections of the observations $\mathbf{x}$ in the feature subspace, i.e., $\mathbf{x}_I$, and obtain precise estimate of the real mixing matrix, $\mathbf{A}$.

Step 4'. the final estimate of the real component vector is $\mathbf{A}^{-1}\mathbf{x}$.

Simulations and Heterogeneity Correction Experiments

A large number of experiments for testing and comparison are performed, including simulation experiment and application of PICA to the resolution of partial volume correction for real microarray data set. In all of our simulation experiments, each element in mixing matrix is a uniformly distributed random variable ranged in [0,1] and mixing matrix has independent elements. The symbol "s.i." in our following figures indicates the indices of selected features by C-space feature selection method. In the following, we exemplify only two simulation results because of limitation of the space.

Our aim is to separate two 242×242 images, coin image and its transpose, from two mixtures of them. Evidently, the component sources are dependent again. The PICA is performed with the number of features selected to be , i.e., 1000 out of 242×242"=58564 features are selected. Much better estimation performance for components is again obtained by comparison of the closeness between Figure 9 (e) and 9 (f) and the closeness between Figure 9 (b) and Figure 9 (f)

For separating the mixture of tire image and coin image (both are in the size of 205×232), feature selection parameter M is set to M= 3000, i.e., the mixture images with 30000 out of 205×232=47560 selected features are fed to PICA algorithm. The performance of component estimates from PICA and its comparison to that from ICA can be seen in Figure 10 and Figure 11. The scatter plot of the features in component space sketched in Figure 6.3 also shows the much better performance of PICA than that of ICA on estimates of the components. In practice, the correlation coefficient of the pure tire and coin image is -0.1843, while the correlation coefficient of them with only selected features becomes only 0.00092433. This means that C-space feature selection with an appropriate choice of parameter M do functions of decorrelation, which leads to components represented in the corresponding feature subspace to be much more independent, while independence is a prerequisite of applying basic ICA for getting better recoveries.

9(c) and 9 (d), and 10 (c) and 10 (d) help us to have a better understanding on how PICA works. The observations with only selected features (pixels) and the estimates of components (source images) with these features (pixels) are shown in Figure 9(c) and 9 (d), and 10 (c) and 10 (d), where the pixels with the lowest intensities (black pixels) are unselected ones. It is seen that the features (pixels) which are selected are nearly the overlap pixels of the objects in two source images, while the features (pixels) which could be though of as common background of the source images are not selected by the feature selection algorithm.

The effectiveness of our PICA method was also tested on NCI data for partial volume correction (PVC) problem. The Source and mixed signals are given in Figure 13 respectively. The original 2308 genes were ranked according to their separable power to the two classes and the top 64/128 features (genes) are selected for ICA analysis. The original expression levels (reflected by intensities of the array image) of the top 64/128 features, mixed expression levels and the separation result are shown in Figure 13 (a) and (b) and (c) respectively (the upper part corresponds to the situation of 64 genes, the lower part corresponds to the situation of 128 genes). By comparison of Figure 13 (a) and Figure 13 (c), the blind source separation by PICA recovered the original array almost perfectly. Many other independent trials using other gene sets reached a similar result.

The scatter plots of genes in component space are shown in Figure 14 for the number of selected genes being $M = 64,128,2308$ from top to down and for the case when all of the $M = 2308$ genes (the bottom scatter plot in Figure 14) are selected. Our experiment indicates that the scatter plot that the estimates of the sources are almost the real sources for $M \leq 256$. This indicates that the statistically independent genes were retained for $M \leq 256$. That the estimates of the sources are very far from the real sources for too large M, say for $M \leq 256$, indicates that if too large number of genes are retained for PICA, some dependent part of the sources are also retained, which will mislead direction searching in the following-up ICA analysis. Therefore, the number of features selected is an important parameter and problem dependent in application of PICA to real world problems.

Heterogeneity Correction with Non-Matrix Factorization (NMF)

In this section, we propose a tissue heterogeneity correction approach based on non-negative matrix factorization (NMF) by noticing that the microarray profiles are non-negative in gene expression levels. NMF is a newly developed approach to factorize a non-negative matrix into two non-negative matrices, and has found wide applications in feature extraction (Lee & Seung, 1999; Guilamet & Vitria, 2001, 2003; Li et al., 2004; Liu & Zheng, 2004). Assume that V is a n x m non-negative matrix showing m samples in n dimensional space. NMF is to factorize V into WH where W and H are n x r and r x m non-negative matrix respectively such that some cost function is the minimum. The two typical cost functions are $F_1 = \sum_i \sum_u [V_{iu} \log(WH)_{iu} - (WH)_{iu}]$ and $F_2 = \sum_i \sum_u (V_{iu} - (WH)_{iu})^2$. The non-negative constraints of NMF model lead to a part-based representation because they allow only additive, not subtractive, combinations. NMF is already successful in learning facial parts and semantic topics (Lee & Seung, 1999). Also, Lee and Seung proved the convergence of NMF algorithms for the above two cost functions respectively (Daniel & Seung, 2001).

Matrix Factorization and Blind Source Separation with Non-Negative Matrix Factorization

BSS is a very active topic recently in signal processing and neural network fields (Hyvarinen et al., 2001; Hoyer & Hyvärinen, 2000). It is an approach to recover the sources from their combinations (observations) without any understanding of how the sources are mixtured. For a linear model, the observations are a linear combination of the sources, i.e., $X=AS$, where S is an $r \times n$ matrix indicating r source signals in n dimensional space, X is an $m \times n$ matrix showing observations in dimensional space, and A is an $m \times r$ mixing matrix. Therefore, BSS is a matrix factorization, to factorize observation matrix V into mixing matrix A and source matrix S.

An NMF problem is: given a non-negative $n \times m$ matrix V, find non-negative $n \times r$ and matrix factors W and H such that the difference measure between V and WH is the minimum according to some cost function, i.e.:

Figure 15. Dependent source separation with NMF: (a) Real sources; (b) observations; (c) recovered sources with ICA method; (d) recovered sources with NMF; (e) cost function vs. iterations for NMF

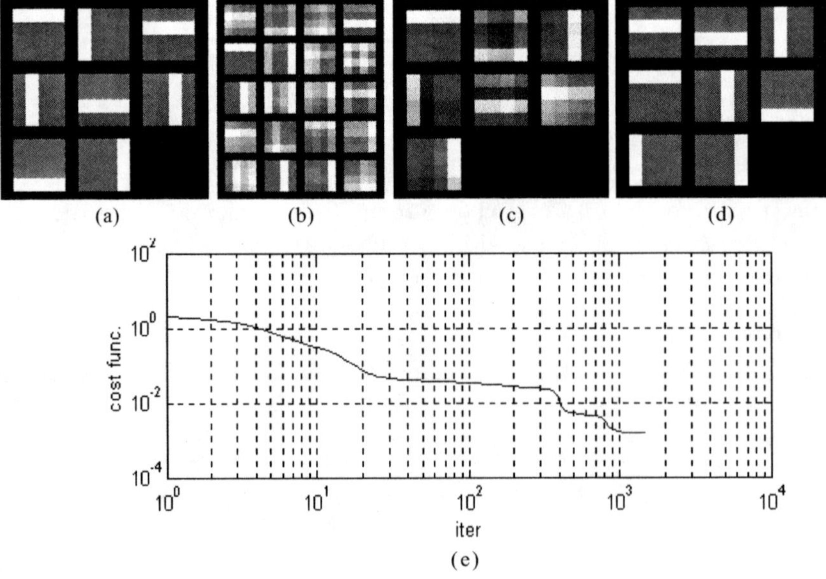

$$V \approx WH \tag{25}$$

NMF is a method to obtain a representation of data using non-negative constraints. These constraints lead to a part-based representation because they allow only additive, not subtractive, combinations of the original data. For the ith column of equation (25), i.e., , where v_i and hi is the ith column of V and H, the ith datum (observation) is a non-negative linear combination of the columns of $(W_1, W_2, ..., W_r)$, while the combinatorial coefficients are the elements of . Therefore, the columns of , i.e., $v_i = Wh_i$, can be viewed as the basis of the data V when V is optimally estimated by its factors.

If $(W_1, W_2, ..., W_r)$ are considered to be r sources/basis of the observation space spanned from the columns of V, notice that NMF does not constrain any statistical properties on the sources. However, we will show that they have some association: they should be of

Figure 16. Partial Volume Correction (PVC) with ICA and NMF: (a) BSS results with ICA and NMF, and (b) the scatter plots of the recovered sources

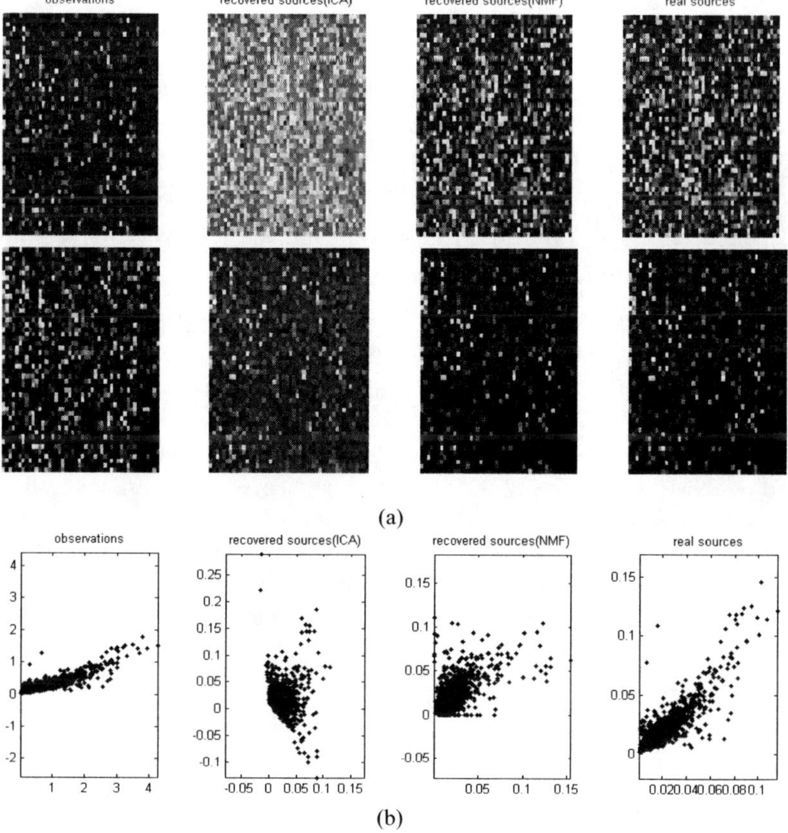

linear independence (as the basis of a space) which will induce some degree statistical independence (i.e., statistical dependence) in statistical meanings. Denote w_i to be a random variable where each element in W_i is its realization. Notice that there may be two parts of correlation between observations. One is from the model, where the observations are linear combinations of sources, which leads to first-order statistical original dependence between observations, and another, which comes from the original dependent sources themselves, may exist according to the sources. It is clear that the task of BSS is to decorrelate the observations such that the recovered sources has the lowest origin-based correlation coefficient:

$$\gamma(w_i, w_j) = \frac{E(w_i w_j)}{\sqrt{E(w_i^2)E(w_j^2)}} = \frac{W_i' W_j}{\|W_i\| \cdot \|W_j\|} \quad (26)$$

rather than the center-based one

$$\gamma'(w_i, w_j) = \frac{E[(w_i - \overline{w_i})(w_j - \overline{w_j})]}{\sqrt{E[(w_i - \overline{w_i})^2]E[(w_j - \overline{w_j})^2]}}$$

Consider the case that W_i and W_j are orthogonal, which is an extreme case of linear independence between W_i and W_j. The correlation coefficient $\gamma(w_i, w_j)$ becomes 0, indicating that W_i and W_j are first-order origin-based statistically independent. In contrast to linear independence between W_i and W_j, let us consider the case where W_i and W_j are linearly dependent, i.e., $W_i = kW_j$, where is a constant. We have $\gamma(w_i, w_j) = 1$, indicating that W_i and W_j are totally first-order origin-based statistically dependent.

In fact, the linear independence between the columns of themselves corresponds to some degree first-order origin-based statistical independence (i.e., statistical dependence), i.e., the correlation coefficient $\gamma(W_i, W_j) \in [0,1)$. This becomes our fundamental understanding that NMF can help to find the basis of the data space which are some degree first-order origin-based statistically independent (i.e., statistical dependent). From this understanding, NMF can be used for decorrelating observations to recover the sources that are not necessarily statistically independent and/or Gaussian distributed. This is the reason why our experiments with NMF in the fourth section worked well for non-negative BSS even when there is statistically linear dependence between the sources. Therefore, when NMF is used for non-negative BSS, the ICA transpose model, i.e.:

$$X^T = S^T A^T \quad (27)$$

is used, and the transpose of the factored W will be the recovered source matrix S.

In our following experiments, the NMF algorithm presented in Lee and Seung (1999), i.e.,

$$\begin{cases} W_{ia} \leftarrow W_{ia} \sum \frac{V_i}{(WH)_i} H_a \\ W_{ia} \leftarrow \frac{W_{ia}}{\sum_j W_{ja}} \\ H_a \leftarrow H_a \sum_i W_{ia} \frac{V_i}{(WH)_i} \end{cases} \tag{28}$$

is used. The convergence of the algorithm is proved in Daniel and Seung (2001).

Blind Source Separation and Heterogeneity Correction Experiments

We did a lot of simulation experiments on non-negative BSS with NMF. The BSS simulation experiments are divided into three groups: statistically independent / dependent sources, statistically independent Gaussian distributed sources, and linear bar separation problem (dependent sources) respectively. In the following, only the linear bar separation is illustrated because of the limitation of the space, even though good results were obtained for the other group of simulation experiments. Notice that in our simulation, each element of mixing matrix is a random number that follows [0,1] uniform distribution, and fastICA algorithm (Foldiak, 1990) is used when ICA method is applied for BSS. Then we will illustrate our NMF method for heterogeneity correction of the real world microarray data.

The linear bar problem (Foldiak, 1990) is a blind separation of bars from their combinations. Eight feature images (sources) sized 4×4 including four vertical and four horizontal bar images are randomly mixtured to form 20 observation images. It is very clear that there exist highly statistical correlation between sources. For example, the background of the images is the same for all of the sources. Therefore, ICA does not work well. Figure 15 demonstrates our result of the recovered sources from NMF method compared with that from ICA method. It is seen from Figure 15 that only seven sources rather than eight were found with ICA method, and the recovered seven sources are very far from the real sources, while the recovered eight sources from NMF are very close to the real sources, indicating that NMF is effective in separating sources where the sources are highly dependent to each other.

The effectiveness of our NMF method was also tested on NCI data for heterogeneity correction. Two Sources (samples from two diseases respectively), two mixed signals, recovered sources from ICA and the ones from NMF are given in Figure 16 (a) respectively. Notice that the true sources are determined, in our present case, by separately profiling the pure cell lines that provide the ground truth of the gene expression profiles from each cell populations. In our clinical case, we use laser-capture microdissection (LCM) technique to separate cell populations from real biopsy samples. By comparison of the second and the third column in Figure 16 (a), the blind source separation by NMF method recovered the original array very well. Many other independent trials using other gene sets reached a similar result.

Our experiments also showed that the NMF algorithm sometimes does not converge to desired solutions. For example, we mixed three independent Gaussian distributed signals with a linear model and got three observation signals. However, the real sources are much less likely to be well-recovered by the application of NMF algorithm to these observations. To our knowledge, it seems that there are two main reasons for NMF to converge to undesired solutions. One is that the basis of a space may not be unique theoretically, and therefore separate runs of NMF may lead to different results. Another reason may come from the algorithm itself that the cost function sometimes gets stock into local minimum during its iteration, the same situation as that in ICA algorithm. From the scatter plots shown in Figure 16 (b), it is clear by comparison of the third column (the scatter plot of the recovered sources) and the fourth column (the scatter plot of the real sources) that the NMF decorrelates the origin-based correlation between observations to recover sources without consideration of whether such correlation in observations comes from the linear combination of the sources or from the sources themselves. This also tells us that the NMF method presented here for non-negative BSS still has room to be improved.

Summary

In this chapter we present material dealing with the jointly discriminatory gene selection for molecular signature decomposition. It is appropriate that we now reflect over that material and ask: how useful is jointly discriminatory gene selection? The answer to this question, of course, depends on the application of interest.

If the main objective is to achieve high classification rate for tissue sample classification and good generalization performance for cancer prediction, the use of jointly discriminatory gene selection offers a useful learning procedure, and a compact and intrinsic representation for good data compression while preserving as much information about the inputs as possible for classification and prediction of cancer. Firstly, it can both compress the microarray data and preserve the separable information for classification of tissue samples. Secondly, since the separable information for classification of tissue samples has its own intrinsic dimensions which is less than the dimensionality of the sample space due to the huge dimensionality of the space and a relatively small number of tissue samples in the space, a representation of a tissue sample based on the selected jointly discriminatory genes is preferable and more compact for classification, which will result in higher classification rate, compared to their original microarray gene profiles of the tissue samples. The more compact and intrinsic representation of the tissue samples by their expression levels of the selected jointly discriminatory genes may make the classifier to be more effective and efficient, i.e., the classifier easier to design both in structure and parameters, and with better generalization potentials.

Heterogeneity correction of gene microarray profiles is another challenging problem that we tackled in this chapter, which is also related to jointly discriminatory gene selection. For heterogeneity correction to get a molecular decomposition of the microarray profiles at hand, independent component analysis is not applicable due to the fact that the sources are mostly statistically dependent, which disabbays the independence assumption of the basic ICA model. In order for ICA to be still applicable for heterogeneity correction of gene microarray profiles, functionally independent genes needs to be selected. Due to the difficulty of finding functionally independent genes, jointly discriminatory genes are selected instead. By noticing that jointly discriminatory genes are in some sense, the genes which are most independent in the function of classification of cancer, a novel idea, first selecting jointly discriminatory genes and then performing ICA on microarray profiles with the expression levels of only selected genes, is presented. Accordingly, separation of tissue samples and heterogeneity correction of tissue samples via jointly discriminatory gene selection are compactly correlated in some technical point of view.

References

Anil, K., Robert, P. R., & Mar, J. (2000). Statistical pattern recognition: A Review. *IEEE Trans. on Pattern Analysis and Machine Intelligence*, *22*(1), 4-37.

Cardoso, J. F. (1998). Multidimensional independent component analysis, In *Proc. IEEE Int. Conf. on Acoustics, Speech and Signal Processing (ICASSP '98)*. Seattle, WA.

Choi, K. (2002). Input feature selection for classification problems. *IEEE Trans. on Neural Networks*, *13*(1), 143-159.

Choudrey, R. & Roberts, S. (2003). Variational mixture of bayesian independent component analyzers. *Neural Computation, 15*, 213-252.

Chung, S. H., Park, C. S., & Park, K. S. (2005). Application of independent component analysis (ICA) method to the Raman spectra processing. *Proceedings of SPIE*, *5702*, 168-172.

Daniel, D. L., & Seung, H. S. (2001). Algorithms for non-negative matrix factorization. *Advances in Neural Information Processing Systems, 13*, 556-562.

Déniz, O., Castrillón, M., & Hernández, M. (2003). Face recognition using independent component analysis and support vector machines. *Lecture Notes in Computer Science*, *2091*, 59.

Dudoit, S., Fridlyand, J., & Speed, T. P. (2002). Comparison of discrimination methods for the classification of tumors using gene expression data. *J. Am. Stat. Assoc*, *97*(457), 77-87.

Foldiak, P. (1990). Forming sparse representations by local anti-Hebbian learning. *Biological Cybernetics*, *64*, 165-170.

Guillamet, D., Vitrià, J. (2003). Evaluation of distance metrics for recognition based on non-negative matrix factorization. *Pattern Recognition Letters*, *24*(9-10), 1599-1605.

Guillamet, D., & Vitria, J. (2001). Unsupervised learning of part-based representations. *Lecture Notes in Computer Science, 2124*, 700.

Guillamet, D., & Vitrià, J. (2003). Evaluation of distance metrics for recognition based on non-negative matrix factorization. *Pattern Recognition Letters, 24*(9-10), 1599-1605.

Guillamet, M. B., & Vitria, J. (2001). Weighted Non-negative matrix factorization for local representations. *Proceedings of 2001 Computer Vision and Pattern Recognition.* Retrieved from http://citeseer.nj.nec.com/context/2095821/0

Guyon, I., Weston, J., Barnhill, S., & Vapnik, V. (2002). Gene selection for cancer classification using support vector machines. *Machine Learning, 46*(3), 389-422.

Haykin, S. (1999). *Neural Networks: A comprehensive foundation.* Prentice Hall Inc.

Herrero, J., Valencia, A., & Dopazo, J. (2001). A hierarchical unsupervised growing neural network for clustering gene expression patterns. *Bioinformatics, 17*(2), 126-136.

Hoyer, P. O., & Hyvärinen, A. (2000). Independent component analysis applied to feature extraction from color and stereo images. *Network: Computation in Neural Systems, 11*(3), 191-210.

Hyvärinen, A. (1999). Sparse Code Shrinkage: Denoising of nongaussian data by maximum likelihood estimation. *Neural Computation, 11,* 1739-1768.

Hyvärinen, A., Hoyer, P. O., & Inki, M. (2001). Topographic independent component analysis. *Neural computation, 13*, 1527-1558.

Hyvärinen, A., Karhunen, J., & Oja, E. (2001). *Independent component analysis.* New York: John Wiley.

Hyvärinen, A., & Hoyer, P. O. (2000). Emergence of phase and shift invariant features by decomposition of natural images into independent feature subspaces. *Neural Computation, 12*(7), 1705-1720.

Hyvärinen, A., & Oja, E. (2000). Independent Component Analysis: Algorithms and applications. *Neural Networks, 13*, 411-430.

Khan, J., Wei, J. S., Ringner, M., Saal, L. H., Lananyi, M., Westermann, F., et al. (2001). Classification and diagnostic prediction of cancers using gene expression profiling and artificial neural networks. *Nature Medicine, 7*(6), 673-679.

Lee, D., & Seung, H. S.(1999). Learning the parts of objects by non-negative matrix factorization. *Nature, 401*, 788.

Li, Y. Q., Cichocki, A., & Amari, S (2004). Analysis of sparse representation and blind source separation. *Neural Compupation, 16*, 1193-1234.

Liu, W. X., Zheng, N. N. (2004). Non-negative matrix factorization based methods for object recognition. *Pattern Recognition Letters, 25*(8), 893-897.

Loog, M., Duin, & R. P. W. (2001). Multiclass Linear fimension teduction by weighted pairwise fisher criteria. *IEEE trans. on Pattern Analysis and Machine Intelligence, 23*(7), 762-766.

Lotlikar, R., & Kothari, R. (2000). Fractional-step dimensionality reduction. *IEEE Trans. on Pattern Analysis and Machine Intelligence, 22*(8), 623-627.

Lu, C. S., Huang, S. K., Sze, C. J., & Liao, H. Y. M. (2000). Cocktail watermarking for digital image protection. *IEEE Trans. on Multimedia*, *2*(4), 209-224.

Pudil, P., & Novovicova, J. (1998). Novel methods for subset selection with respect to problem knowledge. *IEEE Intelligent Systems*, *13*(2), 66-74

Ristaniemi, T. (2000). *Synchronization and blind signal processing in CDMA systems*. Unpublished doctoral dissertation, University of Jyvaskyla, Jyvaskyla, Finland.

Ristaniemi, R., & Joutsensalo, J. (2000). Advanced ICA-based receivers for DS-CDMA systems. *Proceedings of IEEE Int. Conference on Personal, Indoor, and Mobile Radio Communications (PIMRC'00)*. London.

Tenenbaum, J. B., Silva, V. D., & Langford, J. C. (2000). A global geometricframework for nonlinear dimensionality reduction. *Science*, *290*(22), 2319-2325.

Toronen, P. (2004). Selection of informative clusters from hierarchical cluster Tree with Gene Classes. *BMC Bioinformatics*, *5*(1), 32.

Trunk, G. V. (1979). A Problem of Dimensionality: A simple example. *IEEE Trans. on Pattern Analysis and Machine Intelligence*, *1*(3), 306-307.

Wang, Y., Zhang, J., Huang, K., Khan, J., & Szabo, Z. (2002). *Proc. IEEE Intl. Symp. Biomed. Imaging* (pp.457-460). Washington, DC.

Wang, Y., Zhang, J., Wang, Z., Clarke, R., & Khan, J. (2001). *Gene selection by machine learning in microarray Studies* (Tech. Rep. No. CUA01-018). United States, Washington, DC.. The Catholic University of America.

Xiong, M., Fang, X., & Zhao, J. (2001). Biomarker identification by feature wrappers. *Genome Research* (see www.genome.org), *11*, 1878-188.

Zhang, J., Wei, L., & Wang, Y. (2003). Computational decomposition of molecular signatures based on blind source separation of non-negative dependent sources with NMF. In *2003 IEEE 13th Workshop on Neural Networks for Signal Processing (NNSP'03)*, Toulouse, France (pp. 409- 418).

Chapter XI

A Haplotype Analysis System for Genes Discovery of Common Diseases

Takashi Kido, HuBit Genomix, Inc., Japan

Abstract

This chapter introduces computational methods for detecting complex disease loci with haplotype analysis. It argues that the haplotype analysis, which plays a major role in the study of population genetics, can be computationally modeled and systematically implemented as a means for detecting causative genes of complex diseases. In this chapter, the author provides a review of issues on haplotype analysis and proposes the analysis system which integrates a comprehensive spectrum of functions on haplotype analysis for supporting disease association studies. The explanation of the system and some real examples of the haplotype analysis will not only provide researchers with better understanding of current theory and practice of genetic association studies, but also present a computational perspective on the gene discovery research for the common diseases.

Introduction

In recent years, much attention has been focusing on finding causative genes for common diseases in human genetics. (Badano & Katsanis, 2002; Daly, 2001; Fan & Knapp, 2003; Gabriel et al., 2002) These findings of causative genes of common diseases including diabetes, hypertension, heart disease, cancer, and mental illness are expected to be opening doors for realizing new diagnoses and drug discoveries. A promising approach for the gene discovery on common diseases is to statistically examine genetic association between the risk of common diseases and DNA variations in human populations. While single nucleotide polymorphisms (SNPs), the most common genetic variation, are widely used for this genetic association study, haplotypes, the combination of closely linked SNPs on a chromosome, has been shown to have pivotal roles in the study of the genetic basis of disease (Clark, 2004; Niu, 2004; Schaid, 2004). The main purpose of this report is to provide a comprehensive review of haplotype analysis in genetic association studies on complex, common diseases and provide the computational framework which enables us to carry out successful high-throughput genome-wide association study.

In addition to the review of the recent developments of haplotype analysis, the author presents the design, implementation, and application of a haplotype analysis system for supporting genome-wide association study. While there are some useful tools or programs available for haplotype analysis (Kitamura et al., 2002; Niu, Zin, Zu, & Liu, 2002; Sham & Curtis, 1995; Stephens, Smith, & Donnelly, 2001), little work has been reported for a comprehensive analysis pipleline for large-scale and high-throughput SNPs screening which fully integrate these functions. HAPSCORE (Zhang, Rowe, Struewing, & Buetow, 2002) is one of the few examples of those pipeline systems; however, it does not include some analysis functions such as automatic linkage disequilibrium (LD) block partitioning and disease association analysis tools. In this report, the author presents a system, LDMiner (Higashi et al., 2003), which represents the pioneer pipeline system that integrates a comprehensive spectrum of functions related to haplotype analysis. This report introduces the details of LDMiner and shows some examples of haplotype analysis with LDMiner, which helps to explain the theory and practice on population-based association study for common diseases.

Background

Genetic Variations and Common Diseases

The progress on human genome science is opening doors for the discovery of new diagnostics, preventive strategies, and drug therapies for common complex diseases including diabetes, hypertension, heart disease, cancer, and mental illness. Analysis of human genome primarily focuses on variations in the human DNA sequence, since these differences can affect the potential risk of disease outbreaks or the effectiveness of a drug treatment of the diseases.

A common method for determining the genetic differences between individuals is to find single nucleotide polymorphisms (SNPs). A SNP is defined as a DNA sequence variation referring to an alteration of a single nucleotide (A, T, G, C). SNPs represent the most common genetic variations. In fact, there are millions of SNPs in the human genome (Kruglyak & Nickerson, 2001), and it is estimated that there will be on average one SNP every 1,000 base pairs. SNPs are caused when nucleotides replicate imperfectly or mutate. Although most of these SNPs have no ostentatious impacts on the survival of the species, certain SNPs may confer beneficial effects allowing species to evolve and to adapt to new environments more successfully, while certain others may be detrimental. These SNPs are passed on from generation to generation. After hundreds of years, some SNPs become established in the population.

A number of instances are known for which a particular nucleotide at a SNP locus (i.e., a particular SNP allele) is associated with an individual's propensity to develop a disease. For example, it has been reported that functional SNPs in the lymphotoxin-alpha gene were associated with susceptibility to myocardial infection by means of a large-scale case-control association study using 92,788 SNP markers (Ozaki et al., 2002). There have also been a number of reports that show some SNPs in certain genes can determine whether a drug can treat a disease more effectively in individual with certain genotypes compared to those who do not carry such SNPs. For example, Cummins et al. (2004) reported that there is a strong association in Han Chinese between a genetic marker, the human leukocyte antigen HLA-B*1502, and Stevens-Johnson syndrome induced by carbamazepine, a drug commonly prescribed for the treatment of seizures. These SNPs are often linked to the causative genes, but may not be themselves proved to be causative. These are often called surrogate markers for the disease. These disease-associated SNPs are expected to be validated by a number of investigations including biological experiments and epidemiological studies in the next several years.

Why Study Haplotypes?

A haplotype is a set of closely linked alleles (SNPs) inherited as a unit. For example, let's assume that there is a gene containing three SNPs. We now represent two alleles of each SNP with A and B (A is minor allele). The number of possible combinations of the three SNPs is eight (i.e. A-A-A, A-A-B, A-B-A, A-B-B, B-A-A, B-A-B, B-B-A, B-B-B). However, the total number of common (i.e. frequency > 5%) haplotypes in the population usually converge to a number less than eight, for example, three (A-A-A, A-B-A, B-B-B). Also, the number of common haplotypes varies depending on the chromosome regions. The recent empirical findings in human genetic studies show that genomic sequence is comprised of parts; cold spots (LD blocks) having much less variations than other regions and hot spots having more variations (Cardon & Abecasis, 2003). Some recombination hot spots tend to include a large number of haplotypes, while cold spots (or LD blocks) include only small number of haplotypes.

Haplotypes play important rules for searching the causative sites associated with several common diseases. As Clark (2004) discussed the role of haplotypes in candidate gene studies, there are three primary reasons for considering the haplotype configuration, which are listed as follows.

- **Biological function:** Haplotypes may be defined as functional units of genes; the protein products of the candidate genes may depend on particular combinations of amino acid. For example, ApoE is a protein whose function is influenced by a pair of polymorphic amino acids (Fullerton et al., 2000). The two-site haplotypes best describe the functional differences of carious ApoE protein isoforms rather than those of individual SNPs in the gene.
- **Tracing past evolutionary history:** The variation in populations is inherently structured into haplotypes. The haplotypes cover the informative small segments on ancient ancestral chromosome that may harbor a disease locus. Therefore the information on haplotype structure is useful for localizing and for tracing the past historical events on causative sites. For example, we can trace the evolutionary events by classifying haplotypes into clusters on the basis of the idea of population genetic theory, which gives us a useful insight on the origin of the disease. (Kido et al., 2003; Templeton, 1995)
- **Statistical power advantage:** The haplotype-based analysis, which combines the information of adjacent SNPs into multi-locus haplotypes, may increase statistical power of conventional SNP-based analysis to detect disease associated sites. Bader (2001) showed that haplotype-based association tests can have greater power than SNP-based association tests in the case when the disease locus has multiple disease-causing alleles.

Linkage Disequilibrium and Haplotype Analysis

The case-control study, which compares a group of people with disease to a similar group of people without disease, is commonly used in population-based association study design. If we have a strong hypothesis about the special causative alleles, causal alleles can be directly evaluated. For common complex diseases, however, strong hypotheses about the specific causal alleles are generally not available, so the "indirect" association approach has become widespread. In "indirect" association studies, we expect that the actual causative allele is not genotyped but might be located near another marker that is genotyped.

LD is an important concept for indirect association study, which is defined as the lack of independence among different polymorphisms. That is, alleles A and B are said to be in LD if the frequency of pairs of alleles AB is not equal to the product of the frequencies of A and B alleles. In LD analysis, we assume that causative alleles at tightly linked markers will remain in significant LD for extended periods of time from ancestors. Various statistical metrics can be used to summarize the extent of pairwise LD between two linked markers, such as D' and r^2.

One of the important questions for indirect association studies is "how much LD is needed to detect a disease allele using its nearby genetic markers?" Krina and Cardon (2004) discussed the complex interplay among factors that influence allelic association. They outlined the four parameters that affect an odds ratio test of association with a single SNP: (1) the odds ratio of true disease-causing SNP, (2) LD between markers and the causal SNP, (3) the marker allele frequency, and (4) the disease allele frequency.

Ideally a disease allele can be detected provided that the extent of LD between the marker and the causative allele is high and the marker allele frequency matches with the allele frequency of the disease allele.

Since haplotype analysis captures regional LD information of SNPs, characterization of LD patterns is important. Characterization of LD patterns across the human genome is at present an area of highly active research. International HapMap project gathers information on LD structure of variation in human populations (The International HapMap Consortium, 2003). This will facilitate genome-wide association analysis and the search for the genetic determinations of complex diseases.

Computational Algorithms for Haplotype Analysis

Defining LD Blocks

The recent finding in human genetics shows that human genome is broken up into regions of weak LD (called hot spots) and strong LD (called cold spots or LD blocks). Strong LD within a genomic region implies that most of the variation of that region can be captured by just a few informative SNPs, called tag SNPs. It is hoped that the regions of strong LD will facilitate the discovery of genes related to common diseases through association studies.

There are a variety of proposed empiric definitions for haplotype blocks. For example, Gabriel et al. (2002) developed a descriptive approach to comparing block boundaries across different populations. They examined adjacent pairs of SNPs in at least two populations, and labeled SNP pairs as concordant or discordant, based on the strength of linkage disequilibrium between two populations. They then calculated the percentage of concordant pairs among all pairs to define the block boundaries. Liu et al. (2004) propose several similarity measures to compare haplotype boundaries across populations. They found that haplotype block boundaries vary among populations and the definition of haplotype blocks is highly dependent on the density of SNPs and the method used to find blocks. The understanding of the similarities and differences of haplotype blocks between different populations will be important for future genome association studies.

Inferring Haplotype Phases

In most cases, haplotypes are not read directly when multiple human SNPs are genotyped, but must be inferred from unphased genotype data. Although a spectrum of molecular haplotyping methods such as single-molecule dilution, long-range allele-specific PCR were developed, these methods are not widely used because of their low-throughput performance and important technical problems remain unresolved. Therefore a number of algorithms for inferring haplotypes from unphased genotype data have been developed.

Niu et al. (2004) provides a good review of recent computational methods of population-based haplotype inference methods. Clark's algorithm is based on the principle of maximum parsimony which resolves the haplotypes by starting with identifying all unambiguous haplotypes (all homozygotes and single-site heterozygotes) and repeatedly adding a new haplotype to the resolved haplotype. (Clark, 1990) Although Clark's algorithm is a relatively straight-forward procedure, the algorithm does not give unique solutions, because the phasing results are dependent on the order of genotypes that need to be phased. Expectation-maximization (EM) algorithm estimates population haplotype probabilities based on maximum likelihood, finding the values of the haplotype probabilities which optimize the probability of the observed data, based on the assumption of Hardy-Weinberg equilibrium (HWE). (Excoffer & Slatkin, 1995) Actually, EM algorithm is based on solid statistical theory and are quite effective as simulation studies demonstrate that its performance is not strongly affected by the departures from HWE. (Niu et al., 2004). Some disadvantages of EM algorithm, however, are that the iteration may lead to locally optimal point when there are many distinct haplotypes, and it cannot handle a large number of loci. Stephens et al. (2001) proposed a coalescence-based Markov-chain Monte Carlo (MCMC) approach: a pseudo-Gibbs sampler (PGS) for reconstructing haplotypes from genotype data. This algorithm is implemented as program named PHASE. Niu et al. introduced a divide-conquer-combine algorithm: partition-ligation (PL), which can handle a large number of loci. This algorithm is implemented as programs named Haplotyper (Bayesian implementation) and PLEM (EM implementation)

Although there is an ongoing debate in regard to whether the best haplotype phasing algorithm exits, different algorithms perform differently in different populations with their own strengths and limitations. Niu et al. demonstrates that different algorithms have different degrees of sensitivity to various extends of population diversities and genotyping error rates using empirical simulations based on real genome data sets.

Statistical Advantages and Methods for Haplotype-Based Disease Association

Haplotype analyses can sometimes provide greater power than single-marker analyses for genetic disease associations, due to the ancestral structure captured in the distribution f haplotypes. Whether haplotype analyses outperform single locus analyses varies depending on the assumptions about the number of disease causing SNPs, the amount of LD among SNPs from the marker and disease causing SNPs. Schaid (2004) reviews the recent works on the statistical advantages of haplotypes. Bader (2001) reports that when the set of measured SNPs includes causative SNPs, single-locus tests are more powerful than haplotype-based tests when the number of causative SNPs is less than the number of haplotypes. In contrast, haplotype analyses can be more powerful than single-locus analyses when the SNPs are in LD with a causative diallelic locus. (Akey & Xiong, 2001) Both single-locus analyses and haplotype analyses lose power when there are multiple alleles at a causative locus, but comparatively speaking haplotype analyses lose less power. In this situation, the power advantage for haplotype analyses is greatest when the marker alleles are not in strong LD with each other, yet in strong LD with the causative alleles.

Schaid provides a comprehensive review on how to evaluate the association of haplotypes with human traits. When haplotypes are directly observed, then statistical methods can be applied to compare the frequencies of haplotypes between cases and controls. If the haplotype- phases are unknown, haplotype frequencies for the cases and controls can be estimated by EM algorithm. One of the common methods for haplotype-based disease association is to use likelihood ratio statistic test (LRT) which tests the equality of haplotype frequencies between cases and controls (Zhao & Sham, 2002). Although LRT is a practical method, one of the disadvantages of LRT is that it may not be adequate when there are many haplotypes, it lacks adjustment for environmental covariates. In order to solve this problem, Schaid introduces regression models for haplotypes in which haplotypes are treated as categorical covariates with other environmental covariates. Schaid also reviews recent developments of statistical regression methods for haplotype analyses.

A Haplotype System Analysis

LDMiner: The Pipeline System for SNP Analysis

We have developed a system named LDMiner that supports disease association studies to detect genes that may cause complex diseases (Higashi et al., 2003). The main functions of LDMiner are as follows:

1. LDMiner is a total analysis system for SNPs, LD blocks haplotypes, and disease association.
2. LDMiner automatically combines SNPs located in the vicinity with respect to each other on a chromosomal region and constructs LD block, and accordingly haplotypes within the block using the EM algorithm (Kitamura et al., 2002).
3. LDMiner automatically assesses the association of haplotypes and diplotypes between cases and controls. By combining SNPs together into LD blocks, we can improve the statistical power for association study.
4. LDMiner visualizes the analyzed results along with a genome viewer. The main viewer displays the genomic structure and is linked to another main viewer showing the in-depth analysis result. These viewers allow the user to easily check and make interpretations of the results.

LD analysis and some advanced association studies can be efficiently performed using LDMiner with handy tools for eliminating the inadequate data and so on. Consequently, the number of SNPs the system can analyze is about 30 to 50 times higher than by the standard manual procedures per unit of time (Higashi et al., 2003).

Figure 1. Analysis pipeline with LDMiner

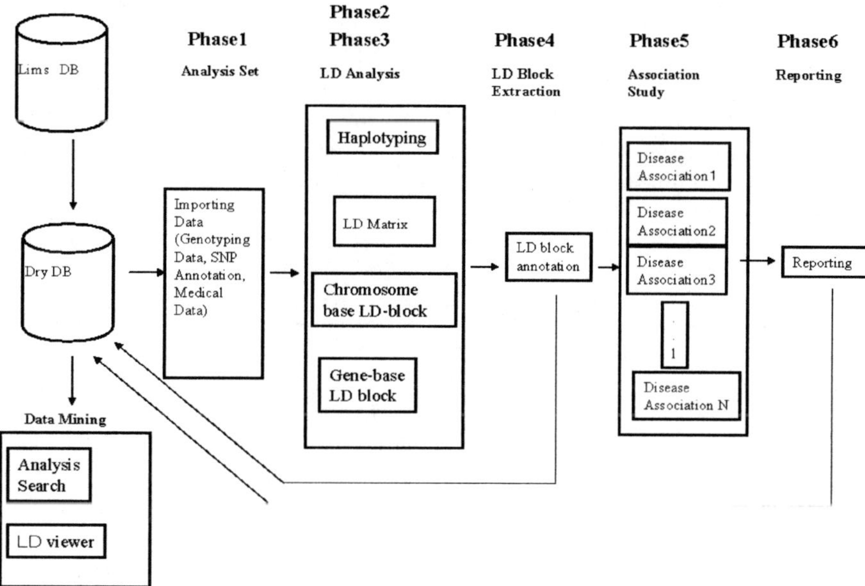

The system design of the LDMiner analysis pipeline is shown in Figure 1. In phase 1, we import genotype and medical data with SNP annotation for the analysis. In phases 2 and 3, we examine the extents of LDs among SNPs within the same genes or the same chromosomal regions as well as to estimate haplotypes and diplotypes using EM algorithm. In phase 4, LD blocks are automatically constructed by a sliding-window approach in searching for strong LD regions as discretized LDblocks. In phase 5, we perform several association tests for either individual SNPs or haplotypes within LD blocks. In phase 6, all association results are automatically annotated with SNPs and LD blocks information in order to easily retrieve and compare. These analysis results are visualized by linking the genome browser which maps LD status into genome structure.

Algorithms

Haplotyping

Given the genotypes of the individuals in a sample, the frequencies of the haplotypes in the population can be estimated. There are primarily three categories of algorithms for the inference of haplotype phases of individual genotype data; these categories are Clark's algorithm (Clark, 1990), the EM algorithm (Kitamura et al., 2002) and Bayesian algorithm (Stephens et al., 2001). Niu et al. reports the performance comparison of these algorithms and conclude that EM algorithm is shown to be accurate and practical. Furthermore they proposed PL algorithm, which is an extended version of a stochastic sampling algorithm incorporating the idea of "divide-conquer-combine", implemented the program called HAPLOTYPER which is applicable to a large number of SNPs. Although our system can incorporate several haplotyping programs as module, we mainly use EM algorithm implemented as LDSupport programs by Kamatani et al. (2002) and then determine the posterior probability distribution of diplotype configuration (diplotype distribution).

Automatic LD Block Extraction

We define the LD block based on the idea that SNPs within each LD block are tightly linked, which means that most of the pairwise LD measurement of two SNPs within the LD block is high, for example D' ≥ 0.8 Because our major purpose of LD block extraction is to improve the performance of haplotype association analysis, the number of SNPs in the LD block should be limited in order to keep down the degree of freedom of the association study. For example, based on our experience, three to five SNP markers are appropriate for our association study when the majority of the pairwise LD values across all SNPs within the same LD block is strong enough. Otherwise, if we select too many SNPs for the LD block, the power of the multi-locus association test becomes smaller because of the inflation of the number of haplotypes as well as the fact that the posterior probability of the diplotype determination becomes smaller. For these reasons, we propose the window-sliding method for LD block construction where the maximum number of SNPs in the LD block is always lower than predefined window size. One of the

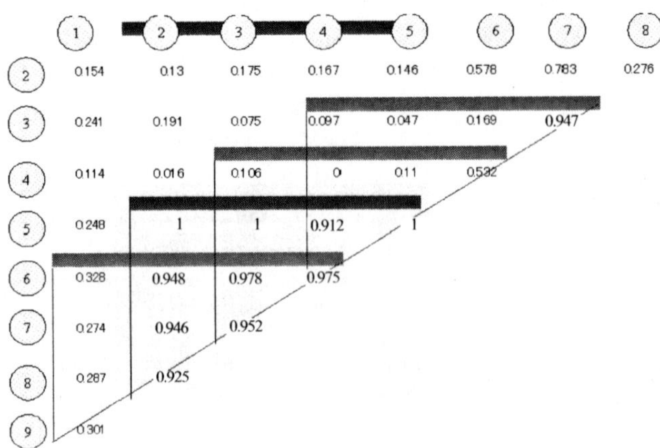

Figure 2. Window-sliding method for LD block extraction

examples of the window-sliding method is shown in Figure 2. In this example, SNP2, 3, 4, 5 were extracted to be within a single LD block because all pairwise LDs among the four SNPs is strong enough (D' > 0.9). Although D' is the quite popular measure of LD, appropriate LD measures and parameters should be selected depending on the situation. LDminer provides a variety of LD measures and parameters for flexibly defining the LD block. We implement this method for both of gene-base and chromosome-base LD block extraction.

Disease Association for Multi-Locus Haplotypes

After building LD blocks, our system automatically performs both single-locus tests and haplotype /displotype association tests. In haplotype / diplotype association tests, we have options to select several multi-locus association tests such as Log Likelihood Test (LRT) (Zhao & Sham, 2002), Monte Carlo tests. (Sham & Curtis, 1995)

LRT is the nonparametric tests for homogeneity in haplotype / diplotype frequencies between cases and controls. For example, the fast EH program (FEHP) developed by Zhao and Sham can be used for this purpose. In their statistics, the maximum log-likelihood was calculated for cases alone, for controls alone, and for cases and controls pooled altogether. Denoting these maximum log-likelihoods as $\ln L_{case}$, $\ln L_{control}$ and $\ln L_{case+control}$, LRT statistic was defined as $2(\ln L_{case} + \ln L_{control} - \ln L_{case+control})$, which is asymptotically chi-squared with the degree of freedom of, where is the number of haplotypes / diplotypes. Then we obtain asymptotic p-values as well as empirical p-values by the use of permutation procedures.

In Monte Carlo tests, some programs are well known such as CLUMP by Sham and Curtis (1995), which outputs the results of four different association tests from T1 to T4. In order to estimate the odds ratio (OR: a measure of the degree of association which estimates the relative risk of the disease in case-control study) and 95%CI for each haplotype/ diplotype, we perform the Fisher's exact test for each haplotype by constructing a 2 × 2 table which consists of the number of haplotype holders /others in the column and the number of case/control in the row.

Viewer for Genome Structure and Tools for Analysis Interpretation

LDMiner also has a supplicated vitalization tool (Higashi et al., 2003). The main viewer displays the genomic structure and is linked to another main viewer showing the in-depth analysis result. As real causative polymorphisms are likely to be located in the vicinity of the analyzed polymorphisms rather than themselves, it is important to examine the surrounding genes and polymorphisms. For this purpose, we also proposed the algorithm on haplotype classification for detecting disease-associated sites (Kido et al., 2003). These viewers and tools allow the user to easily check and make an interpretation of the results.

Figure 3. Selected SNPs and detected LD block for Caucasian, African American and Japanese

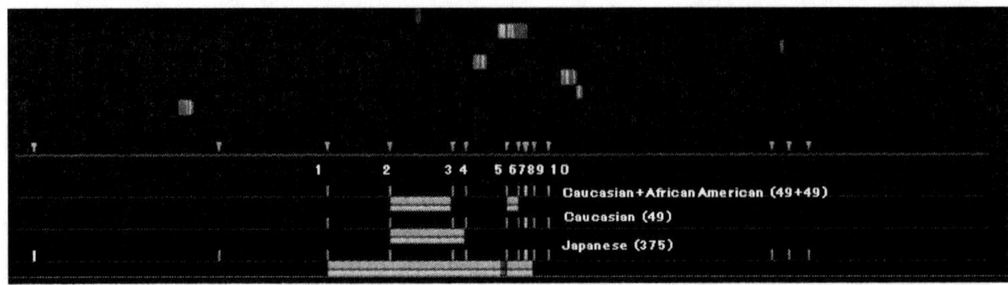

Examples of Haplotype Analyses

Experimental Setting: Genotype Data Used

To evaluate our methods, we performed a simulation study based on observed SNPs from genes around NPHS1 in chromosome 19 regions with a Japanese population of 375 independent individuals, a Caucasian population of 49 individuals and an African-American population of 49 individuals. This enables us to base our study on real patterns of LD within those genes. Some genes were genotyped in both exonic and intronic regions. We chose a subset of 10 SNPs mainly from NPHS1 genes denoted by 1 to 10 as shown in Figure 3.

Experimental Setting for the Simulation Study for Association Study

In order to examine whether the multi-locus association tests of LD0blocks have more statistical power than that of single-locus marker and how LD status influences the power of association, we evaluated it with simulated data sets. We designed a disease model for typical common disease described in report by Jannot, Essioux, Reese, & Clerget-Darpoux (2003) and phenotypic trait distributions were simulated under the samples of Japanese 375 (case 148, control 227) individuals, keeping our SNP typing information. In this model, we assumed that SNP 7, which allele frequency is 0.068, is the causative with prevalence rate of 0.08. We then compared the multi-locus haplotype tests with single-locus tests.

Single-Locus Analysis

The results of the single-locus analysis of allelic and genotype association are shown in Figure 4. As we designed, the causative SNP7 shows significant difference between

Figure 4. P-value distribution of the single-locus analysis and LD mapping

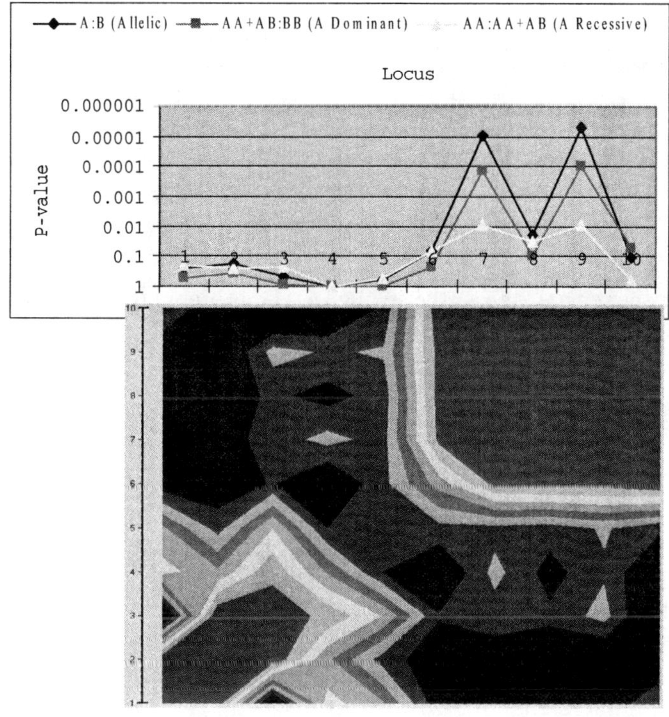

Figure 5. P-value distribution of the haplotype analysis and comparison with those of single-locus analysis

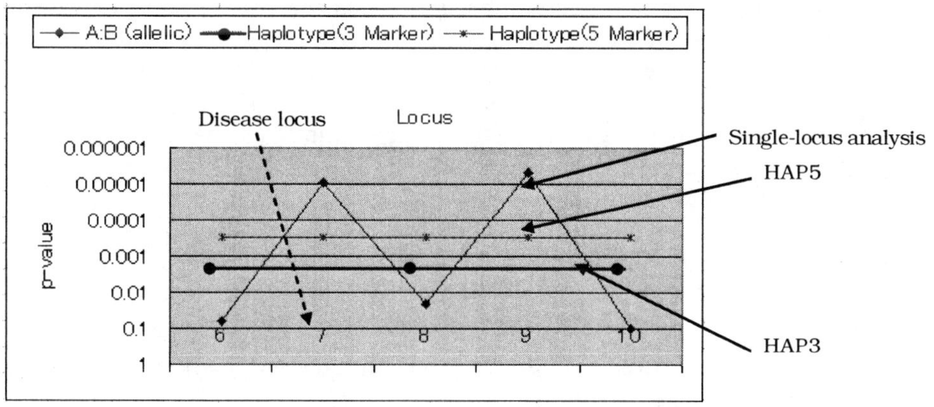

case and control allele frequencies with a p-value is less than 0.00001. SNP9 also has a low p-value in allelic association. This is because LD between SNP7 and SNP9 is strongly high (|D'|=0.978) and the discrepancy between SNP7 and SNP9 frequencies is very small (SNP7 minor allele frequency [MAF]=0.068, and SNP9 MAF = 0.064). LD theory introduced by Zondervan and Cardon (2004) estimates that the expected OR of locus 9 is 2.06 and this fits with our simulation result. However, the SNP6, SNP8, SNP10 are also in the strong LD with locus, their p-values are not significant. This can be explained by the discrepancy between the disease SNP (locus 7) and marker SNP frequency is not small. The MAF of SNP6 is 0.135, that of SNP8 is 0.120 and that of SNP10 is 0.056. This case shows that common variant marker which has high allele frequency can not detect the effect of row variant causative SNP even though they are in the same LD block.

In LD mapping, the red part shows the strong LD (D' is higher than 0.8) and the blue part shows weak LD. Read parts can be extracted as LD block.

Haplotype Analysis

Selecting SNPs for Haplotype Analysis

First we selected 5 SNPs - SNP6, SNP7 (causative), SNP8, SNP9, SNP10 that are in the strong LD block as show in Figure 4. The average of|D'| in this LD block is 0.964. We then select 3 SNPs - SNP6, SNP8 SNP10. This is because, minor allele frequencies of SNP7 and SNP9 are small (less than 0.1) and they are usually considered as inappropriate SNP markers. Here, an interesting question for us is whether haplotype analysis consisting of three SNPs can detect the causative region or not, although all three SNPs are not positive in single-locus analysis.

Results of Haplotype Association: Single-Locus Analysis vs. Haplotype Analysis

The p-value distributions of the haplotype association and comparisons with those of single-locus analysis are shown in Figure 5. The HAP5 consisting of five SNPs has a significant overall p-value by FEHP program; empirical p-value is less than 0.001. However it is not higher than those of single-locus analysis at SNP7 and SNP9. Interestingly, HAP3 consisting of three SNPs has also significant overall p-value, which is less than 0.01. (The empirical p-value by FEHP program with 20000 permutations is 0.0022, and the asymptotic p-value is 0.0007. In Figure 3, we plot the empirical p-value of FEHP.) Furthermore this p-value is higher than those of single-locus analysis at SNP6, SNP8 and SNP10. This is a good example to show the advantage of haplotype analysis; although three SNPs in HAP3 are not positive in single-locus analysis, haplotype analysis can detect the significant difference in the HAP3 block.

Table 1 shows the results of haplotype analysis with three SNP markers (HAP3). Only six haplotypes are estimated by EM algorithm, major haplotype 'BBB' dominates 85.8 % over the population. For the major haplotype 'BBB', the difference of frequencies

Table 1. The results of haplotype analysis with 3 SNP markers

Haplotype ID	Sequence	Control Case	Frequency in case/control	Posterior probability distribution	Frequency	2x2 Fisher's p-value	Odds ratio	Odds ratio 95% confidence interval
0	Missing	0%(0) 0%(2)			0	0.0	0.0	null
1	Haplotype[BBB]	87%(400) 82%(243)			0.838	0.083636	0.694286	0.458361 - 1.051644
2	Haplotype[AAB]	3%(15) 11%(33)			0.065	2.2E-5	3.745946	1.996388 - 7.028751
3	Haplotype[AAA]	6%(29) 3%(10)			0.052	0.091979	0.522133	0.250532 - 1.088175
4	Haplotype[ABB]	1%(9) 1%(4)			0.017	0.775473	0.689815	0.210468 - 2.260884
5	Haplotype[ABA]	0%(2) 0%(1)			0.004	1.0	0.780069	0.070413 - 8.642005
6	Haplotype[BAB]	0%(1) 0%(1)			0.003	1.0	1.563574	0.097414 - 25.096648

between case and control is not significant with a p-value obtained by Fisher's exact test of 0.085. The frequency of the minor haplotype 'AAB' - 6.5 %, is significant (p-value is less than 0.0001). It works as a risk haplotype, such that its frequency among cases (11%) is much higher than in controls (3%). The Fisher's p-value of 'AAB' is less than 0.0001 and the OR is 3.75 and its 95% confidence interval is 1.99 – 7.03. Because the posterior probability for determining diplotype is almost 100% (represented by blue bar) or more than 99% (represented by green bar) in most haplotypes, we can validate that haplotype estimation is mostly reasonable.

Diplotype Analysis

We also perform three-locus diplotype association as well as haplotype association study. This is done by determining individual's diplotype after estimating haplotype by EM algorithm (Kitamura et al., 2002). We only choose them from which posterior probability is more than 80%, otherwise we rule out from our analysis. The results are shown in Table 2. Minor diplotype [1:2] which represents the combination of haplotype1 ('BBB') and haplotype2 ('AAB') works as risk diplotype (Fisher's p-value is 0.005, OR=2.65). The diplotype also works as risk diplotype in our analysis (Fisher's p-value is 0.008, OR=17), although it is in the rare cases.

Summary

The main purpose of this chapter was to review the roles of haplotype analysis for association study on complex common diseases and provide the computational framework that enables to realize high throughput genome-wide association study. The motivation, background and computational methods for finding causative genes by use

Table 2. The result of diplotype analysis with 3 SNP markers

Diplotype ID	Control Case	Frequency in control/case	Posterior probability distribution	Frequency	2x2 Fisher's p-value	Odds ratio	Odds ratio 95% onfidence interval
Missing	0%(0) 0%(1)				null	null	null
Diplotype[1:1]	77%(176) 71%(105)				0.270774	0.756652	0.470371 - 1.217172
Diplotype[1:2]	6%(15) 15%(23)				0.005039	2.655285	1.335409 - 5.279683
Diplotype[1:3]	10%(25) 4%(6)				0.020673	0.348000	0.139146 - 0.870335
Diplotype[1:4]	2%(6) 2%(3)				1.000000	0.776224	0.191075 - 3.153330
Diplotype[1:5]	0%(1) 0%(0)				1.000000	0.517634	0.020944 - 12.793105
Diplotype[1:6]	0%(1) 0%(1)				1.000000	1.565517	0.097154 - 25.226514
Diplotype[2:2]	0%(0) 3%(5)				0.008690	17.763251	0.974761 - 323.702877
Diplotype[3:3]	0%(2) 1%(2)				0.645199	1.569444	0.218627 - 11.266481
Diplotype[4:4]	0%(1) 0%(0)				1.000000	0.517634	0.020944 - 12.793105
Diplotype[4:5]	0%(1) 0%(1)				1.000000	1.565517	0.097154 - 25.226514

of haplotype and LD information have been reviewed. A Haplotype Analysis System, LDMiner, has been introduced with some real examples of association studies. Although there have been significant advances in this area, further developments of computational methods for the genome-wide association studies and the detection of complex gene-gene interactions have been pursued. (Meng, Zaykin, Zu, Wanger, & Ehm, 2003; Ritchie et al., 2001; Toivonen et al., 2000) It is hoped that this report will entice further research and developments in bioinformatics for finding causative genes.

Acknowledgment

We would like to thank Dr. Tianhua Niu at Harvard University for useful discussions and comments.

References

Akey J., & Xiong, M. (2001). Haplotype vs single marker linkage disequilibrium tests: What do we gain? *Eur J Hu Genet, 9*, 291-300.

Badano, J. L, & Katsanis, N. (2002). Beyond Mendel: An evolving view of human genetic disease transmission. *Nature Genetics*, *3*, 779-789.

Bader, J. S. (2001). The relative power of SNPs and haplotype as genetic markers for association tests. *Pharmacogenomics*, *2*(1), 11-24

Cardon, L. R, & Abecasis, G.R. (2003). Using haplotype blocks to map human complex trait loci. *Trends Genet.*, *19*, 135-140.

Clark, A. G. (1990). Inference of haplotypes from PCR-amplified samples of diploid populations. *Mol. Biol. Evol.*, *7*, 111-122.

Clark, A. G. (2004). The role of haplotypes in candidate gene studies. *Genetic Epidemiology*, *27*, 321-333.

Cummins, J. M., Rago, C., Kohli, M., Kinzer, K. W., Lengauer, C., & Vogelstein, B. (2004). A marker for Stevens-Hohnson Syndrome. *Nature, 428*, 486.

Excoffier, L., & Slatkin, M. (1995). Maximum-likelihood estimation of molecular haplotype frequencies in a diploid population. *Mol. Biol. Evol.*, *12*, 921-927.

Fan, R., & Knapp, M. (2003). Genome association studies of complex diseases by case-control designs. *Am .J .Hum. Genet.*, *72*, 850-868.

Fullerton, S. M., Clark A. G, Weiss, K. M, Nickerson, D. A., Taylor, S. L., Stengard, J. H., et al. (2000). Apolipoprotein E variation at the sequence haplotype level: Implications for the origin and maintenance of a major human polymorphism. *Am. J. Hum. Genet.*, *67*, 881-900.

Gabriel, S. B., Schaffner, S. F., Nguyen, H., Moore, J. M., Roy, J., Blumenstiel, B., et al. (2002). The structure of haplotype blocks in the human genome. *Science, 296*, (pp. 2225-2229).

Higashi, Y., Higuchi, H., Kido, T., Matsumine, H., Baba, M., Morimoto, T., et al. (2003). SNP analysis system for detecting complex disease associated sites. *The Proceeding of IEEE Computational Systems Bioinformatics* (CSB2003), 450-451.

Hung, S. L., Chung, W. H., Liou, L. B., Cgy, C. C., Lin, M., Huang, H. P., et al. (2005). HLA-B*5801 allele as a genetic marker for severe cutaneous adverse reactions caused by allopurinol. *PNAS*, *102*, 11, 4134-4139.

International HapMap Project. (n.d.). Retrieved from http://www.hapmap.org/

The International HapMap Consortium. (2003) The International HapMap Project. *Nature*, *18*, 426(6968), 789-96.

Jannot, A. S., Essioux, L., Reese, M. G., & Clerget-Darpoux, F. (2003). Improved use of SNP information to detect the role of genes. *Genetic Epidemiology, 25*, 158-167.

Kido, T., Baba, M., Matsumine, H., Higashi, Y., Higuchi, H., & Muramatsu, M. (2003). Haplotype pattern mining & classification for detecting disease associated site. *The proceeding of IEEE Computational Systems Bioinfomratics* (CSB2003), (pp. 452-453).

Kitamura, Y., Moriguchi, M., Kaneko, H., Morisaki, H., Morisaki, T., Toyama, K., & Kamatani, N. (2002). Determination of probability distribution of diplotype configuration (diplotype distribution) for each subject from genotypic data using the EM algorithm. *Annals of Human Genetics*, *66*, 183-193.

Kruglyak, L., & Nickerson, D. A. (2001 March). Variation is the spice of life. *Nature Genetics., 27*(3), 234-6.

Liu, N., Sawyer, S. L., Mukherjee, N., Pakstis, A. J., Kidd, J. R., Kidd, K. K., et al. (2004). Haplotype block structures show significant variation among populations. *Genetic Epidemiology, 27*, 385-400.

Meng, Z., Zaykin, D. V., Zu, C. F., Wanger, M., & Ehm, M. G. (2003). Selection of genetic markers for association analysis, using linkage disequilibrium and haplotypes. *Am. J.Hum. Genet., 73*, 115-130.

Niu, T. (2004). Algorithms for Inferring Haplotypes. *Genetic Epidemiology, 27*, 334-347.

Niu, T., Zin, Z. S., Zu, X., & Liu, J. S. (2002). Bayesian haplotype inference for multiple linked single-nucleotide polymorphisms, *Am. J. Hum. Genet., 70*, 157-169.

Ritchie, M. D, Hahn, L. W., Roodi, N., Dupont, W. D., Parl, F. F., & Moore, J. H. (2001). Multifactor dimensionality reduction reveals high-order interactions among estrogen metabolism genes in sporadic breast cancer. *Annals of Human Genetics, 69*, 138-147.

Schaid, D. J. (2004). Evaluating associations of haplotypes with traits. *Genetic Epidemiology, 27*, 348-364.

Sham, P. C., & Curtis, D. (1995). Monte Carlo tests for associations between disease and alleles at highly polymorphic loci. *Annals of Human Genetics, 59*, 97-105.

Stephens, M., Smith, N. J., & Donnelly, P. (2001). A new statistical method for haplotype reconstruction from population data. *Am. J. Hum . Genet., 68*, 978-989.

Toivonen, H. T. Onkamo, P., Vasko, K., Ollikainen, V., Sevon, P., Mannila, H., Herr, M., & Kere, J. (2000). Data mining applied to linkage disequilibrium mapping. *Am. J. Hum. Genet., 67*, 133-145.

Zhang, J., Rowe, W. L., Struewing, J. P., & Buetow, K. H. (2002). HapScope: A software system for automated and visual analysis of functionally annotated the allelic association. *Nature Genetics, 5*, 89-100.

Zhao, J. H. (2004). Haplotype block structures show significant variation among populations. *Genetic Epidemiology, 27*, 385-400.

Zhao, J. H., & Sham, P.C. (2002). Faster haplotype frequency estimation using unrelated subjects. *Hum Hered., 53*, 36-41.

Chapter XII

A Bayesian Framework for Improving Clusterng Accuracy of Protein Sequences Based on Association Rules

Peng-Yeng Yin, National Chi Nan University, Taiwan
Shyong-Jian Shyu, Ming Chuan University, Taiwan
Guan-Shieng Huang, National Chi Nan University, Taiwan
Shuang-Te Liao, Ming Chuan University, Taiwan

Abstract

With the advent of new sequencing technology for biological data, the number of sequenced proteins stored in public databases has become an explosion. The structural, functional, and phylogenetic analyses of proteins would benefit from exploring databases by using data mining techniques. Clustering algorithms can assign proteins into clusters such that proteins in the same cluster are more similar in homology than those in different clusters. This procedure not only simplifies the analysis task but also enhances the accuracy of the results. Most of the existing protein-clustering algorithms compute the similarity between proteins based on one-to-one pairwise sequence alignment instead of multiple sequences alignment; the latter is prohibited due to expensive computation. Hence the accuracy of the clustering result is deteriorated. Further, the traditional clustering methods are ad-hoc and the resulting clustering often converges to local optima. This chapter presents a Bayesian framework for improving clustering accuracy of protein sequences based on association rules. The experimental results manifest that the proposed framework can significantly improve the performance of traditional clustering methods.

Introduction

One of the central problems of bioinformatics is to predict structural, functional, and phylogenetic features of proteins. A protein can be viewed as a sequence of amino acids with 20 letters (which is called the primary structure). The explosive growth of protein databases has made it possible to cluster proteins with similar properties into a family in order to understand their structural, functional, and phylogenetic relationships. For example, there are 181,821 protein sequences in the Swiss-Prot database (release 47.1) and 1,748,002 sequences in its supplement TrEMBL database (release 30.1) up to May 24, 2005. According to the secondary structural content and organization, proteins were originally classified into four classes: α, β, $\alpha+\beta$, and α/β (Levitt & Chothia, 1976). Several others (including multi-domain, membrane and cell surface, and small proteins) have been added in the SCOP database (Lo Conte, Brenner, Hubbard, Chothia, & Murzin, 2002). Family is a group of proteins that share more than 50% identities when aligned, the SCOP database (release 1.67) reports 2630 families.

Pairwise comparisons between sequences provide good predictions of the biological similarity for related sequences. Alignment algorithms such as the Smith-Waterman algorithm and the Needleman-Wunsch algorithm and their variants are proved to be useful. Substitution matrices like PAMs and BLOSUMs are designed so that one can detect the similarity even between distant sequences. However, the statistical tests for distant homologous sequences are not usually significant (Hubbard, Lesk, & Tramontano, 1996). Pairwise alignment fails to represent shared similarities among three or more sequences because it leaves the problem of how to represent the similarities between the first and the third sequences after the first two sequences have been aligned. It is suggested in many literatures that multiple sequence alignment should be a better choice. While this sounds reasonable, it causes some problems we address here. The most critical issue is the time efficiency. The natural extension of the dynamic programming algorithm from the pairwise alignment to the multiple alignment requires exponential time (Carrillo & Lipman, 1988), and many problems related to finding the multiple alignment are known to be NP-hard (Wang & Jiang, 1994). The second issue is that calculating a distance matrix by pairwise-alignment algorithm is fundamental. ClustalW (Thompson, Higgins, & Gibson, 1994) is one of the most popular softwares for multiple-alignment problems. It implements the so-called progressive method, a heuristic that combines the sub-alignments into a big one under the guidance of a phylogenetic tree. In fact, the tree is built from a pre-computed distance matrix using pairwise alignment.

Many protein clustering techniques exist for sorting the proteins but the resulting clustering could be of low accuracy due to two reasons. First, these clustering techniques are conducted according to homology similarity, thus a preprocessing of sequence alignment should be applied to construct a homology proximity matrix (or similarity matrix). As we have mentioned, applying multiple sequence alignment among all proteins in a large data set is prohibited because of expensive computation. Instead, an all-against-all pairwise alignment is adopted for saving computation time but it may cause deterioration in accuracy. Second, most of the traditional clustering techniques, such as hierarchical merging, iterative partitioning, and graph-based clustering, often converge to local optima and are not established on statistical inference basis (Jain, Murty, & Flynn, 1999).

This chapter proposes a Bayesian framework for improving clustering accuracy of protein sequences based on association rules. With the initial clustering result obtained by using a traditional method based on the distance matrix, the strong association rules of protein subsequences for each cluster can be generated. These rules satisfying both minimum support and minimum confidence can serve as features to assign proteins to new clusters. We call the process to extract features from clusters the alignment-less alignment. Instead of merely comparing similarity from two protein sequences, these features capture important characteristics for a whole class from the majority, but ignore minor exceptions. These exceptions exist due to two reasons: the feature itself or the sequence itself. For the first reason, the feature being selected could be inappropriate and thus causes exceptions. Or, there does not exist a perfect feature that coincides for the whole class. The second reason is more important. The sequence causing the exception may be pre-classified into a wrong cluster; therefore, it should be re-assigned to the correct one. The Bayes classifier can provide optimal protein classifications by using the *a priori* feature information through statistical inference. As such, the accuracy of the protein clustering is improved.

The rest of this chapter is organized as follows. The background reviews existing methods relevant to protein clustering and the motivations of this chapter. The third section presents the ideas and the theory of the proposed method. The fourth section gives the experimental results with a dataset of protein sequences. The final section concludes this chapter.

Background

Related Works

Many clustering techniques for protein sequences have been proposed. Among them, three main kinds of approaches exist, namely the hierarchical merging, iterative partitioning, and graph-based clustering. All of these methods use a pre-computed similarity matrix obtained by performing pairwise alignments on every pair of proteins. In the following we briefly review these approaches.

Figure 1. Summary of hierarchical merging clustering

```
        create.cluster(p);
Repeat
        Find clusters x, y such that similarity(x, y) is maximal;
        If similarity(x, y)>cutting_off_threshold then merge.cluster(x, y);
        Otherwise, terminate;
```

- **Hierarchical Merging:** The hierarchical merging clustering (Yona, Linial, & Linial, 1999, 2000; Sasson, Linial, & Linial, 2002) starts with a partition that takes each protein as a separate cluster, and then iteratively merges the two clusters that have the highest similarity from all pairs of current clusters. The similarity between two clusters is derived from the average similarity between the corresponding members. Thus, the hierarchical merging procedure forms a sequence of nested clusterings in which the number of clusters decreases as the number of iterations increases. A clustering result can be obtained by specifying an appropriate cutting-off threshold, in other words, the iterative merging procedure progresses until the maximal similarity score between any two clusters is less than the cutting-off threshold. The algorithm for hierarchical merging clustering is summarized in Figure 1.
- **Iterative Partitioning:** The iterative partitioning method (Jain & Dubes, 1988; Guralnik & Karypis, 2001; Sugiyama & Kotani, 2002) starts with an initial partition of k clusters. The initial partition can be obtained by arbitrarily specifying k proteins as cluster centers then assigning each protein to the closest cluster whose center has the most similarity with this protein. The next partition is obtained by computing the average similarity between each protein and all members of each cluster. The partitioning process is iterated until no protein changes its assignment between successive iterations. Although the iterative partitioning method has some variants like ISODATA (Ball & Hall, 1964) and *K-means* (McQueen, 1967), their general principles can be described as shown in Figure 2. A post-processing stage can be added to refine the clusters obtained from the iterative partitioning by splitting or merging the clusters based on intra-cluster and inter-cluster similarity scores (Wise, 2002).
- **Graph-based Clustering:** These methods (Tatusov, Koonin, & Lipman, 1997; Enright & Ouzounis, 2000; Bolten, Schliep, Schneckener, Schomburg, & Schrader, 2001) represent each protein sequence as a graph vertex, and every pair of these vertices are connected by an edge with a label on it. The label denotes the similarity score between the two proteins represented by the vertices connecting to the corresponding edge. A partition of the graph can be generated by cutting off the edges whose labels are less than a specified similarity threshold, and each connected component of vertices corresponds to a cluster of proteins since the

Figure 2. Summary of iterative partitioning method

Select *k* proteins as initial cluster centers
Repeat
    Compute the average similarity between each protein and each cluster;
    Generate a new partition by assigning each protein to its closest cluster;
Until no proteins change assignment between successive iterations

similarity scores between proteins from the same component are higher than those between proteins from different components. The general idea of the graph-based clustering is outlined in Figure 3. A post-refining process can be conducted by using the graph-based clustering result as the input to a cluster-merging algorithm which iteratively merges the nearest neighboring clusters if the relative entropy decreases with the merging of the clusters (Abascal & Valencia, 2002).

Motivations

The protein clustering result obtained by using the above mentioned approaches could be of low accuracy. This is partly due to the reason that these clustering methods use only pairwise sequence alignment information and partly because these clustering techniques often converge to local optima. Since it is computationally prohibitive to derive multiple sequence alignment information, an alternative is to calculate the statistics from the sequences directly by using data mining techniques. More precisely, the *association rules* between the amino acids in the protein sequences are mined within each cluster. Then the rules satisfying minimum confidence can serve as salient features to identify each cluster. Moreover, matching each protein sequence with the association rule provides a good estimate of the *a priori* probability that the protein satisfies the rule. The statistical inference can compensate the accuracy inadequacy of the clustering

Figure 3. Summary of graph-based clustering

Represent each input sequence as a vertex, and every pair of vertices are connected by an edge.
Label each edge with the similarity score between the two connected vertices.
Create a partition of the graph by cutting off the edges whose similarity scores are less than a specified threshold.

result due to the pairwise alignment and the local clustering technique. Therefore, we propose to improve the clustering accuracy by using the Bayes classifier with the conditional probabilities of association rules with each cluster and the alignment scores between protein sequences and these rules.

Methods

With the assistance of association rule mining and Bayes classifier, our system improves the clustering accuracy of traditional protein-clustering methods. The system overview is shown in Figure 4. First, a traditional protein-clustering approach (either one of the hierarchical merging, iterative partitioning, or graph-based clustering methods) is performed to obtain an initial clustering result of the input protein sequences. In general, the traditional protein-clustering approach consists of three steps as shown in the upper grey box: (1) perform the local alignment (such as BLAST with scoring matrix of BLOSUM 62) between each pair of protein sequences, (2) construct a distance matrix from the raw distance scores (such as the E-values produced by BLAST) of the local alignment, and (3) apply the clustering method with the distance matrix to get the clustering result.

Second, the refining process (as shown in the lower grey box) is performed such that the clustering result is improved. The refining process also consists of three steps: (1) generate strong association rules which satisfy minimum confidence for each cluster, (2) perform sequence rule matching between each protein sequence and each association rule based on local alignment, and (3) adapt cluster membership of each protein sequence by using Bayes classifier and its matching score. The proposed refining process improves the clustering result based on statistical inference. The association rules used are statistically confident and the reassignment of protein sequences using Bayes classifier satisfies the maximum *a posteriori* criterion.

We now present the refining process in details.

Sequence Association Rule Generation

Association rule mining has been intensively used for finding correlation relationships among a large set of items and has delivered many successful applications, such as catalog design, cross marketing, and loss-leader analysis (Han & Kamber, 2001). Association rule mining finds significant associations or correlation relationships among a large repository of data items. These relationships, represented as rules, can assist the users to make their decisions. Traditionally, association rule mining works with unordered itemset, that is, the order of the items appearing in the itemset does not matter (Agrawal, Imielinski, & Swami, 1993). However, the order of the amino acids appearing in a protein sequence reserves important phylogeny information among homologies and should be taken into account. The *sequence Apriori algorithm* (Agrawal & Srikant, 1994, 1995) which adapts the classical association rule mining algorithm for sets of sequences can be used for this purpose.

Figure 4. System overview

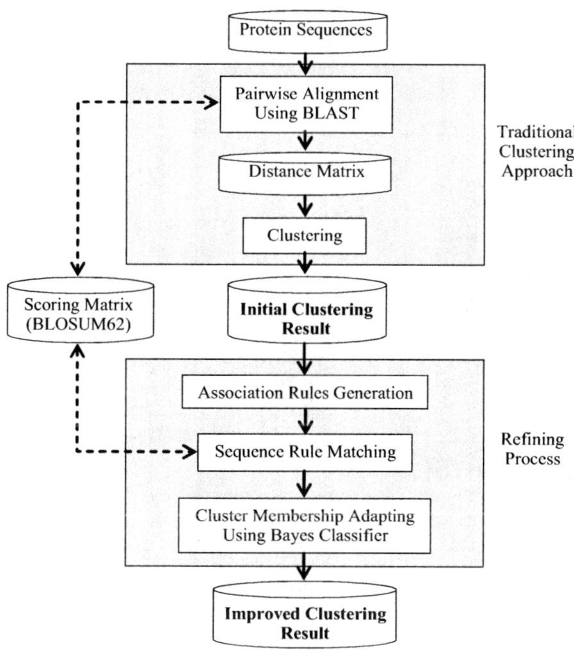

Given a cluster C of protein sequences, the algorithm finds sets of sequences that have support above a given *minimum support*. The *support* of a sequence X with respect to C, denoted by $\text{support}_C(X)$, is the number of sequences in C that contain X as subsequence. A sequence is called *frequent* if its support is greater than the given *minimum support*. To expedite the search for all frequent sequences, the algorithm enforces an iterative procedure based on the a priori property which states that any subsequence of a frequent sequence must be also frequent. It starts with the set of frequent 1-sequences which are the frequent sequences of length 1 and uses this set to find the set of frequent 2-sequences, then the set of frequent 2-sequences is used to find the set of frequent 3-sequences, and so on, until no more longer frequent sequences can be found. As the number of frequent sequences can be extremely large for a cluster of protein sequences having length of hundreds of amino acids, we retain the frequent sequences that are not contained in longer ones and restrict the search for frequent sequences of length 5 to 10. The reduced set of frequent sequences is sufficient for deriving similarity statistics in homology since longer frequent sequences of length more than 10 usually have lower support and a number of short frequent sequences within the specified range of lengths that the longer ones contain as subsequences can still be reserved.

With the set of found frequent sequences, we generate sequence association rules as follows. For a given frequent sequence X, it can be divided into two disjoint subsequences A and B with their position information attached. For example, let X be '*abcde*', one of the possible divisions could be A = 'a_c_e' and B = '_b_d_'. A sequence association rule is of the form $A \Rightarrow B$ where both A and B contain at least one amino acid. A is called the rule *antecedent* and B is called the rule *consequent*. A sequence association rule is *strong* if it has a confidence value above a given *minimum confidence*. The confidence of a rule $A \Rightarrow B$ with respect to protein cluster C, denoted by $\text{confidence}_C(A \Rightarrow B)$, is defined as:

$$\text{confidence}_C(A \Rightarrow B) = \frac{\text{support}_C(A \cup B)}{\text{support}_C(A)}, \tag{1}$$

where $A \cup B$ denotes the supersequence that is properly divided into A and B. We generate all strong sequence association rules for each cluster of protein sequences.

Sequence Association Rule Matching

In order to determine the cluster membership of each protein, we need to propose a measure which estimates the possibility that the evolution of a protein follows a particular association rule. Here we propose a rule matching scheme which is analogous to local sequence alignment without gaps. Given a sequence t and a strong association rule r, we compute the alignment score between t and r according to a substitution matrix with the constraint that no gap is allowed and the rule *antecedent* cannot be substituted in order to detect most similarities. For example, suppose we use the substitution matrix as shown in Figure 5 (a). Let the sequence be *'adabdacd'* and the association rule be *'da_d'* ⇒ '_ _*d*_', there are two possible alignments as illustrated in Figure 5 (b) and the best alignment score without gaps is 11.

Figure 5. An illustrative example of sequence association rule matching: (a) A substitution matrix, (b) Two possible alignments between the sequence and the association rule

	a	b	c	d
a	2	-1	2	0
b	-1	3	-2	1
c	2	-2	2	-1
d	0	1	-1	4

(a)

Sequence:	*adabdacd*	Score
Association rule:	*dadd*	4+2+1+4=11
Association rule:	*dadd*	4+2 -1+4= 9

(b)

Assume that we obtain a set of strong association rules $\Re = \{r_1, r_2, ..., r_n\}$ from the procedure of sequence association rule generation, the probability that the evolution of a protein t follows a particular association rule r_i can be estimated by:

$$p(r_i \mid t) = \frac{w(r_i, t)}{\sum_{h=1}^{n} w(r_h, t)}, \quad (2)$$

where $w(r_i, t)$ is the alignment score between r_i and t. As such, we can use the probabilities $(p(r_1 \mid t), p(r_2 \mid t), ..., p(r_n \mid t))$ as the feature values of protein t, and determine its cluster membership by using the Bayes classifier.

Cluster Membership Adapting by Using Bayes Classifier

The Bayes classifier is one of the most important techniques used in data mining for classification. It predicts the cluster membership probabilities based on statistical inference. Studies have shown that the Bayes classifier is comparable in performance with decision tree and neural network classifiers (Han & Kamber, 2001). Herein we propose to predict the cluster membership of each protein by using the Bayes classifier with association rules.

Let the initial clustering result consists of k clusters, denoted by $\Im = \{C_1, C_2, ..., C_k\}$, from which a set of n strong association rules is derived. The *a priori* probability $p(C_i)$ that a protein belongs to cluster C_i can be calculated by counting the ratio of C_i in size to the whole set of proteins. The condition probability $p(r_i \mid C_j)$ that a protein satisfies association rule r_i given that this protein is initially assigned to cluster C_j is estimated by the average probability $p(r_i \mid t)$ for any $t \in C_j$. The conditional probability $p(r_i \mid t)$ that the evolution of protein t follows association rule r_i is estimated by using Equation (2). We then use the naïve Bayes classifier to assign protein t to the most probable cluster C_{Bayes} given by:

$$C_{Bayes} = \arg\max_{C_j \in \Im} \left\{ \prod_{i=1}^{n} \left(p(r_i \mid C_j) p(r_i \mid t) \right) p(C_j) \right\}. \quad (3)$$

In theory, the naïve Bayes classifier makes classification with the minimum error rate.

Experimental Results

We validate our method by using protein sequences selected from SCOP database (release 1.50, Murzin, Brenner, Hubbard, & Chothia, 1995; Lo Conte et al., 2002) which is a protein classification created manually. SCOP provides a hierarchy of known protein folds and their detailed structure information. We randomly select 1189 protein sequences from SCOP and these sequences compose 388 protein clusters according to manual annotations on structure domains. The mean length of these protein sequences is 188 amino acids. Our method is implemented in C++ programming language and the experiments are conducted on a personal computer with a 1.8 GHz CPU and 512 MB RAM.

Accuracy Measures

To define the accuracy measures of protein classification, some notations are first introduced. For every pair of protein sequences, the predicted classification and the annotated classification have four possible combinations. *TP* (true positive) is the number of pairs predicted in the same domain given that they are in the same SCOP domain, *TN* (true negative) is the number of pairs predicted with different domains given that they are in different SCOP domains, *FP* (false positive) is the number of different SCOP-domain pairs that are predicted in the same domain, and *FN* (false negative) is the number of SCOP-domain pairs that are predicted in different domains. Two accuracy measures, namely *sensitivity* (S_n) and *specificity* (S_p), are defined as follows:

$$S_n = \frac{TP}{TP + FN} \tag{4}$$

and $$S_p = \frac{TP}{TP + FP}. \tag{5}$$

Sensitivity is the proportion of SCOP-domain pairs that have been correctly identified, and specificity is the proportion of pairs predicted in the same domain that are actually SCOP-domain pairs. Sensitivity and specificity cannot be used alone since perfect sensitivity can be obtained if all the pairs are predicted in the same domain, and specificity is not defined if all the pairs are predicted in different domains. To compare the performance between two competing methods, a sensitivity vs. specificity curve is usually used for evaluation. Or alternatively, the mean of sensitivity and specificity values can be used as a unified measure.

Performance Evaluation

We first evaluate the clustering performance using the traditional methods, in particular, we have implemented the hierarchical merging method (Sasson et al., 2002), the K-means algorithm (modified from Guralnik & Karypis, 2001, by changing the feature space to pair-wise E-value), and the graph-based clustering (Bolten et al., 2001). Various clustering thresholds are specified to obtain the performances at different specified numbers of

Figure 6. Performances of traditional clustering methods, (a) hierarchical merging clustering, (b) K-means clustering, (c) graph-based clustering

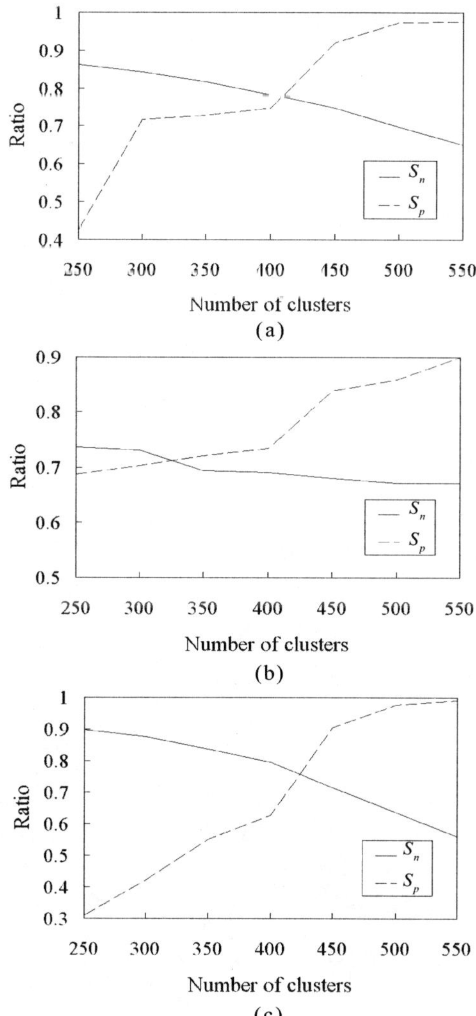

Figure 7. Performance improvement for specificity vs. sensitivity, (a) hierarchical merging clustering, (b) K-means clustering, (c) graph-based clustering

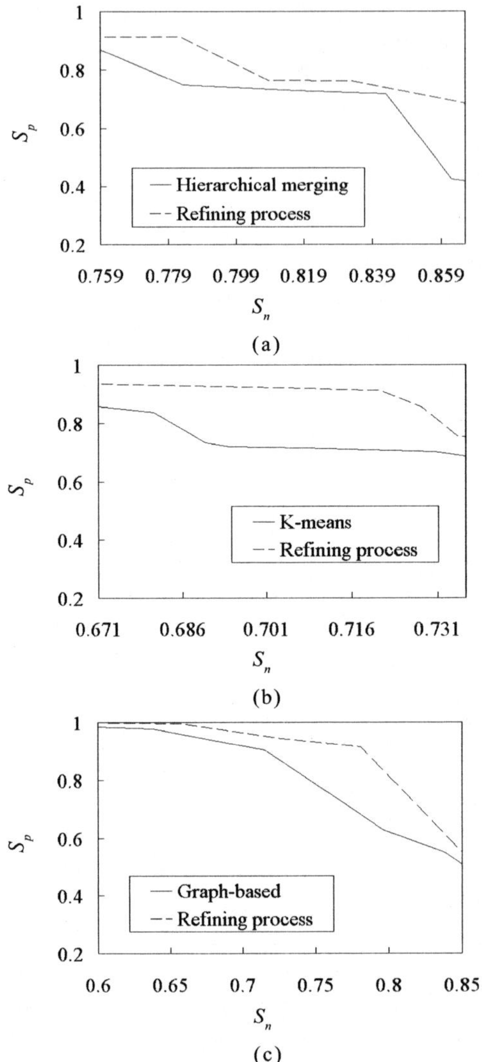

clusters. Figures 6 (a) - 6 (c) show the variations of sensitivity and specificity as the number of clusters increases for hierarchical merging, K-means, and graph-based clustering, respectively. These curves are intuitive since the mean size of clusters is smaller if the partition with more clusters is obtained and, in general, the smaller the cluster-mean size is, the lower the sensitivity is, but the higher the specificity is. If we take the accuracy values obtained when the number of clusters is equal to the true number (388), the hierarchical merging method with $S_n = 0.79$ and $S_p = 0.73$ is superior to the K-

Figure 8. Performance improvement for the unified measure, (a) hierarchical merging clustering, (b) K-means clustering, (c) graph-based clustering

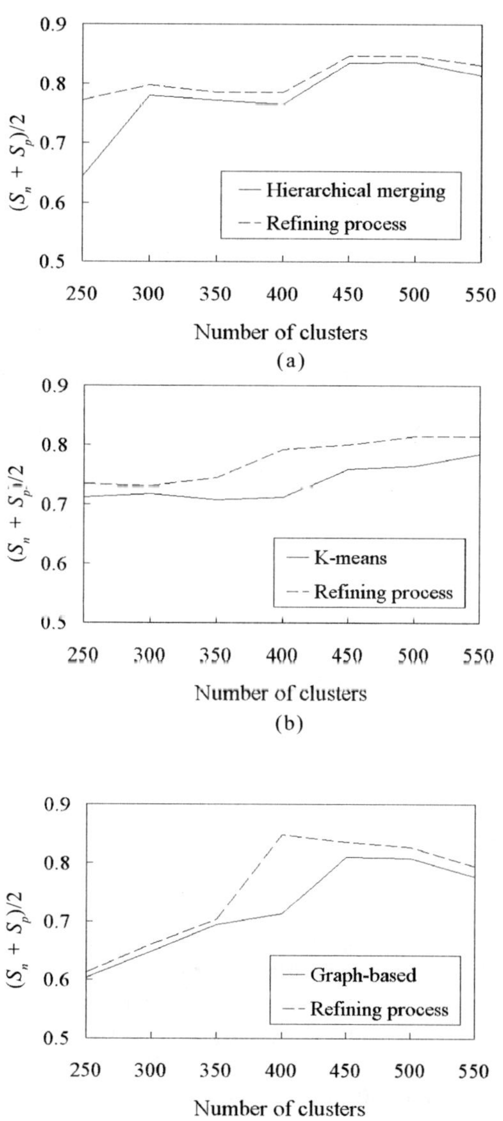

means algorithm which produces $S_n = 0.69$ and $S_p = 0.72$, while the graph-based clustering has medium performance ($S_n = 0.80$ and $S_p = 0.63$).

Next we evaluate the improvement for clustering accuracy due to our Bayesian framework. From the clustering results obtained by traditional clustering methods, we apply the sequence association rule mining to generate strong association rules. Each protein

Table 1. The computation time used by each component of the proposed method

	CPU Time (second)	Percentage
Distance Matrix Computation	162	35%
Traditional Clustering	64	14%
Association Rules Generation	157	33%
Rule Matching and Bayesian Classification	87	18%

sequence is matched with these rules and updates its cluster membership by using the Bayes classifier in order to improve the accuracy. Because the sensitivity values obtained by hierarchical merging, K-means, and graph-based clusterings with the true cluster number (388) are 0.79, 0.69, and 0.80, respectively, we compute the specificity improvement for these methods within a range of sensitivity close to these values. Figures 7 (a) - 7 (c) show the sensitivity vs. specificity curve for illustrating the improvement achieved. It is observed that the proposed refining process can significantly improve the specificity of the traditional methods. The average improvements in Sp are 0.09, 0.11, and 0.06 for hierarchical merging, K-means, and graph-based clustering, respectively. We also compute the accuracy improvement by using the unified measure of $(S_n + S_p)/2$ as shown in Figs. 8 (a) - 8 (c). The average improvements over various specified numbers of clusters are 0.03, 0.04, and 0.03 for hierarchical merging, K-means, and graph-based clustering, manifesting the robustness of the proposed framework.

Table 1 shows the incurred computation time (in seconds) by each component of the proposed framework and the corresponding percentage to the whole for our collective database. We observe that the distance matrix computation using pairwise sequence alignment and the association rules generation using sequence Apriori algorithm are the most time-consuming components, and they consume 35% and 33% of the total computation time, respectively. Further, the computation time for the refining process involving the last two components of Table 1 is about half of the whole time needed. That is, for our collective database, the proposed framework provides a considerable amount of accuracy improvement (as shown in Figures 7-8) for the traditional protein-clustering algorithms by doubling the computation time. In general cases, the time proportion for the refining process diminishes when the number of sequences in the database increases. Suppose there are m sequences in the database, and for simplicity, assume that their average length is r. The time complexity for calculating a distance matrix is $\Theta(m^2r^2)$, for performing traditional clustering is $\Omega(m^2)$, and for generating the association rules of fixed length is $\Theta(mr)$. Rule matching and Bayesian classification depends on two factors: the number of features extracted and the number of clusters classified (and remember that each rule is within a fixed length). The former can be regarded as a constant once the lengths of rules and the alphabet (which is 20 for amino acids) are fixed to constants. Therefore a rough calculation asserts that it takes $\Theta(mrc)$ where c is the number of clusters. As a summary, the time complexity for each component is $\Theta(m^2r^2)$, $\Omega(m^2)$, $\Theta(mr)$, and $\Theta(mrc)$, respectively. Parameter r, the average length of peptide sequences, is around 300 and can be regarded as a constant. Parameter c grows mildly as the number of sequences increases. As a result, when the number of sequences in a database grows larger, the percentage of the computing time on the refining process becomes smaller.

Summary

Protein sequence clustering is useful for structural, functional, and phylogenetic analyses. As the rapid growth in the number of sequenced proteins prohibits the analysis using multiple sequences alignment, most of the traditional protein-clustering methods derive the similarity among sequences from pairwise sequence alignment. In this chapter, we have proposed a Bayesian framework based on association rule mining for improving the clustering accuracy using existing methods. A selective dataset from SCOP has been experimented and the result manifests that the proposed framework is feasible. The main features of the proposed framework include:

- The proposed framework improves the accuracy of a given initial clustering result which can be provided by any clustering methods. Therefore, the user is still able to choose a particular clustering method which is suited to his/her own analysis of the final result.
- From the initial clustering result, the sequence association rules among amino acids are mined. These rules represent important relationships for the sequences belonging to the same cluster.
- The association rules serve as features for classification of proteins. Using Bayes classifier, the classification error can be minimized based on statistical inference.

Future research is encouraged in expediting the computation for distance matrix and association rule generation in order to extend the application of the proposed framework to large protein databases such as SWISSPROT.

References

Abascal, F., & Valencia, A. (2002). Clustering of proximal sequence space for the identification of protein families. *Bioinformatics, 18*(7), 908-921.

Agrawal, R., Imielinski, T., & Swami, A. (1993). Mining association rules between sets of items in large databases. *Proceedings of ACM-SIGMOD, International Conference on Management of Data*, Washington, DC (pp. 207-216).

Agrawal, R., & Srikant, R. (1994). Algorithms for mining association rules. *Proceedings of the 20th International Conference on Very Large Databases*, Santiago, Chile (pp. 487-499).

Agrawal, R., & Srikant, R. (1995). Mining sequential patterns. *Proceedings of the International Conference on Data Engineering (ICDE)*, Taipei, Taiwan.

Ball, G. H., & Hall, D. J. (1964). Some fundamental concepts and synthesis procedures for pattern recognition preprocessors. *Proceedings of International Conference on Microwaves, Circuit Theory, and Information Theory*, Tokyo.

Bolten, E., Schliep, A., Schneckener, S., Schomburg, D., & Schrader, R. (2001). Clustering protein sequences—structure prediction by transitive homology. *Bioinformatics, 17*(10), 935-941.

Carrillo, H., & Lipman, D. (1988). The multiple sequence alignment problem in biology. *SIAM Journal on Applied Mathematics, 48*, 1073-1082.

Enright, A. J., & Ouzounis, C. A. (2000). GeneRAGE a robust algorithm for sequence clustering and domain detection. *Bioinformatics, 16*(5), 451-457.

Guralnik, V., & Karypis, G., (2001, August 26). A scalable algorithm for clustering protein sequences. *Proceedings of Workshop on Data Mining in Bioinformatics,* San Francisco (pp. 73-80). ACM Press.

Han, J., & Kamber, M., (2001). *Data mining: Concepts and techniques*. San Francisco: Morgan Kaufmann Publishers.

Hubbard, T. J., Lesk, A. M., & Tramontano, A. (1996). Gathering them into the fold. *Nature Structure Biology, 4*, 313.

Jain, A. K., & Dubes, R. C. (1988). *Algorithms for clustering data*. Englewood Cliffs, NJ: Prentice Hall.

Jain, A. K., Murty, M. N., & Flynn, P. J. (1999). Data clustering: A review. *ACM Computing Surveys, 13*(3), 264-323.

Levitt, M., & Chothia, C. (1976). Structural patterns in globular proteins. *Nature, 261*, 552-558.

Lo Conte, L., Brenner, S. E., Hubbard, T. J. P., Chothia, C., & Murzin, A. G. (2002). SCOP database in 2002; refinements accommodate structural genomics. *Nucleic Acids Research, 30*, 264-267.

McQueen, J.B. (1967). Some methods of classification and analysis of multivariate observations. *Proceedings of Fifth Berkeleys Symposium on Mathematical Statistics and Probability* (pp. 281-297).

Murzin A. G., Brenner S. E., Hubbard T., & Chothia C. (1995). SCOP: A structure classification of proteins database for the investigation of sequences and structures. *Journal of Molecular Biology, 247*, 536-540.

Sasson, O., Linial, N., & Linial, M. (2002). The metric space of proteins—comparative study of clustering algorithms. *Bioinformatics, 18*(1), S14-S21.

Sugiyama, A., & Kotani, M. (2002). Analysis of gene expression data by self-organizing maps and k-means clustering. *Proceedings of IJCNN* (pp. 1342-1345).

Tatusov, R. L., Koonin, E. V., & Lipman, D. J. (1997). A genomic perspective on protein families. *Science, 278*, 631-637.

Thompson, J. D., Higgins, D. G., & Gibson, T. J. (1994). CLUSTAL W: Improving the sensitivity of progressive multiple sequence alignment through sequence weighting, position-specific gap penalties and weight matrix choice, *Nucleic Acids Research, 22*, 4673-4680.

Wang, L., & Jiang, T. (1994). On the complexity of multiple sequence alignment. *Journal of Computational Biology, 1*, 337-348.

Wise, M. J. (2002). The POPPs: clustering and searching using peptide probability profiles. *Bioinformatics, 18*(1), S38-S45.

Yona, G., Linial, N., & Linial, M. (1999). ProtoMap: Automatic classification of protein sequences, a hierarchy of protein families, and local maps of the protein space. *Proteins, 37*, 360-378.

Yona, G., Linial, N., & Linial, M. (2000). ProtoMap: Automatic classification of protein sequences, a hierarchy of protein families, and local maps of the protein space. *Nucleic Acids Research, 28*, 49-55.

Chapter XIII

In Silico Recognition of Protein-Protein Interactions: Theory and Applications

Byung-Hoon Park, Oak Ridge National Laboratory, USA
Phuongan Dam, University of Georgia, USA
Chongle Pan, University of Tennessee, USA
Ying Xu, University of Georgia, USA
Al Geist, Oak Ridge National Laboratory, USA
Grant Heffelfinger, Sandia National Laboratories, USA
Nagiza F. Samatova, Oak Ridge National Laboratory, USA

Abstract

Protein-protein interactions are fundamental to cellular processes. They are responsible for phenomena like DNA replication, gene transcription, protein translation, regulation of metabolic pathways, immunologic recognition, signal transduction, etc. The identification of interacting proteins is therefore an important prerequisite step in understanding their physiological functions. Due to the invaluable importance to various biophysical activities, reliable computational methods to infer protein-protein interactions from either structural or genome sequences are in heavy demand lately. Successful predictions, for instance, will facilitate a drug design process and the reconstruction of metabolic or regulatory networks. In this chapter, we review: (a) high-throughput experimental methods for identification of protein-protein interactions, (b) existing databases of protein-protein interactions, (c) computational approaches to predicting protein-protein interactions at both residue and protein levels, (d) various statistical and machine learning techniques to model protein-protein interactions, and (e) applications of protein-protein interactions in predicting protein functions. We also discuss intrinsic drawbacks of the existing approaches and future research directions.

Introduction

Protein-protein interactions are one of the most ubiquitous and fundamental phenomena in all cellular activities including DNA transcription and regulation of metabolic or signaling pathways. Uncovering the mechanisms that govern protein-protein interactions, therefore, has been the subject of great interest in the post-genomic era. Emerging genome-wide high-throughput experimental techniques (Ho et al., 2002; Ito et al., 2001; Uetz et al., 2000; Zhu et al., 2001) are generating an increasing amount of data that provide significant insights into the mechanisms underlying protein-protein interactions. In parallel, new computational approaches are being introduced to infer novel interactions or to reconstruct complex interactions such as functional modules and protein complexes from experimental data.

In a narrow sense, a protein-protein interaction refers to a physical binding between two or more proteins. Such a physical interaction can be categorized based on the composition of the complex, the function of the complex versus that of a monomer, the binding affinity of subunits in the complex, the duration of the complex formation, or interactions between specific functional groups. Accordingly, protein-protein interaction can be homo-oligomeric (between identical subunits), such as a dimer of a transcription factor, or hetero-oligomeric (between different subunits), such as the transcription machinery (RNA polymerase). A physical interaction can also be classified based on the duration of the contact. It can be transient as in the case of an enzyme-substrate contact or ligand-receptor interaction, or it can be stable as in the case of the formation of an actin filament or an antigen-antibody interaction. Furthermore, protein-protein interactions are often characterized as interactions between specific functional groups. Examples include interactions between SH_2 domain and a phosphorylated tyrosine residue or between two PDZ domains.

In the field of bioinformatics, protein-protein interactions are understood in a much broader sense. They not only include physical interactions but also embrace functional associations between proteins at different times or locations. This new interpretation includes two proteins that appear in the same pathway, although they are not physically in contact with each other. Such functional interactions can be defined in many ways, including proteins that are functioning in the same pathway, co-expressed under the same conditions, co-evolved or co-localized in the same region of genomes, or even co-present in the same literature abstracts (von Mering et al., 2005).

Computational methods typically address physical or functional interactions at the levels of the protein and its component residues. The former is an effort to determine the potential interaction between two proteins given their structural or sequence information. In contrast, studies of protein interactions at the residue level focus on identifying specific residues that interact with other residue(s) in different proteins. Computational methods utilize various techniques to predict protein interactions at both levels, including simple correlation analysis, sophisticated statistical assessments, and machine-learning approaches for classification.

In this chapter, we review *in silico* approaches to recognize protein-protein interactions in a wide spectrum. We first introduce high-throughput experimental methods for sifting putative physical interactions, followed by a discussion of a series of computational methods targeted at protein and residue levels. Applications of protein interactions, protein function and complex predictions are described within the context of protein-protein interaction networks. We conclude the chapter with the discussion of some future directions and needs for computational approaches to characterizing protein-protein interactions.

Experimental Methods and Databases of Protein-Protein Interactions

Experimental protein-protein interaction data are an invaluable resource to infer biological phenomena. Traditionally, experimental methods for protein-protein interaction identification are low-throughput, or focusing on individual cases. The procedures include some degree of purification of one or more protein complexes followed by (1) identification of the complex components using a wide-range of techniques, or (2) direct detection of protein-protein interactions by NMR or crystallography. In practice, purification of a protein complex largely depends on physical characteristics of the complex as mass, ionic charge, hydrophobicity, or its affinity to a substrate such as lectin or metal ions (Doonan, 1996). After the protein purification step, the subunits (component proteins) are identified by their mass (mass spectrometry), their amino acid sequence (sequencing of the peptides), or their affinity to specific antibodies generated against them in previous studies. These low-throughput procedures are costly, time-consuming, and do not scale.

With the advent of "proteomic" approaches, several high-throughput experimental methods have emerged. Different methods can probe different types of protein interactions, but all suffer from generating false positives and false negatives to some extent. Since most protein-protein interaction databases, the foundation of many computational approaches, are populated with data obtained through these methods, it is important to understand their strengths and weaknesses, as well as the types of interactions they can discover. The yeast two-hybrid assay (Fields & Song, 1989; Tucker, Gera, & Uetz, 2001; Uetz, 2002; Uetz et al., 2000), phage display (Tong et al., 2002), affinity purification (Gavin et al., 2002) (Aloy & Russell, 2002), and protein microarray (chip) (Zhu et al., 2001), four of the most widely used experimental methods are depicted in Figure 1.

Figure 1. Four high-throughput experimental methods to sift protein-protein interactions

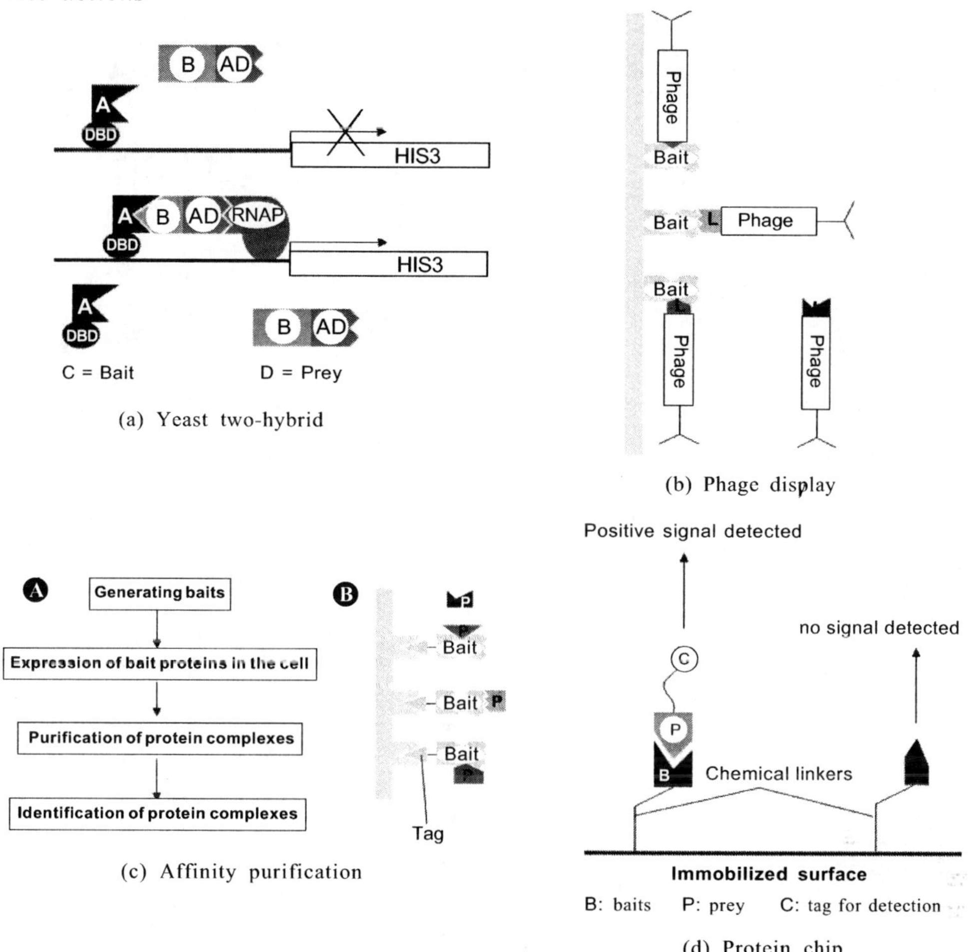

(a) Yeast two-hybrid

(b) Phage display

(c) Affinity purification

(d) Protein chip

Protein-protein interaction data, whether produced by high-throughput methods or retrieved from literature, are invaluable resources for computational methods. There exists a number of publicly accessible protein interaction databases systematically maintained to serve researchers. Built on meticulously designed relational schemas, most provide extensive search capabilities via nice user interfaces. Retrieval of related data in various formats is readily available, simplifying large-scale comparative analysis of protein-protein interactions. Therefore, understanding the characteristics of these databases is the first step in pursuing protein interaction inference.

Every year new databases dedicated for protein interactions are introduced. In particular, a comprehensive list of protein-protein interaction databases and their URLs is published in the first issue of Nucleic Acids Research journal each year. These protein-protein interaction databases are currently in the category of Metabolic and Signaling Pathways/Intermolecular interactions and signaling pathways (Galperin, 2005). From the

Table 1. A list of widely used protein-protein interaction databases (Note that names of databases are shown in acronyms)

Database	URL	Data	Source	Comments
DIP (Xenarios et al., 2005)	http://dip.doe-mbi.ucla.edu	44,482 interactions 17,173 proteins	Experiments	Vast majority of interactions are from eukaryotes. Only *Helicobacter pylori* and *Escherichia coli* are prokaryotes.
BIND (Alfarano et al., 2005)	http://bind.ca	134,886 interactions 2,525 complexes 8 pathways	Experiments Scientific Literature	Object-oriented design. An interaction is the basic unit, and complexes and pathways are represented in terms of interactions.
MIPS (Mewes et al., 2002)	http://mips.gsf.de	11,200 interactions 1,050 complexes	Experiments	Cross-reference for functional classification, subcellular localization, and EC numbers are available. Widely used for benchmark data.
MINT (Zanzoni et al., 2002)	http://cbm.bio.uniroma2.it/mint	43,273 interactions 18,549 proteins	Experiments Scientific Literatures	Special emphasis on proteomes from mammalian organisms
iPfam (Finn, Marshall, & Bateman, 2004)	http://www.sanger.ac.uk/Software/Pfam/iPfam/	2,828 domain interactions	PDB, Pfam	Collection of domain-domain interactions found in PDB (Pfam pairs).
InterDOM (Ng, Zhang, & Tan, 2003)	http://InterDom.lit.org.sg	30,037 domain interactions	Gene Fusion events DIP, BIND, PDB Scientific Literature	Data are integrated by assigning higher confidence to domain interactions that are independently derived from different data sources and methods.
IntACT (Hermjakob et al., 2004)	http://www.bioinf.man.ac.uk/ resources/ interactpr.shtml	39,715 interactions and complexes	Scientific Literatures	On-going collaboration between 9 bioinformatics centers including EBI, MPI-MG, SIB, SDU,UB1, CNB-CSIC, HUJI, GSK, and MINT
PDB (Deshpande et al., 2005)	http://www.pdb.org	29,101 molecular structures	X-ray crystallography or NMR spectroscopy	Comprehensive resources on biomolecular structures, including proteins and protein complexes.
STRING (Mering et al., 2003)	http://string.embl.de/	Predicted interactions in 444,238 genes from 110 species	Experiments Genomic Context Scientific literature Predicted interactions	Database of known and predicted interactions. Both direct and indirect interactions are listed. Predictions in terms of COG orthologs are also available.
PREDICTOME (Mellor, Yanai, Clodfelter, Mintseris, & DeLisi, 2002)	http://predictome.bu.edu	300,505 predicted interactions	44 microbial genomes Experiments Predicted interactions	Nice tool for visualizing the predicted functional associations among genes and proteins in many different organisms.
LiveDIP (Duan, Xenarios, & Eisenberg, 2002)	http://dip.doembi.ucla.edu/ldip.html	35 types of chemical modification 408 interactions	Extension of DIP	Describes protein interactions by protein states and state transitions.

list, 29 out of 719 databases introduced in the year of 2005 are databases of protein interactions (http://www3.oup.co.uk/nar/database/cap/). We list 11 widely used databases in Table 1. Besides listed databases, some specialized databases include MHC-Peptide Interaction Database — a highly curated database for sequence-structure-function information on MHC-peptide interactions, PDZBase — a database of protein-protein interactions involving PDZ domains, PINdb — a database of protein-protein interactions in the nucleus of human and yeast, POINT — a database of predicted human protein-protein interactome, ProNIT — a protein-nucleic acid interaction database and InterPare – a database of interacting residues for all protein-protein interfaces in the PDB.

Interface Site Prediction: Protein Interaction at Residue Level

When two or more proteins physically interact, the residues on a protein surface that are actually in contact with the residues of the other proteins are called protein interface sites. The prediction of protein interface sites is an effort to identify residues that constitute such contact regions from either primary sequence or tertiary structural information. Identification of protein interface sites is of great importance in areas like protein docking (Wodak & Mendez, 2004) that suffers from the astronomical size of the search space of possible conformations.

Definition and Properties of Interface Sites

Protein interface sites are defined in terms of either the distances between residues or the reduction of Accessible Surface Area (ASA). For example, if the distance between any heavy atoms (or between C_α or C_β atoms) is less than a certain threshold, such as 5 Å (Zhou & Shan, 2001), the residue pair is classified as an interface site. In some other cases, the threshold is set in terms of absolute or relative reduction (percentage of ASA loss upon complexation) of ASA, such as 1 $Å^2$ (Bahadur, Chakrabarti, Rodier, & Janin, 2003; Jones & Thornton, 1995), or 5% (Caffrey, Somaroo, Hughes, Mintseris, & Huang, 2004). Over the last decades, a number of attempts have been made to explain interface sites in terms of residue propensity (Ofran & Rost, 2003b), hydrophobicity (Bahadur et al., 2003; Jones & Thornton, 1995, , 1996; Korn & Burnett, 1991), shape complementarities (Jones & Thornton, 1996), and in other terms. However, the underlying principles that govern protein interface sites are not yet fully understood.

Computational Approaches to Interface Sites Prediction

Most computational approaches are based on the assumption that residue type propensities, computed from multiple sequence alignment of orthologs, are not random in areas immediately surrounding interface sites.

Since, for each residue in a known protein complex, the label (interface, surface, or core) can be easily obtained, machine learning methods like artificial neural network or support vector machines are widely used to predict interface sites. Specifically, for each residue, amino acid propensities of either spatial or sequence neighbors are transformed into a feature vector. Occasionally, additional information such as ASA is appended (Zhou & Shan, 2001). If spatial location is considered when building the feature vector, the resulting classifier is only valid for predicting proteins with known structures (Fariselli, Pazos, Valencia, & Casadio, 2002; Zhou & Shan, 2001). In that sense, a classifier constructed from sequence neighbors has wider applications (Ofran & Rost, 2003b; Yan, Dobbs, & Honavar, 2004).

The overall accuracy in identifying interface sites is still far from satisfactory. To boost performance, a more sophisticated classification framework has recently been introduced, where predictions from multiple SVMs are merged by a weighted scheme (Park, Munavalli, Geist, & Samatova, 2004) or integrated by a Bayesian classifier (Yan et al., 2004). While such efforts will help produce a slightly more accurate classifier, the fundamental limitation of computational approaches can only be resolved when a more appropriate set of complexes and more descriptive feature vectors becomes available. Focusing on specific subsets of interface sites is one such effort (Ofran & Rost, 2003a, 2003b).

Computational Approaches to Predicting Protein-Protein Interactions

While the vast amount of high-throughput experimental data gives us new opportunities to compile rich sources of valuable genome-wide interactomes, we are still far behind in understanding fundamental principles that govern protein-protein interactions. Furthermore, high-throughput data is considered to be error prone, and the coverage of data is very limited to simpler organisms such as *Saccharomyces cerevisiae* and *Hellicobacter pylori*. Computational approaches play a complementary role to experimental methods by predicting novel interactions, assessing reliabilities of the existing interactions, and building mathematical models that may explain protein interactions through transforming biological intuition into mathematical formulations.

Figure 2. Genomic context-based methods are based on the assumption that functionally related proteins are encoded by genes that are regulated or evolved together: Three different approaches attempt to uncover such evidences through comparative genome analysis, (a) Phylogenenetic profile captures pairs of genes based on their co-occurrences pattern across genomes, (b) gene neighborhood identifies pairs of genes in vicinity, and (c) gene fusion examines pairs of genes that exist individually in some organisms, but as a fused gene in other organisms

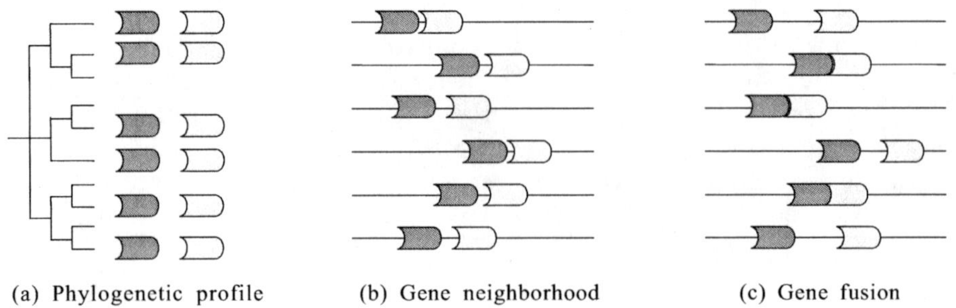

Genomic Context-Based Methods

Genomic context-based methods are based on the assumption that functionally related proteins are encoded by genes that are co-regulated or co-evolved. These genes are believed to leave detectable signatures over multiple genomes that can be identified by comparative analysis. Different methods seek to uncover such signatures in different ways (see Figure 2). Gene fusion methods (Enright, Iliopoulos, Kyrpides, & Ouzounis, 1999; Marcotte, Pellegrini, Thompson, Yeates, & Eisenberg, 1999; Sali, 1999) search for two genes that exist individually in some genomes, but as a fused gene in other genomes. Typically, a non-overlapping side-by-side BLAST search is performed to identify fused genes (Enright & Ouzounis, 2001; Huynen, Snel, Mering, & Bork, 2003; Mellor, Yanai, Clodfelter, Mintseris, & DeLisi, 2002; Suhre & Claverie, 2004). The gene neighborhood method considers the conserved proximity between two genes for a potential interaction (Dandekar, Snel, Huynen, & Bork, 1998; Overbeek, Fonstein, D'Souza, Pusch, & Maltsev, 1999). Whereas the above two methods consider spatial co-location in various genomes, phylogenetic profiling (Pellegrini, Marcotte, Thompson, Eisenberg, & Yeates, 1999)

Figure 3. Two methods for predicting interacting proteins: (a) Given two proteins, a pair of organism-aligned MSAs is built (b) from two mirror phylogenetic trees A and B that are constructed from MSAs in (a), each pair of distances is mapped to 2-D plane. A high correlation indicates potential interaction (c) both correlations of intra and inter residue pairs are computed, and three distributions are compared to determine potential interaction

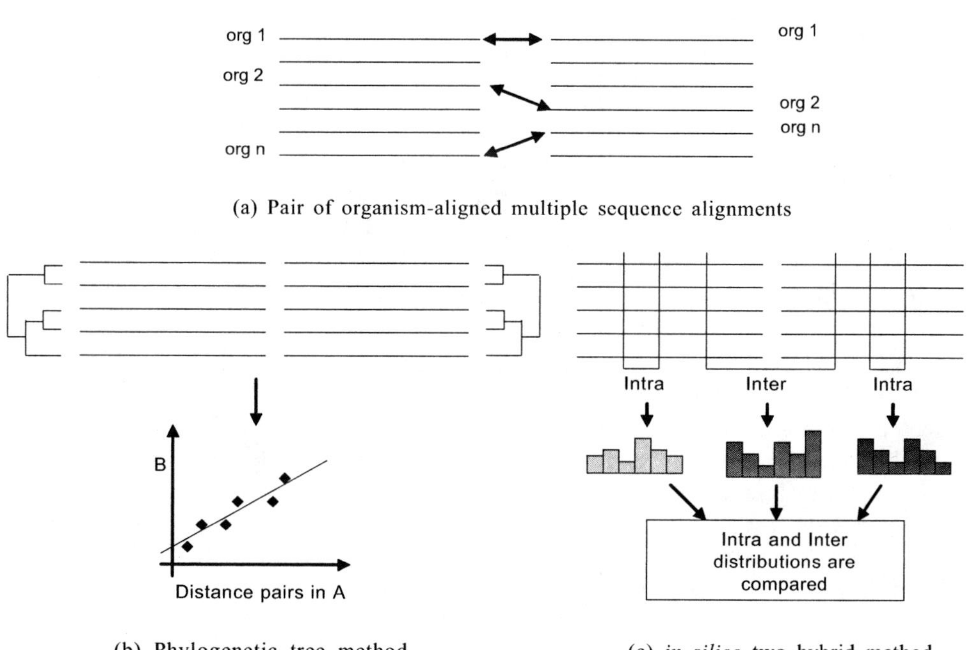

examines temporal co-appearance as an evidence of potential interaction. In practice, two proteins are predicted to interact, if their orthologs are found to have correlated appearances (Matthews et al., 2001) over a set of genomes. In a further refinement, phylogenetic trees (Pazos & Valencia, 2001) can be used to quantify the degree of co-evolution rather than co-existence between two genes. For two genes, two mirror phylogenetic trees are constructed from organism-aligned multiple sequence alignments of associated protein families. Then all pairs of distances measured from the same path in each tree are measured. Two genes (or proteins accordingly) are predicted to interact if high correlation is observed between all pairs of distances. The *in silico* two hybrid (i2h) method {Pazos, 2002 #129}, the distributions of correlations between inter-protein positions and correlations between intra-protein positions are compared. Based on the difference, the i2h predicts a potential interaction between the proteins. Detailed procedures are shown in Figure 3.

A genomic context-based approach suggests that genomic associations between genes reveal functional associations between them; the stronger the genomic association, the stronger the functional association (Snel, Bork, & Huynen, 2002). For this reason, genomic associations cover a relatively wide range of protein interactions including physical and functional interactions. However, its applicability is somewhat limited for several reasons; it requires sufficient number of completely sequenced genomes, it is only applicable to prokaryotes, and not many proteins follow the principle of genomic associations (Brun et al., 2003).

Domain-Domain Interactions

Protein domains are evolutionarily conserved segments of an amino acid sequence. Biologically, they are often responsible for the structural formation or functional behavior of proteins, and are frequently involved (either directly or indirectly) in intermolecular interactions. For this reason, there have been attempts to search for pairs of protein domains whose coexistence can account for putative protein-protein interactions. In essence, most approaches focus on assessing statistical significances of domain pairs found in high-throughput experimental data. For example, let P_i and P_j be interacting proteins that include domains $\{a,b\}$ and $\{c,d\}$, respectively. Then the interaction may be explained by some of 4 possible domain interactions: (a,c), (a,d), (b,c), and (b,d). The main objective is, through statistical assessments, to identify which domain pairs are indeed responsible for the interaction.

Sprinzak and Margalit (2001) consider protein interactions in terms of InterPro (Mulder et al., 2003) signature pairs whose co-occurrences are more frequent than what is expected at random. The Protein Interaction Classification by Unlikely Profile Pair (PICUPP) introduces a more reliable method that assesses the interaction probability of a domain pair by bootstrapping (Park et al., 2003). Another probabilistic approach that utilizes the scale-free property (see Section 6 for details) of protein interactions (Gomez, Noble, & Rzhetsky, 2003) has been proposed that considers not only the number of times the domain pair is observed, but also the number of times it is found in non-interacting proteins. The Interaction Domain Profile-Pair (IDPP) method (Wojcik & Schachter, 2001) represents protein interactions as interactions between domain clusters. Whereas most approaches aim to estimate the likelihood of each domain pair individually, a global optimization approach that simultaneously estimates probabilities of all domain-domain interactions using Expectation-Maximization (EM) algorithm has been proposed (Deng, Mehta, Sun, & Chen, 2002). As an integrative approach, Ng, Zhang, and Tan (2003) combined evidences of interacting Pfam domain pairs obtained from various data sources. Each domain pair was given a number of odd ratio scores (observed/expected) which were integrated using an unweighted scheme.

Classification Methods

The most commonly used classifier for predicting interacting protein pairs is the Support Vector Machine (SVM) (Bock & Gough, 2001, 2003; Martin, Roe, & Faulon, 2005). SVM originated from statistical learning theory (Vapnik, 1998) and has been widely accepted in other bioinformatics areas, such as homology detection (Leslie, Eskin, Weston, & Noble, 2003; Saigo, Vert, Ueda, & Akutsu, 2004), gene classification (Brown et al., 2000), prediction of RNA-binding proteins (Han, Cai, Lo, Chung, & Chen, 2004), protein folding recognition (Ding & Dubchak, 2001), prediction of subcellular localization (Park & Kanehisa, 2003), and prediction of protein secondary structure (Hua & Sun, 2001). The main challenge in applying SVM to predict protein-protein interactions is to employ an appropriate encoding of protein-protein interactions in a vector space. This primarily involves the choice of biochemical (or biological) properties that can differentiate interacting from non-interacting pairs.

Most classification methods assume the absence of structural information, i.e. only a pair of primary sequences is available. Therefore, physiochemical properties associated with a primary sequence such as hydrophobicity, charge, and surface tension (Bock & Gough, 2001) are often considered. To generate a feature vector, the values for each residue in a sequence are tabulated in a fixed order. Note, however, this type of transformation suffers from two drawbacks. First, the transformation of proteins of various lengths into a fixed length vector can be problematic. Second, if v_i and v_j denote the transformed feature vectors for protein *i* and protein *j*, respectively, the construction of a feature vector for protein pair (v_i,v_j) is not readily obvious. An early SVM-based method simply merged the two vectors; the final feature vector for a protein pair became either v_i+v_j or v_j+v_i, where + denotes concatenation of the two vectors. Therefore, direct mapping from physiochemical properties to a vector space needs to be reexamined.

Despite the fact that the classification approach has been proven to be successful in many other areas, and an early attempt applying this method to protein-protein interaction prediction achieved a certain level of success (Bock & Gough, 2001), its practical applicability is still in question mainly due to the absence of negative instances. Technically, there exists no experimental method that can detect pairs of proteins that do not interact at all. In practice, randomly chosen pairs of proteins substitute for negative instances (Bock & Gough, 2003; Martin et al., 2005). Occasionally, pairs of synthetically generated proteins are used for the same purpose (Bock & Gough, 2001). However, their biological validity is difficult to estimate. Thus, the creation of negative instances based on true biological principles (subcellular localization, for example) is much anticipated.

Inference from Imperfect Data

Although the advances in high-throughput experimental methods make exhaustive inspections for all potential interactions possible in some smaller organisms, there remain significant discrepancies in data produced by different methods. In the *Saccharomyces cerevisiae* proteome, for example, roughly 10,000 interactions are identified from four different methods, yet the overlaps are quite limited (see Table 2). Also, the total number of interactions currently identified is only one third of what is predicted to exist (von Mering et al., 2002). Although the small overlap may be partially due to differences in the types of interactions that an experimental method can discover, it mainly comes from the lack of reproducibility of the methods as being discussed in the yeast two-hybrid system. This unreliable characteristic of high-throughput datasets calls for (1) computational assessment of data quality and (2) robust approaches to infer from such imperfect data.

In principle, data quality can be assessed by validation utilizing high-quality benchmark data. However, due to the limited availability of such a reliable dataset (von Mering et al., 2002), quality is often examined by considering additional properties found in different information sources. For example, Deane et al. introduce two measures: the Expression Profile Reliability Index (EPR Index) and the Paralogous Verification Method (PVM) (Deane, Salwinski, Xenarios, & Eisenberg, 2002). The EPR assesses the quality of protein interaction datasets by comparing mRNA expression profiles of interacting protein pairs in the dataset to those of known interacting and non-interacting protein pairs. The PVM predicts that two proteins potentially interact if their paralogs also interact. Further discussions on data quality assessment can be found in (Salwinski & Eisenberg, 2003).

Two approaches exist for inferences from imperfect data. In the first approach, the most reliable subset of high-throughput data is sought by either taking the intersection of multiple data sources (Bader & Hogue, 2002; von Mering et al., 2002), performing

Table 2. Intersections of protein interactions identified in different high-throughput experimental datasets. Values in bold denote the number of interactions in each database, and others show overlaps between databases that are referenced by corresponding column and row. Small-scale DIP denotes the reliable small-scale studies that are listed in DIP database (Salwinski et al., 2004). Every pair of high-throughput datasets shows very limited agreements. However, all datasets identified many interactions in small-scale studies. The data in this table are borrowed from (Salwinski & Eisenberg, 2003).

	Ito *et al.*	**Uetz *et al.***	**Gavin *et al.***	**Ho *et al.***	**Small-scale DIP**
Ito *et al.*	**4363**	186	54	63	442
Uetz *et al.*		**1403**	54	56	415
Gavin *et al.*			**3322**	198	528
Ho *et al.*				**3596**	391

statistical assessments (Sprinzak, Sattath, & Margalit, 2003), analyzing network topology (Goldberg & Roth, 2003; Saito, Suzuki, & Hayashizaki, 2003), or by taking the combination of these three (Bader, Chaudhuri, Rothberg, & Chant, 2004). The second approach takes the opposite direction: integration of multiple data sources through sophisticated probabilistic approaches. For example, STRING (Huynen et al., 2003) combines predictions from multiple methods, which use different types of data, by assigning a unique weight to each method based on its benchmark result against a common reference set. The Probabilistic Interactome (PI) (Jansen et al., 2003) takes a step further in merging multiple data sources. Like STRING, PI also assesses the weight of each data source. For this, two Bayesian networks are constructed from high-throughput experimental data and other genomic features, respectively. Then, a final Bayesian network (a simple naïve Bayes classifier) is built based on probabilities estimated from the two networks. Alternatively, multiple data sources can be selectively integrated. Based on structural risk minimization, Bock and Gough introduced a framework that iteratively brings in experimental data sets for different organisms in such a way that the prediction accuracy through cross-validation is improved (Bock & Gough, 2003). It is a growing consensus that the integration of multiple data sources can significantly facilitate inference of protein interactions. In the near future, more diverse data sources are expected to be integrated along this direction (Salwinski & Eisenberg, 2003).

Characterization of Protein Interaction Networks and Prediction of Protein Functions

The responses of a cell to environmental conditions are governed by complex intermolecular interactions. For example, in a metabolic pathway, hundreds of proteins are interconnected through various biochemical reactions. Thanks to large-scale, high-throughput identification of protein-protein interactions and various computational techniques, genome-scale protein interaction networks are available, where both direct and indirect interactions between proteins are represented. However, identifying the functional role of a protein or cellular functional units through analysis of interaction networks has become a major challenge in proteomics. Most approaches follow three steps: (1) the construction of genome-wide protein-protein interaction networks in terms of mathematical representation, i.e. a graph, where nodes are proteins, and edges are interactions between them (Ito et al., 2001; Uetz et al., 2000); (2) the determination of graph properties such as cliques, shortest paths, etc., and (3) functional characterization of proteins and protein complexes based on these properties. In this section, we describe how a protein-protein interaction network is utilized to infer biological complexes and protein functions.

Network Properties

A number of network models have been suggested to characterize protein-protein interaction networks. Each model lays a foundation to identify protein function and functional modules within its framework. The most widely accepted models are scale-free networks (Barabasi & Albert, 1999; Fraser, Hirsh, Steinmetz, Scharfe, & Feldman, 2002; Hoffmann & Valencia, 2003; Jeong, Mason, Barabasi, & Oltvai, 2001; Wuchty, 2002) and small-world networks (Goldberg & Roth, 2003; Watts & Strogatz, 1998). Whereas the scale-free model delineates the global connectivity of a network, the small-world model rather focuses on elucidating the local connectivity. The former suggests that there exist very few highly connected nodes following the power-law distribution. It implies that such highly connected proteins (often called hubs) participate in numerous metabolic reactions, thus explaining the robust self-organizing behavior of a protein interaction network. Another view, the small-world model implies that the shortest path between any pair of proteins tends to be small, and the network is full of densely connected local neighborhoods. The two models are not in conflict; rather, they complement each other (Snel et al., 2002). In fact, it has recently reported that the yeast co-expression network can be successfully modeled as a small-world, scale-free network (van Noort, Snel, & Huynen, 2004).

An alternative view, which is based on the assumption that cellular functionality can be divided into a set of modules, is also widely used. This modular network model (Brun et al., 2003; Rives & Galitski, 2003) suggests that protein interaction networks consist of several densely interconnected functional modules that are connected by few cross-talks. However, such a modular structure implies that most nodes roughly have the equal edge degrees, which is against the scale-free nature. There has been an attempt to resolve this conflict by devising a hierarchical modular network, where modules constitute higher level modules (Ravasz, Somera, Mongru, Oltvai, & Barabasi, 2002). However, the disparity between these two models still remains.

Protein Function Prediction through Network Analysis

Protein function prediction through network analysis is based on the notion of "guilt-by-association" (Oliver, 2000). It implies that a protein performs its function through interactions with other proteins that have the same cellular function. Hence, large complex protein interaction networks are now analyzed in detail to associate proteins of known functions with hitherto hypothetical or uncharacterized proteins, and to extract cellular processes in which a number of interacting proteins are involved. For example, given a whole proteome that consists of both known and unknown proteins, functional annotations for the unknowns can be determined by finding the best configurations that maximize the presence of the same functional association among interacting proteins (Vazques, Flammini, Maritan, & Vespignani, 2003).

The graph-theoretic properties of a protein interaction network facilitate functional classifications of proteins. Following the scale-free network model, highly connected proteins (hubs) are identified to account for viability of the cell (Jeong et al., 2001).

Figure 4. Three illustrative network topologies. In scale-free networks (a), a few highly connected nodes are linked with others. In small-world networks (b), any node can be reached by a small number of intermediate nodes. In hierarchical modular networks (c), a node belongs to a module, which in turn belongs to a higher module in the hierarchy.

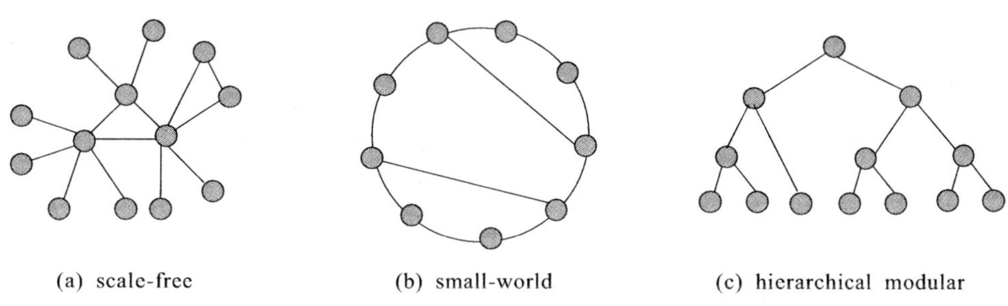

Biologically, such proteins are confirmed to evolve very slowly (Fraser et al., 2002). In contrast, several studies show that clustering the interaction network can isolate a number of functional modules, demonstrating the modular property of protein interaction networks (Brun et al., 2003; Rives & Galitski, 2003). Such modularity is particularly evident in networks with a high clustering coefficient that indicates how densely the neighbors of a protein are interconnected. In summary, it has been shown that some graph-theoretic properties are successfully mapped to cellular functional units; signaling pathways, important proteins, and protein complexes are described in terms of cliques, articulation nodes, subgraphs, etc. (Bader & Hogue, 2002; Przulj, Wigle, & Jurisica, 2004; Watts & Strogatz, 1998).

Protein Complex Prediction

Unlike a functional module, which consists of proteins that bind to (or interact with) each other at different times and locations to perform a certain cellular process, a protein complex is comprised of multiple proteins (subunits) that physically interact to form a single molecular structure. Though computational approaches to predict protein complexes are typically done via docking simulations with high resolution structures available, many recent studies demonstrate that protein complexes can also be detected by analyzing protein-protein interaction maps. For example, Spirin and Mirny (2003) studied highly connected subgraphs of meso-scale (5-25 proteins) clusters and confirmed that many of these clusters correspond to protein complexes. A probabilistic approach also reported that a strong interaction highly correlates with a potential membership in the same complex (Asthana, King, Gibbons, & Roth, 2004), where a strong interaction is represented as an edge with high probability in the network.

Computational methods for predicting protein functions and complexes inevitably deal with imperfect data. Protein interaction networks are mostly constructed from experimental data that are susceptible to errors and noise. Although we expect more extensive and accurate interaction data due to the advancement of recent mass spectrometry during the last decade (Janin & Seraphin, 2003; Sobott & Robinson, 2002; Yates III, 2000), attempts to appropriately integrate various data sources will also help encompass the coverage and enhance the integrity of the networks (Hoffmann & Valencia, 2003; Lee, Date, Adai, & Marcotte, 2004). For example, many signatures of protein complex membership have been observed in other data sources including gene expression data and subcellular localization data. Therefore, systematic integration of various genomic data with the current protein-protein interaction maps is much anticipated (Jansen, Lan, Qian, & Gerstein, 2002).

Summary

Ultrascale computing and high-throughput technologies, such as genomics, transcriptomics, proteomics and metabolomics have been generating an avalanche of data that offers new opportunities to build detailed models of molecular interaction networks including physical, chemical, and biological processes at increasing levels of detail. However, there still remain growing challenges of interpreting and integrating, in a biologically meaningful way, the flux of complex and noisy data thus produced. As a result, false-positive and false-negative ratios of current computational methods are high. Also the coverage of such predictions, both within the same organism and among differing organisms, is still very limited.

We foresee several directions in future computational efforts to remedy the current limitations. First, integration of heterogeneous data of complementary methods (microarray data, for example) into a single data mining framework will be in great demand. This is particularly useful for systematic biological inference through well-defined data mining techniques given the absence of strong direct evidences. For example, the growth of identified protein complexes is outpaced by that of other experimental data. However, given the importance of structural information for understanding how interactions are associated with functions, an integrating effort to sift indirect inference is challenging, yet much anticipated. Second, with a constantly increasing volume of experimental data, the most reliable benchmark data set needs to be materialized through highly robust statistical assessment methods. This data will greatly facilitate the development of new computational frameworks by providing a reliable and universal test-bed. In contrast to integration efforts, we expect more diverse and more specialized computational methods to emerge. For example, in protein interface site prediction, mate-specific methods will appear. Instead of trying to identify interface residues in a protein given a blind assumption that the protein interacts with other proteins, the method will be refined to locate interacting residues with respect to a specific partner. Finally, representing/reconstructing protein-protein interaction networks with spatial and temporal dimensions will help unveil particular functional blocks that constitute other biological networks including metabolic/signaling pathways and regulatory networks.

Acknowledgments

This work is funded in part or in full by the US Department of Energy's Genomes to Life program (http://doegenomestolife.org/) under the project "Carbon Sequestration in Synechococcus sp.: From MolecularMachines to Hierarchical Modeling." The work of N.F.S. is partially sponsored by the Laboratory Directed Research and Development Program of Oak Ridge National Laboratory. P.D. and Y.X. are partially supported by the National Science Foundation (#NSF/DBI-0354771, #NSF/ITR-IIS-0407204).

References

Alfarano, C., Andrade, C. E., Anthony, K., Bahroos, N., Bajec, M., Bantoft, K., et al. (2005). The biomolecular interaction network database and related tools 2005 update. *Nucleic Acids Research, 33*(suppl_1), D418-424.

Aloy, P., & Russell, R. B. (2002). The third dimension for protein interactions and complexes. *Trends in Biochemical Sciences, 27*(12), 633-638.

Asthana, S., King, O. D., Gibbons, F. D., & Roth, F. P. (2004). Predicting protein complex membership using probabilistic network reliability. *Genome Res., 14*(6), 1170-1175.

Bader, G., & Hogue, C. (2002). Analyzing yeast protein-protein interaction data obtained from different sources. *Nature Biotechnology, 20*(10), 991-997.

Bader, J. S., Chaudhuri, A., Rothberg, J. M., & Chant, J. (2004). Gaining confidence in high-throughput protein interaction networks. *Nature Biotechnology, 22*(1), 78-85.

Bahadur, R. P., Chakrabarti, P., Rodier, F., & Janin, J. (2003). Dissecting subunit interfaces in homodimeric proteins. *Proteins, 53*(3), 708-719.

Barabasi, A. L., & Albert, R. (1999). Emergence of scaling in random networks. *Science, 286*(5439), 509-512.

Bock, J. R., & Gough, D. A. (2001). Predicting protein-protein interactions from primary structure. *Bioinformatics, 17*(5), 455-460.

Bock, J. R., & Gough, D. A. (2003). Whole-proteome interaction mining. *Bioinformatics, 19*(1), 125-134.

Brown, M. P. S., Grundy, W. N., Lin, D., Cristianini, N., Sugnet, C. W., Furey, T. S., et al. (2000). Knowledge-based analysis of microarray gene expression data by using support vector machines. *PNAS, 97*(1), 262-267.

Brun, C., Chevenet, F., Matin, D., Wojcik, J., Guenoche, A., & Bernard, J. (2003). Functional classification of proteins for the prediction of cellular function from a protein-protein interaction network. *Genome Biology, 5*(R6), 1465-6906.

Caffrey, D. R., Somaroo, S., Hughes, J. D., Mintseris, J., & Huang, E. S. (2004). Are protein-protein interfaces more conserved in sequence than the rest of the protein surface? *Protein Sci, 13*(1), 190-202.

Dandekar, T., Snel, B., Huynen, M., & Bork, P. (1998). Conservation of gene order: a fingerprint of proteins that physically interact. *Trends in Biochemical Sciences, 23*(9), 324-328.

Deane, C. M., Salwinski, L., Xenarios, I., & Eisenberg, D. (2002). Protein interactions: Two methods for assessment of the reliability of high-throughput observations. *Mol Cell Proteomics*, M100037-MCP100200.

Deng, M., Mehta, S., Sun, F., & Chen, T. (2002). Inferring domain-domain interactions from protein-protein-interactions. *Genome Res., 12*(10), 1540-1548.

Deshpande, N., Addess, K. J., Bluhm, W. F., Merino-Ott, J. C., Townsend-Merino, W., Zhang, Q., et al. (2005). The RCSB protein data bank: a redesigned query system and relational database based on the mmCIF schema. *Nucleic Acids Research, 33*(suppl_1), D233-237.

Ding, C. H. Q., & Dubchak, I. (2001). Multi-class protein fold recognition using support vector machines and neural networks. *Bioinformatics, 17*(4), 349-358.

Doonan, S. (1996). Protein purification protocols. General strategies. *Methods Mol Biol, 59*, 1-16.

Duan, X. J., Xenarios, I., & Eisenberg, D. (2002). Describing biological protein interactions in terms of protein states and statetransitions : THE LiveDIP DATABASE. *Mol Cell Proteomics, 1*(2), 104-116.

Enright, A. J., Iliopoulos, I., Kyrpides, N. C., & Ouzounis, C. A. (1999). Protein interaction maps for complete genomes based on gene fusion events. *Nature, 402*, 86.

Enright, A. J., & Ouzounis, C. A. (2001). Functional associations of proteins in entire genomes by means of exhaustive detection of gene fusions. *Genome Biology, 2*, 341-347.

Fariselli, P., Pazos, F., Valencia, A., & Casadio, R. (2002). Prediction of protein-protein interaction sites in heterocomplexes with neural networks. *Eur J Biochem, 269*(5), 1356-1361.

Fields, S., & Song, O. (1989). A novel genetic system to detect protein-protein interactions. *Nature, 340*(6230), 245-256.

Finn, R. D., Marshall, M., & Bateman, A. (2004). iPfam: visualisation of protein-protein interactions in PDB at domain and amino acid resolutions. *Bioinformatics*, bti011.

Fraser, H. B., Hirsh, A. E., Steinmetz, L. M., Scharfe, C., & Feldman, M. W. (2002). Evolutionary Rate in the Protein Interaction Network. *Science, 296*(5568), 750-752.

Galperin, M. Y. (2005). The Molecular Biology Database Collection: 2005 update. *Nucleic Acids Research, 33*(suppl_1), D5-24.

Gavin, A. C., Bosche, M., Krause, R., Grandi, P., Marzioch, M., Bauer, A., et al. (2002). Functional organization of the yeast proteome by systematic analysis of protein complexes. *Nature, 415*(6868), 141-147.

Goldberg, D. S., & Roth, F. P. (2003). Assessing experimentally derived interactions in a small world. *PNAS, 100*(8), 4372-4376.

Gomez, S. M., Noble, W. S., & Rzhetsky, A. (2003). Learning to predict protein-protein interactions from protein sequences. *Bioinformatics, 19*(15), 1875-1881.

Han, L. Y., Cai, C. Z., Lo, S. L., Chung, M. C. M., & Chen, Y. Z. (2004). Prediction of RNA-binding proteins from primary sequence by a support vector machine approach. *RNA, 10*(3), 355-368.

Hermjakob, H., Montecchi-Palazzi, L., Lewington, C., Mudali, S., Kerrien, S., Orchard, S., et al. (2004). IntAct: An open source molecular interaction database. *Nucleic Acids Research, 32*(90001), D452-455.

Ho, Y., Gruhler, A., Heilbut, A., Bader, G. D., Moore, L., Adams, S. L., et al. (2002). Systematic identification of protein complexes in Saccharomyces cerevisiae by mass spectrometry. *Nature, 415*(6868), 180-183.

Hoffmann, R., & Valencia, A. (2003). Protein interaction: same network, different hubs. *Trends in Genetics, 19*(12), 681-683.

Hua, S., & Sun, Z. (2001). A novel method of protein secondary structure prediction with high segment overlap measure: Support vector machine approach. *Journal of Molecular Biology, 308*(2), 397-407.

Huynen, M. A., Snel, B., Mering, C. v., & Bork, P. (2003). Function prediction and protein networks. *Current Opinion in Cell Biology, 15*(2), 191-198.

Ito, T., Chiba, T., Ozawa, R., Yoshida, M., Hattori, M., & Sakaki, Y. (2001). A comprehensive two-hybrid analysis to explore the yeast protein interactome. *PNAS, 98*(8), 4569-4574.

Janin, J., & Seraphin, B. (2003). Genome-wide studies of protein-protein interaction. *Curr Opin Struct Biol, 13*(3), 383-388.

Jansen, R., Lan, N., Qian, J., & Gerstein, M. (2002). Integration of genomic datasets to predict protein complexes in yeast. *Journal of Structural and Functional Genomics, 2*(2), 71-81.

Jansen, R., Yu, H., Greenbaum, D., Kluger, Y., Krogan, N. J., Chung, S., et al. (2003). A Bayesian networks approach for predicting protein-protein interactions from genomic data. *Science, 302*(5644), 449-453.

Jeong, H., Mason, S. P., Barabasi, A. L., & Oltvai, Z. N. (2001). Lethality and centrality in protein networks. *Nature, 411*, 41.

Jones, S., & Thornton, J. M. (1995). Protein-protein interactions: a review of protein dimer structures. *Prog Biophys Mol Biol, 63*(1), 31-65.

Jones, S., & Thornton, J. M. (1996). Principles of protein-protein interactions. *Proc Natl Acad Sci USA, 93*(1), 13-20.

Korn, A. P., & Burnett, R. M. (1991). Distribution and complementarity of hydropathy in multisubunit proteins. *Proteins, 9*(1), 37-55.

Lee, I., Date, S. V., Adai, A. T., & Marcotte, E. M. (2004). A probabilistic functional network of yeast genes. *Science, 306*(5701), 1555-1558.

Leslie, C., Eskin, E., Weston, J., & Noble, W. (2003). Mismatch string kernels for SVM protein classification. *Advances in Neural Information Processing Systems, 15*, 1441-1448.

Marcotte, E. M., Pellegrini, M., Thompson, M., Yeates, T., & Eisenberg, D. (1999). A combined algorithm for genome-wide prediction of protein function. *Nature, 402*(23), 25-26.

Martin, S., Roe, D., & Faulon, J. L. (2005). Predicting protein-protein interactions using signature products. *Bioinformatics, 21*(2), 218-226.

Matthews, L. R., Vaglio, P., Reboul, J., Ge, H., Davis, B. P., Garrels, J., et al. (2001). Identification of potential interaction networks using sequence-based searches for conserved protein-protein interactions or "interologs". *Genome Res., 11*(12), 2120-2126.

Mellor, J. C., Yanai, I., Clodfelter, K. H., Mintseris, J., & DeLisi, C. (2002). Predictome: A database of putative functional links between proteins. *Nucleic Acids Research, 30*(1), 306-309.

Mering, C. v., Huynen, M., Jaeggi, D., Schmidt, S., Bork, P., & Snel, B. (2003). STRING: A database of predicted functional associations between proteins. *Nucleic Acids Research, 31*(1), 258-261.

Mewes, H. W., Frishman, D., Guldener, U., Mannhaupt, G., Mayer, K., Mokrejs, M., et al. (2002). MIPS: A database for genomes and protein sequences. *Nucleic Acids Research, 30*(1), 31-34.

Mulder, N. J., Apweiler, R., Attwood, T. K., Bairoch, A., Barrell, D., Bateman, A., et al. (2003). The InterPro Database, 2003 brings increased coverage and new features. *Nucleic Acids Research, 31*(1), 315-318.

Ng, S.-K., Zhang, Z., & Tan, S.-H. (2003). Integrative approach for computationally inferring protein domain interactions. *Bioinformatics, 19*(8), 923-929.

Ofran, Y., & Rost, B. (2003a). Analysing six types of protein-protein interfaces. *Journal of Molecular Biology, 325*(2), 377-387.

Ofran, Y., & Rost, B. (2003b). Predicted protein-protein interaction sites from local sequence information. *FEBS Letters, 544*(1-3), 236-239.

Oliver, S. (2000). Proteomics: Guilt-by-association goes global. *Nature, 403*(6770), 601-603.

Overbeek, R., Fonstein, M., D'Souza, M., Pusch, G. D., & Maltsev, N. (1999). The use of gene clusters to infer functional coupling. *PNAS, 96*(6), 2896-2901.

Park, B.-H., Munavalli, R., Geist, A., & Samatova, N. F. (2004, April, 2004). *Analysis of Protein Interaction Sites through Separated Data Spaces.* Paper presented at the Bioinformatics Workshop in conjunction with the Fourth SIAM International Conference on Data Mining, Orlando, FL.

Park, B.-H., Ostrouchov, G., Gong-Xin, Y., Geist, A., Gorin, A., & Smatova, N. F. (2003). *Inference of Protein-Protein Interactions by Unlikely Profile Pair.* Paper presented at the SIAM International Conference on Data Mining.

Park, K.-J., & Kanehisa, M. (2003). Prediction of protein subcellular locations by support vector machines using compositions of amino acids and amino acid pairs. *Bioinformatics, 19*(13), 1656-1663.

Pazos, F., & Valencia, A. (2001). Similarity of phylogenetic trees as indicator of protein-protein interaction. *Protein Eng., 14*(9), 609-614.

Pellegrini, M., Marcotte, E. M., Thompson, M. J., Eisenberg, D., & Yeates, T. O. (1999). Assigning protein functions by comparative genome analysis: Protein phylogenetic profiles. *PNAS, 96*(8), 4285-4288.

Przulj, N., Wigle, D. A., & Jurisica, I. (2004). Functional topology in a network of protein interactions. *Bioinformatics, 20*(3), 340-348.

Ravasz, E., Somera, A. L., Mongru, D. A., Oltvai, Z. N., & Barabasi, A.L. (2002). Hierarchical organization of modularity in metabolic networks. *Science, 297*(5586), 1551-1555.

Rives, A. W., & Galitski, T. (2003). Modular organization of cellular networks. *PNAS, 100*(3), 1128-1133.

Saigo, H., Vert, J.P., Ueda, N., & Akutsu, T. (2004). Protein homology detection using string alignment kernels. *Bioinformatics, 20*(11), 1682-1689.

Saito, R., Suzuki, H., & Hayashizaki, Y. (2003). Construction of reliable protein-protein interaction networks with a new interaction generality measure. *Bioinformatics, 19*(6), 756-763.

Sali, A. (1999). Functional links between proteins. *Nature, 402*, 23.

Salwinski, L., & Eisenberg, D. (2003). Computational methods of analysis of protein-protein interactions. *Current Opinion in Structural Biology, 13*(3), 377-382.

Salwinski, L., Miller, C. S., Smith, A. J., Pettit, F. K., Bowie, J. U., & Eisenberg, D. (2004). The database of interacting proteins: 2004 update. *Nucleic Acids Research, 32*(90001), D449-451.

Snel, B., Bork, P., & Huynen, M. A. (2002). The identification of functional modules from the genomic association of genes. *PNAS, 99*(9), 5890-5895.

Sobott, F., & Robinson, C. V. (2002). Protein complexes gain momentum. *Current Opinion in Structural Biology, 12*(6), 729-734.

Spirin, V., & Mirny, L. A. (2003). Protein complexes and functional modules in molecular networks. *PNAS, 100*(21), 12123-12128.

Sprinzak, E., & Margalit, H. (2001). Correlated sequence-signatures as markers of protein-protein interaction1. *Journal of Molecular Biology, 311*(4), 681-692.

Sprinzak, E., Sattath, S., & Margalit, H. (2003). How Reliable are Experimental Protein-Protein Interaction Data? *Journal of Molecular Biology, 327*, 919-923.

Suhre, K., & Claverie, J.M. (2004). FusionDB: a database for in-depth analysis of prokaryotic gene fusion events. *Nucleic Acids Research, 32*(90001), D273-276.

Tong, A. H. Y., Drees, B., Nardelli, G., Bader, G. D., Brannetti, B., Castagnoli, L., et al. (2002). A Combined Experimental and Computational Strategy to Define Protein Interaction Networks for Peptide Recognition Modules. *Science, 295*(5553), 321-324.

Tucker, C. L., Gera, J. F., & Uetz, P. (2001). Towards an understanding of complex protein networks. *Trends Cell Biol, 11*(3), 102-106.

Uetz, P. (2002). Two-hybrid arrays. *Curr Opin Chem Biol, 6*(1), 57-62.

Uetz, P., Giot, L., Cagney, G., Mansfield, T. A., Judson, R. S., Knight, J. R., et al. (2000). A comprehensive analysis of protein-protein interactions in Saccharomyces cerevisiae. *Nature, 403*(6770), 623-627.

van Noort, V., Snel, B., & Huynen, M. A. (2004). The yeast coexpression network has a small-world, scale-free architecture and can be explained by a simple model. *EMBO Reports, 5*(3), 280-284.

Vapnik, V. (1998). *Statistical learning theory*. New York: Wiley Interscience.

Vazques, A., Flammini, A., Maritan, A., & Vespignani, A. (2003). Global protein function prediction from protein-protein interaction networks. *Nature Biotechnology, 21*(6), 697-700.

von Mering, C., Jensen, L. J., Snel, B., Hooper, S. D., Krupp, M., Foglierini, M., et al. (2005). STRING: known and predicted protein-protein associations, integrated and transferred across organisms. *Nucleic Acids Research, 33 Database Issue*, D433-437.

von Mering, C., Krause, R., Snel, B., Cornell, M., Oliver, S. G., Fields, S., et al. (2002). Comparative assessment of large-scale data sets of protein-protein interactions. *Nature, 417*(6887), 399-403.

Watts, D. J., & Strogatz, S. H. (1998). Collective dynamics of 'small-world' networks. *Nature, 393*(6684), 440-442.

Wodak, S. J., & Mendez, R. (2004). Prediction of protein-protein interactions: the CAPRI experiment, its evaluation and implications. *Current Opinion in Structural Biology, 14*(2), 242-249.

Wojcik, J., & Schachter, V. (2001). Protein-protein interaction map inference using interacting domain profile pairs. *Bioinformatics, 17*(90001), 296S-305.

Wuchty, S. (2002). Interaction and domain networks of yeast. *Proteomics, 2*(12), 1715-1723.

Xenarios, I., Salwinski, L., Xiaoqun, J., Higney, P., Kim, S., & Eisenberg, D. (2005). DIP — Database of Interacting Proteins. *Nucleic Acids Research, 33*(Database Issue).

Yan, C., Dobbs, D., & Honavar, V. (2004). A two-stage classifier for identification of protein-protein interface residues. *Bioinformatics, 20*(suppl_1), i371-378.

Yates III, J. R. (2000). Mass spectrometry: From genomics to proteomics. *Trends in Genetics, 16*(1), 5-8.

Zanzoni, A., Montecchi-Palazzi, L., Quondam, M., Ausiello, G., Helmer-Citterich, M., & Cesareni, G. (2002). MINT: a Molecular INTeraction database. *FEBS Letters, 513*(1), 135-140.

Zhou, H.-X., & Shan, Y. (2001). Prediction of protein interaction sites from sequence profile and residue neighbor list. *PROTEINS: Strucure, Function, and Genetics, 44*, 336-343.

Zhu, H., Bilgin, M., Bangham, R., Hall, D., Casamayor, A., Bertone, P., et al. (2001). Global Analysis of Protein Activities Using Proteome Chips. *Science, 293*(5537), 2101-2105.

Chapter XIV

Differential Association Rules: Understanding Annotations in Protein Interaction Networks

Christopher Besemann, North Dakota State University, USA
Anne Denton, North Dakota State University, USA
Ajay Yekkirala, North Dakota State University, USA
Ron Hutchison, The Richard Stockton College of New Jersey, USA
Marc Anderson, North Dakota State University, USA

Abstract

In this chapter, we discuss the use of differential association rules to study the annotations of proteins in one or more interaction networks. Using this technique, we find the differences in the annotations of interacting proteins in a network. We extend the concept to compare annotations of interacting proteins across different definitions of interaction networks. Both cases reveal instances of rules that explain known and unknown characteristics of the network(s). By taking advantage of such data mining techniques, a large number of interesting patterns can be effectively explored that otherwise would not be.

Introduction

Many high-throughput techniques have been developed to identify protein interaction networks in organisms. One conventional definition of interactions within an organism considers whether the protein molecules interact physically. High-throughput techniques to test for such interactions include the popular yeast-two-hybrid method (Ito et al., 2001; Uetz et al., 2000). Another common definition of interactions relies on the concept of genetic interactions in which the lethality of two gene deletions is examined as in (Tong et al., 2001; Tong et al., 2004). Finally, there also exist a number of computer-driven procedures like the identification of domain-fusion interactions. Two genes have a domain fusion interaction in one species if their homologs occur as exons of a single gene in another species (Marcotte et al., 1999; Truong & Ikura, 2003). With so much interaction data available, there is a clear need to develop techniques that are targeted at understanding the role of networks in bioinformatics.

Interactions are not the only type of information available for proteins. Many genes and proteins are studied extensively in model species, such as yeast. A rich supply of experimental data, in the form of annotations, is available for these species and can be studied in much the same way as data associated with objects in other disciplines. These annotations are the "items" of information that belong to particular objects. A popular technique for the study of such items is association rule mining (ARM). ARM has the goal of discovering associations between frequently occurring sets of items. In the prototypical case, ARM is used in the analysis of shopping transactions (Agrawal, Imielinski, & Swami, 1993; Han & Fu, 1995); the technique has also been applied in relational database settings (Dehaspe & Raedt, 1997; Oyama, Kitano, Satou, & Ito, 2002). The combination of network data and protein annotation data is a special case of relational data. In a relational database, one table is used to describe protein annotations and one or more separate tables define the network(s). A relational framework allows analysis beyond statistical comparison of single annotations of interacting proteins (Schwikowski, Uetz, & Fields, 2000). Relational frameworks also support the study of associations involving more complex attribute combinations. Within the same framework, it also becomes possible to contrast different interaction definitions.

This chapter explores a novel ARM algorithm called differential ARM that enables the discovery of differences between interacting proteins and between multiple network types. In the Background section, we describe representative work on protein interactions and introduce key definitions in association rule mining. The next section, Mining Differential Association Rules, defines our goals and the problems to overcome, ending with a description of the differential ARM method. Finally, in the section on Differential Association Rule Experiments, an experiment on real-world yeast interaction data is described.

Background

We begin by discussing related work in the area of protein interaction network analysis. Basic notation for association rule mining is covered and notes on related references in the area of ARM are given.

Protein Interaction Network Analysis

Several papers have analyzed yeast protein interaction networks. The majority of these works have studied isolated networks (Wagner, 2001) or used comparisons of the statistics of multiple networks (Maslov & Sneppen, 2002). In Schwikowski et al. (2000) the authors did some analysis of the yeast physical protein interaction network using "counting" methods to determine percentages of interactions that have certain functional behavior. Most examples of interest are techniques that require some manual intervention. In larger datasets, in which patterns of interest are harder to perceive, automated techniques would greatly assist analysis. Tong et al. (2004) performed statistical analysis on part of the yeast genetic interaction network. In this work, the authors begin with observations of interesting statistics observed within interaction networks, and conclude with clustering analysis and population genetics. Some of the conclusions drawn relate to the density of the network and comparison of how the genetic interaction network predicts parts of the physical interaction network. Again, the emphasis is on a single network and simple statistical comparisons. In Ozier, Amin, and Ideker (2003, 2003) global statistics of a physical and genetic interaction network are compared. ARM helps move beyond these studies.

The first use of association rule mining for protein interaction graphs was documented in Oyama et al. (2002). Oyama et al. use a relational ARM method to study yeast physical protein interactions towards predicting protein-binding site information. The idea of the paper is to combine protein interaction and annotation data into a single table that is mined using traditional ARM techniques. The table has two columns, one for the "left-hand side" protein and the other for the "right-hand side" protein of the interaction. Some problems were observed because many interacting proteins share the same annotations. This behavior is related to what Jensen and Neville (2002) call autocorrelation in the context of classification. Techniques described in this chapter help address the issues; they first appear in Besemann, Denton, Yekkirala, Hutchison, and Anderson (2004).

Association Rule Mining

General descriptions of association rule mining can be found in many papers. It is first described in Agrawal et al. (1993). In this section, we introduce the notation for relational ARM that is used in the later parts of this chapter.

Association rule mining is commonly defined for sets of items. We combine the concept of sets with the relational algebra framework by choosing an extended relational model similar to Han and Fu (1995). Attributes within this model are allowed to be set-valued, thereby violating first normal form in database theory. For more background on relational

Figure 1. Example of a node and an edge table

Node	
ORF	Annotations
YPR184W	*{<cytoplasm >}*
YER146W	*{<cytoplasm >}*
YNL287W	*{< Sensitivity_aaaod >}*
YBL026W	*{< transcription >, < nucleus >}*
YMR207C	*{< nucleus >}*

Edge	
ORF0	ORF1
YPR184W	YER146W
YNL287W	YBL026W
YBL026W	YMR207C

notation, introductory texts such as Elmasri and Navathe (2004) can be reviewed. In this notation $R(A_1,...,A_n)$ represents a relation and A_i is an attribute of the relation (e.g., protein relation containing annotation attributes). Classic ARM assumes that the dataset is represented in a single relation, R(A), with the single attribute, A, being set-valued. The members of the set-valued attribute are items. In our example, the items could be annotations of a protein. A rule in ARM is $A \rightarrow C$ for non-intersecting itemsets A and C. The support of a rule is the percent of transactions (rows in the table) that have ($A \cup C$) and the confidence of the rule is the fraction $\frac{\sup(A \cup C)}{\sup(A)}$ or how often we find "C" given that "A" is present.

The goal of ARM is to find all rules that meet thresholds of minimum support and confidence. In the original problem, a database consisting of only a single relation was considered. Multiple relations are considered by using relational join operations. This approach was taken in Oyama et al. (2002). Multi-relational association rule mining has also been performed in the context of inductive logic programming (Dehaspe & Raedt, 1997). Dehaspe and Raedt address the general task of allowing all relations in the database to be mined using a predicate logic language such as Prolog. Association rule mining has further been extended to structural properties of graphs. The work on graph-based and frequent substructure ARM (Inokuchi, Washio, & Motoda, 2000; Michihiro & Karypis, 2001) differs from relational approaches in that nodes in the graph are assumed to have at most one label.

A difficulty in association rule mining of data associated with nodes in a network is elimination of rules that are not considered interesting. Uninteresting rules can also be a problem in traditional ARM. Work has been done to eliminate redundant rules using the theory of closed sets (Zaki, 2000; Bastide, Pasquier, Taouil, Stumme, & Lakhal, 2000). Measures other than support and confidence can be used such as in Brin, Motwani, Ullman, and Tsur (1997). Finally, efficient methods have been developed to constrain the desired rule output (Srikant, Vu, & Agrawal, 1997; Pei, Han, & Lakshmanan, 2004).

Mining Differential Association Rules

A motivating scenario for the use of differential ARM can be taken from the protein interaction setting. Consider using standard ARM to identify differences between interacting proteins. Assume, for example, that with standard ARM proteins with "transcription" as annotation are associated frequently with proteins that are localized in the "nucleus." This association rule may be due to two other rules: one that associates "transcription" and "nucleus" within a single protein, and others that represent the autocorrelation of "transcription" and/or "nucleus" between interacting proteins. We would not consider this a difference between interacting proteins. Nor would we consider the autocorrelation of "transcription" and/or "nucleus" interesting, or the fact that "transcription" and "nucleus" are related for individual proteins. However, a rule of the same format could still stand for a difference in other cases.

Consider the rule that proteins in the "nucleus" are found to interact with proteins in the "mitochondria." A single protein would not simultaneously be located in the "nucleus" and in the "mitochondria". We therefore assume that the rule highlights a difference between interacting proteins and may identify an instance of compartmental crosstalk. This rule is more interesting to a biologist than the rule relating "nucleus" and "transcription." It is more expressive of the properties of the respective interaction network.

We distinguish between two examples based on our biological background knowledge. One example is an artifact while the other correctly identifies differences in interacting proteins that shed light on the nature of the interaction in question. Two approaches could be taken to make this distinction at the level of the ARM algorithm. We could devise a difference criterion involving correlations between neighboring proteins and/or rules found within individual proteins. Such an approach would not benefit from any of the pruning that has made ARM an efficient and popular technique. The differential ARM algorithm takes an approach that makes significant use of pruning: only those items are considered where each item in a set is unique to only one of the interacting nodes or proteins. The rule associating "transcription" and "nucleus" would only be evaluated on those "transcription" proteins that are not themselves in the "nucleus," and those "nucleus" proteins, that are not themselves involved in "transcription."

There are other reasons why a focus on differences is more effective for association rule mining in networks than a standard application of ARM on joined relations. Traditionally association rule mining is performed on sets of items with no known correlations. Interacting proteins are known to have matching annotations in many cases (Tong et al., 2004; Jensen & Neville, 2002). Using association rule mining on such data, in which items are expected to be correlated may lead to output in which the known correlations dominate all other observations either directly or indirectly. This problem is observed when relational association rule mining is directly applied to protein networks (Oyama et al., 2002; Besemann et al., 2004). Excluding matching items of interacting proteins is therefore advisable in the interest of finding meaningful results. Matching annotations can be studied by simple correlation analysis, in which co-occurrence of an annotation in interacting proteins is tested (Tong et al. 2004). In the presence of such correlations, association rules likely reflect nothing but similarities between interacting proteins.

Figure 2. (Left) Example of a joined table. (Right) Table after application of uniqueness operator

TID	Join		Unique	
1	0.*cytoplasm*	1.*cytoplasm*	NULL	NULL
2	0.*Sensitivity_aaaod*	1.*transcription*, 1.*nucleus*	0.*Sensitivity_aaaod*	1.*transcription*, 1. *nucleus*
3	0.*transcription*, 0.*nucleus*	1.*nucleus*	0.*transcription*	NULL

Differential Association Rule Notation

This section describes notation used to formalize the concept of differential ARM, and the implementation that was used in experiments. We assume a relational framework to discuss differences within and between interaction networks. The concept of a network may suggest use of graph-based techniques. Graph-theory typically assumes that nodes and edges have at most one label. Relational algebra on the other hand has the tools for the manipulation of data associated with nodes and edges. A relational representation of a graph with one type of node requires two relations. One relation for data associated with nodes that we call the node relation, and a second relation that describes the relationship between nodes that we call the edge relation.

Consider a database with a protein node relation. Tuples in the relation have a node identifier attribute and descriptor-set attribute such as ORF and Annotations in table "Node" of Figure 1. The annotations are like items in standard ARM, however in the multi-relational setting a label on the items must identify their node of origin. Edge relations have attributes to point to the two connected nodes as in table "Edge" of Figure 1.

Our goal is to mine relational basis sets constructed from the descriptors (annotations) of connected proteins. A *joined-relation basis set* is the set of tuples needed to perform basic relational ARM. Joined-relation basis sets are formed in multiple steps. Edge and node relations are joined through a conditional join operation. The condition on the join excludes cycles in the graph. Attribute names are changed similar to notation of the attribute renaming function in Elmasri and Navathe (2004) such that the names are unique between related proteins. Attributes are identified by consecutive integers to which we refer to as origin identifiers. Origin identifiers are added to the annotations to identify which protein contributed that annotation. The Join table in Figure 2 gives an example of a basis set prepared as described.

The result for a 2-node setting (pair of interacting proteins) assigns "0" to the attributes of the first protein of the edge relation, and assigns "1" to the attributes for the second protein. This would be similar to the description from Oyama et al. (2002) with the left-hand side protein annotations being "0" and right-hand side protein annotations being "1." Note also, that we have multiple alternatives for joining corresponding to different subgraph structures. As an example, four nodes can form either a straight line (0-1-2-3) or a "T"-shape (0-1-2, 1-3)....

Once the joining has occurred and origin identifiers added to the annotations, we have simple items. A simple item may be "0.nucleus" that indicates the protein at index 0 of the join has the annotation of "nucleus." The "Join" section in Figure 2 shows an example of the product as the result of the operations applied to Tables "Node" and "Edge" from Figure 1.

After the joined-relation basis sets are prepared, the differential portion of the method is applied. This takes place in the form of a uniqueness operator that ensures each descriptor (annotation) is not repeated a tuple from the joined table. Table Unique in Figure 2 shows the results of the unique operation on the joined portion. For example, "cytoplasm" in TID 1 is not unique so it is eliminated from TID 1. The uniqueness operator applies to all set-valued attributes of a joined-relation but alternatives are possible, such as requiring uniqueness only across a subset of edges.

One additional problem occurs when finding patterns involving joins over multiple tables. A pattern that expresses information about a single protein can be found in transactions that involve single proteins, interacting protein pairs, chains of three interacting proteins and so forth. The typical solution is to pick the simplest representation. For a pattern involving a single protein, it is the node table without performing joins. For patterns involving two interacting proteins, we use the table that only involves one interaction table, and so on. This situation is also identified in a report on relational ARM (Cristofor & Simovici, 2001). We formalize this concept by calling an itemset out-of-scope if it does not require all portions of the current join operation, i.e., every node (and edge) from the join must be represented. From Figure 2, this means that TID 1 and TID 3 are out-of-scope.

Summary of differential ARM algorithm:

1. Form basis sets from protein and interaction relations (join relations & set origin identifiers)
2. Process each transaction in the basis set to remove non-unique annotations
3. Remove out-of-scope transactions and itemsets
4. Perform association rule mining over the processed dataset

Differential ARM for Network Comparison

In this section, we go beyond applying the idea of differential ARM to interacting proteins of a single network and look at interacting proteins in multiple networks. If we store each interaction "type" in an attribute of the edge relation then we can perform differential ARM as given.

For network comparison, we use basis sets formed by joining a chain of three nodes, involving both types of network. After the join formula is applied, the other steps of building items and applying the uniqueness operator occur the same way as described earlier in the chapter. Based on this definition, three nodes must interact to allow

Figure 3. Transactions for network comparison

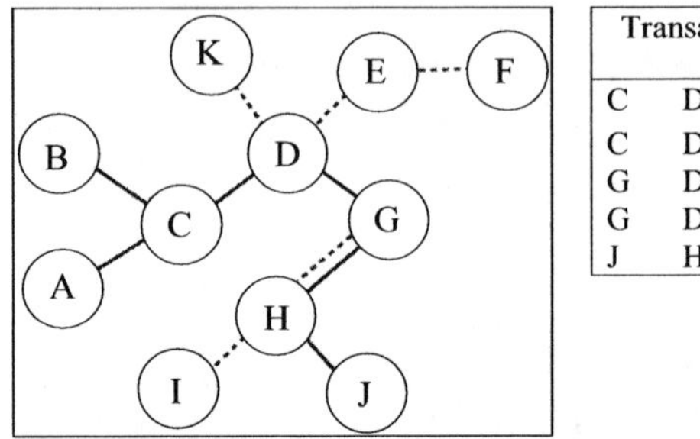

Transactions		
C	D	E
C	D	K
G	D	E
G	D	K
J	H	I

comparison of two networks. This represents the case of node overlap when the first network covers one node, the two networks share a second node, and the second network covers the third node.

An example is seen in Figure 3 where the edge "C, D" (solid) and the edge "D, E" (dashed) are compared by the algorithm, but not the edge "B, C" (solid) because it has no corresponding dashed interaction for either node. Similarly, edge "H, G" causes either a dashed or a solid overlap, so it is not used. When we apply the uniqueness operator to these transactions, we are accepting only those triples that have different, non-overlapping interaction types.

Differential Association Rule Experiments

This section gives an example of applying the differential ARM methods to a bioinformatics dataset. First, we describe the datasets used. We then discuss the results of applying differential ARM on a single-type interaction network and then on a multiple-type interaction network.

Our data consists of a node relation gathered from the Comprehensive Yeast Genome Database at MIPS (Mewes et al, 2002; CYGD, 2004). This node relation represents gene annotation data such as function and localization. The relation contains the ORF identifier as key and the set of annotations related to that ORF as attribute (descriptor set).

It should be noted that the quality of annotations is still in the process of being improved. At this point, there are many proteins with no annotations at all and some annotation categories are only covered by a small subset of proteins. In other cases, only high-level annotations are available. This means that the results are limited to the detail that is provided in the databases. As annotation sources and procedures improve, so will the results of ARM.

We used three different definitions for protein-protein interactions that are represented as undirected edges for yeast: physical, genetic and domain fusion. The physical edge relation was built from the "ppi" table at CYGD where all tuples with type label of "physical" were used. The genetic edge relation was taken from supplemental table S1 of genetic interactions from Tong et al. (2004) where both "Synthetic Sick" and "Synthetic Lethal" entries are used. Our third edge relation was the domain fusion set built from the unfiltered results posted from Truong and Ikura (2003) and Ikura Lab (2004). The set was filtered to reflect only ORFs contained in our node relation from MIPS in order to improve both performance and reliability of results.

Before we discuss the association rule results we produced, we show the efficiency of the approach. Differential ARM outperforms the standard algorithm by three orders of magnitude in the four-node setting and one order of magnitude in the two-node setting. The reduction in the number of rules produced is more significant. The differences between the numbers of rules in differential vs. standard ARM demonstrate how scope and autocorrelation greatly inflate the results. Differential ARM has more than six orders of magnitude fewer rules to interpret. Results of trial experiments are shown in Figure 4. Those orders of magnitude determine the difference between useable and unusable results.

More important than performance concerns are the practical benefits of rules produced by differential ARM. We will first look at an example of a rule that is strong based on the application of a standard ARM algorithm on joined tables but not so if only unique items are considered. Standard ARM on joined tables returns mostly rules that show autocorrelation of items or are out-of-scope. We look at a rule that was mentioned earlier:

$\{0.transcription\} \rightarrow \{1.nucleus\}$ *(support = 0.29% confidence = 28.38%).*

This rule is the consequence of a single-node rule and an autocorrelation rule:

$\{0.transcription\} \rightarrow \{0.nucleus\}$ *(support = 0.70% confidence = 69.59%),*

$\{0.nucleus\} \rightarrow \{1.nucleus\}$ *(support = 5.74% confidence = 29.02%).*

Figure 4. Performance comparison between standard ARM and differential ARM

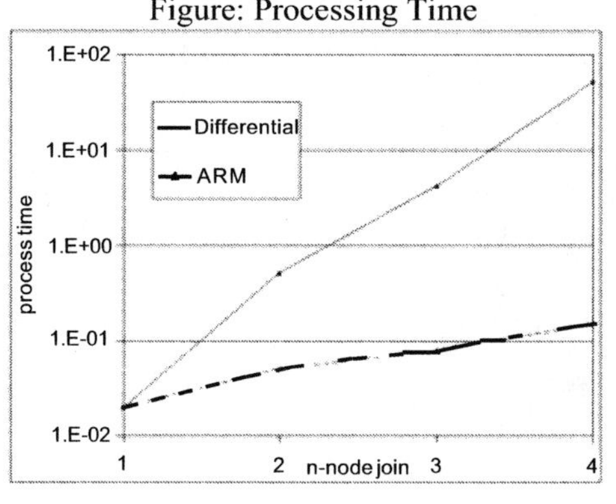

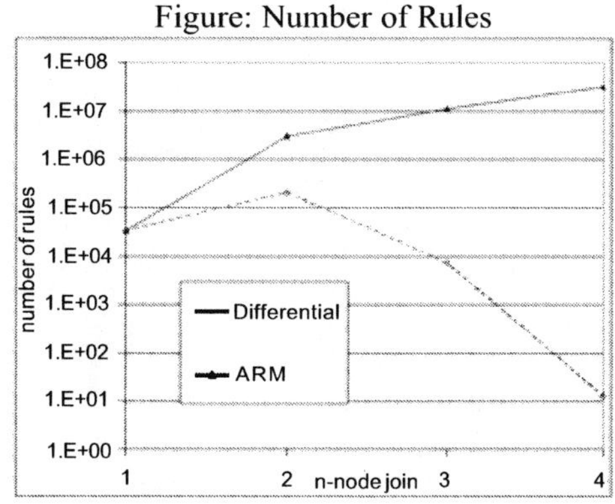

Using the uniqueness operator changes the support of the original rule to 0.02% and a confidence of 2.08%. Strong rules in our data set generally have a support around 2-4% and confidence around 20%. Based on those numbers the rule cannot be considered strong and ranks lower in the differential results. This shows a case where a hypothesis may have been made based on a standard ARM result that the properties of transcription and nucleus are typical for proteins on opposite sides of a physical interaction. Differential ARM results do not support this hypothesis. Differential ARM shows that nucleus is strongly associated itself and proteins co-located in the nucleus interact. Therefore, the differential ARM results correctly exclude this rule.

The following rule was found in the physical network based on differential ARM results:

{1.mitochondria} → {0.cytoplasm} (support = 1.2% confidence = 27.3%).

This rule clearly corresponds to annotations that would not be expected to hold within a single protein but may hold between interacting ones. A protein located in the mitochondria would not have localization cytoplasm. We do expect compartmental crosstalk as studied in Schwikowski et al. (2000) between those two locations. The observation confirms to us that we see rules that are sensible from a biological perspective. We also identify patterns that have not been reported. The pattern *{1.ER, 0.mitochondria} (support 0.21%)* was observed. It was selected because of its comparatively high support and unexpected indication of proteins in the endoplasmic reticulum (ER) physically interacting with proteins in the mitochondria.

We now look at results that derive from the network comparison formalism. Given multiple types of protein-protein interactions we look for significant differences to aid in the understanding of cellular function and as well as the properties and uses of the networks. The networks themselves do not show a significant overlap, in other words, it is common that for any given physical interaction between two proteins there will be no genetic interaction (Tong et al., 2004).

Table 1 shows that even the statistical properties of the networks differ significantly. The average number of interactions of proteins that show at least one interaction varies from 3.55 in the physical network to 44.5 in the domain fusion network. Comparison of annotations across those networks has to compensate for such differences. The process of joining relations and application of the uniqueness operator ensures that each protein

Table 1. Statistics of physical, genetic, and domain fusion networks

	Statistics			
Table	int/orf	max int	#>20	#int
physical	3.55	289	73	14672
genetic	7.88	157	93	8336
domain fusion	44.6	231	305	28040

that is considered for a physical interaction will also be considered for a genetic interaction.

Before looking at details of individual rules, we make some general observations regarding the number of rules we find for different combinations of networks. When comparing physical and genetic networks we found about one order of magnitude more strong rules relating to the physical network compared with the genetic network. Physical interactions also produce the stronger rules when compared with domain fusion networks. That means that the physical network allows the most precise statements to be made. That suggests that physical interactions reflect properties of the proteins better than either of the other two do.

An example of a rule supported by the physical network but not the domain fusion network:

{0.ABC trans fam sign} → {1.ATP/GTP binding site} (support = 0.45% confidence = 90%).

ORF 0 has the motif of an ABC transporter signature that implies it is an ABC transporter coding sequence. ABC transporters have conserved ATP binding domains as the motif in ORF 1 (Falquet et al., 2002). From the rule, we see that these two domains physically interact. However, the domains never occur in a single gene. This supports the observation that the ATP binding domain is found in many other proteins as well (Falquet et al., 2002) and both functions are combined through interactions at the protein level rather than at the genetic level. This observation could warrant further studies or hypothesis on the observed pattern. There are many other examples of rules that allow comparison and contrast of these related networks.

Future Trends

Study of biological networks combined with annotation information is likely to become a topic of increasing importance. Definitions of biological networks exist beyond the ones discussed in this chapter and new ones are added frequently. Important types of networks that have been omitted from the discussion of the chapter are metabolic, regulatory, and signaling pathways. Network-based techniques that are aware of node data are ideally suited as tools that are intermediate in their complexity and scope between large-scale network analysis that ignores anything but connectivity, and small-scale simulations of individual pathways. While data mining is applied to the general scenario of relational databases, little has been specifically done to address the setting of a network with node information. Much can be gained by focusing on this important setting. Networks with node data are ubiquitous from social networks through the structure of the World Wide Web to many areas of bioinformatics. The increasing volume and complexity of such data will require further advances in efficiency and utility.

Summary

This chapter has described issues related to data mining of protein interaction networks in conjunction with information associated with the proteins themselves. Other techniques in statistics and data mining leave significant gaps, including problems in describing differences in the data and analyzing multiple networks. A novel concept, differential association rule mining was introduced, that allows users to discover associations that indicate differences in interacting proteins. The goal of this technique is to highlight differences between items belonging to different interacting nodes or different networks. We demonstrated that such differences would not be identified by application of standard relational ARM techniques. Differential ARM was applied to real-world examples of protein interactions in yeast. Within the results, we recovered association results found in the literature as well as new associations. Overall, the technique offers an efficient and effective method for data exploration in biological and other networks.

References

Agrawal, R., Imielinski, T., & Swami, A. N. (1993). Mining association rules between sets of items in large databases. *Proceedings of the ACM SIGMOD International Conference on Management of Data*, Washington, DC (pp. 207- 216).

Bastide, Y., Pasquier, N., Taouil, R., Stumme, G., & Lakhal, L. (2000). Mining minimal non-redundant association rules using frequent closed itemsets. *Lecture Notes in Computer Science, 1861*(972).

Besemann, C., Denton, A., Yekkirala, A., Hutchison, R., & Anderson, M. (2004). Differential association rule mining for the study of protein-protein interaction networks. *Proceedings of the Workshop on Data Mining in Bioinformatics at KDD*, Seattle, WA.

Brin, S., Motwani, R., Ullman J. D., & Tsur, S. (1997). Dynamic itemset counting and implication rules for market basket data. *Proceedings ACM SIGMOD International Conference on Management of Data*, Tucson, AZ.

Cristofor, L., & Simovici, D. (2001). *Mining association rules in entity-relationship modeled databases*. Technical report, University of Massachusetts Boston.

CYGD. (2004). Retrieved March, 2004 from http://mips.gsf.de/genre/proj/yeast/index.jsp

Dehaspe, L., & Raedt, L. D. (1997). Mining association rules in multiple relations. *Proceedings of the 7th International Workshop on Inductive Logic Programming*, Prague, Czech Republic (pp. 125-132).

Elmasri, R. & Navathe, S. (2004). *Fundamentals of Database Systems*. Boston: Pearson.

Falquet, L., Pagni, M., Bucher, P., Hulo, N., Sigrist, C. J., Hofmann, K., et al. (2002). The prosite database, its status in 2002. *Nucleic Acids Research, 30*, 235-238.

Han, J., & Fu, Y. (1995). Discovery of multiple-level association rules from large databases. *Proceedings of the 21th International Conference on Very Large Data Bases*, Zurich, Switzerland.

Ikura Lab, O. C. I. (2004). *Domain fusion database*. Retrieved March, 2004 from http://calcium.uhnres.utoronto.ca/pi/pubpages/download/index.htm

Inokuchi, A., Washio, T., & Motoda, H. (2000). An apriori based algorithm for mining frequent substructures from graph data. *Proceedings of the European Conference on Principles of Data Mining and Knowledge Discovery*, Lyon, France (pp. 13–23).

Ito, T., Chiba, T., Ozawa, R., Yoshida, M., Hattori, M., & Sakaki, Y. (2001). A comprehensive two-hybrid analysis to explore the yeast protein interactome. *Proc Natl Acad Sci USA, 98*(8), 4569–74.

Jensen, D., & Neville, J. (2002). Autocorrelation and Linkage Cause Bias in Evaluation of Relational Learners. *Lecture Notes in Artificial Intelligence, 2583*, 101-116.

Marcotte, E. M., Pellegrini, M., Ng, H. L., Rice, D. W., Yeates, T. O., & Eisenberg, D. (1999). Detecting protein function and protein-protein interactions from genome sequences. *Science, 285*(5428), 751–3.

Maslov, S., & Sneppen, K. (2002). Specificity and stability in topology of protein networks. *Science, 296*(5569), 910-3.

Mewes, H., Frishman, D., Gldener, U., Mannhaupt, G., Mayer, K., Mokrejs, M., et al. (2002). Mips: A database for genomes and protein sequences. *Nucleic Acids Research, 30*(1), 31–44.

Michihiro, K., & Karypis, G. (2001). Frequent subgraph discovery. *Proceedings of the International Conference on Data Mining*. San Jose, CA (pp. 313–320).

Oyama, T., Kitano, K., Satou, K., & Ito, T. (2002). Extraction of knowledge on protein-protein interaction by association rule discovery. *Bioinformatics, 18*(5), 705–14.

Ozier, O., Amin, N., & Ideker, T. (2003). *Genetic interactions: Online supplement to: Global architecture of genetic interactions on the protein network*. Retrieved January 31, 2005, from http://web.wi.mit.edu/ideker/pub/nbt/

Ozier, O., Amin, N., & Ideker, T. (2003). Global architecture of genetic interactions on the protein network [Letter to the editor]. *Nature Biotechnology, 21*(5), 490-1.

Pei, J., Han, J., & Lakshmanan, L.V. (2004). Pushing convertible constraints in frequent itemset mining. *Data Mining and Knowledge Discovery, 8*, 227–252.

Schwikowski, B., Uetz, P., & Fields, S. (2000). A network of protein-protein interactions in yeast. *Nature Biotechnology, 18*(12), 1242–3.

Srikant, R., Vu, Q., & Agrawal, R. (1997). Mining association rules with item constraints. *Proceedings of the International Conference on Knowledge Discovery and Data Mining*, Newport Beach, CA, (pp. 67–73).

Tong, A. H. Y., Evangelista, M., Parsons, A. B., Xu, H., Bader, G. D., Pag, N., et al. (2001). Systematic genetic analysis with ordered arrays of yeast deletion mutants. *Science, 294*(5550), 2364–8.

Tong, A. H. Y., Evangelista, M., Parsons, A. B., Xu, H., Bader, G. D., Pag, N., et al. (2004). Global mapping of the yeast genetic interaction network. *Science, 303*(5695), 808–815.

Truong, K., & Ikura, M. (2003). Domain fusion analysis by applying relational algebra to protein sequence and domain databases. *BMC Bioinformatics, 4*(16).

Uetz, P., Giot, L., Cagney, G., Mansfield, T. A., Judson, R. S., Knight, J. R., et al. (2000). A comprehensive analysis of protein-protein interactions in Saccharomyces cerevisiae. *Nature, 403*(6770), 623–7.

Wagner, A., & Fell, D. (2001). The small world inside large metabolic networks. *Proc. Roy. Soc. London Series B, 268* (pp. 1803-1810).

Zaki, M. J. (2000). Generating non-redundant association rules. *Proceedings of the ACM SIGKDD International Conference on Knowledge Discovery and Data Mining*, Boston (pp. 34–43).

Chapter XV

Mining BioLiterature: Toward Automatic Annotation of Genes and Proteins

Francisco M. Couto, Universidade de Lisboa, Portugal
Mário J. Silva, Universidade de Lisboa, Portugal

Abstract

This chapter introduces the use of Text Mining in scientific literature for biological research, with a special focus on automatic gene and protein annotation. This field became recently a major topic in Bioinformatics, motivated by the opportunity brought by tapping the BioLiterature with automatic text processing software. The chapter describes the main approaches adopted and analyzes systems that have been developed for automatically annotating genes or proteins. To illustrate how text-mining tools fit in biological databases curation processes, the chapter presents a tool that assists protein annotation. Besides the promising advances of Text Mining of BioLiterature, many problems need to be addressed. This chapter presents the main open problems in using text-mining tools for automatic annotation of genes and proteins, and discusses how a more efficient integration of existing domain knowledge can improve the performance of these tools.

Introduction

Bioinformatics aims at understanding living systems using biological information. The facts discovered in biological research have been mainly published in the scientific literature (BioLiterature) since the 19th century. Extracting knowledge from such a large amount of unstructured information is a painful and hard task, even to an expert. A solution could be the creation of a database where authors would deposit all the facts published in BioLiterature in a structured form. Some generic databases, such as UniProt, collect and distribute biological information (Apweiler et al., 2004). However, different communities have different needs and views on specific topics, which change over time. As a result, researchers do not look only for the facts, but also for their evidence. Before a researcher considers a fact as relevant to his work, he checks the evidence presented by the author, because facts are normally valid only in a specific context. This explains why Molecular Biology knowledge continues to be mainly published in BioLiterature. Another solution is Text Mining, which aims at automatically extracting knowledge from natural language texts (Hearst, 1999). Text-mining systems can be used to identify the following types of information: entities, such as genes, proteins and cellular components; relationships, such as protein localization or protein interactions; and events, such as experimental methods used to discover protein interactions. Bioinformatics tools to collect more information about the concepts they analyze also use Text Mining. For example, information automatically extracted from the BioLiterature can improve gene expression clustering (Blaschke, Oliveros, & Valencia, 2004).

Text Mining of BioLiterature has been studied since the last decade (Andrade & Valencia, 1998). The interest in the topic has been steadily increasing, motivated by the vast amount of publications that curators have to read in order to update biological databases, or simply to help researchers keep up with progress in a specific area. Text Mining can minimize these problems mainly because BioLiterature articles are quite often publicly available. The most widely used BioLiterature repository is MEDLINE, which provides a vast collection of abstracts and bibliographic information. For example, in 2003, about 560,000 citations have been added to MEDLINE. Reading 10 of these documents per day, it would take around 150 years to read all the documents added in 2003. Moreover, the number of new documents added per year increased by more than 20,000 from 2000 to 2003. Hence, text-mining systems could have a great impact in minimizing this effort by automatically extracting information that can be used for multiple purposes and could not possibly be organized by other means.

This chapter starts by providing broad definitions used in Text Mining and describes the main approaches. Then, it summarizes the state-of-the-art of this field and shows how text-mining systems can be used to automatically annotate genes or proteins. Next, the chapter describes a tool designed for assisting protein annotation. Finally, the chapter discusses future and emerging trends and presents concluding remarks.

Text Mining

Text Mining aims at automatically extracting knowledge from unstructured text. Usually the text is organized as a collection of documents, or corpus.

TextMining = NLP + DataMining

Data Mining aims at automatically extracting knowledge from structured data. (Hand, Mannila, & Smyth, 2000). Thus, Text Mining is a special case of Data Mining, where input data is text instead of structured data. Normally, text-mining systems generate structured representations of the text, which are then analyzed by Data Mining tools. The simplest representation of a text is a vector with the number of occurrences of each word in the text (called a bag-of-words). This representation can be easily created and manipulated, but ignores all the text structure. Text-mining systems may also use Natural Language Processing (NLP) techniques to represent and process text more effectively. NLP is a broad research area that aims at analyzing spoken, handwritten, printed, and electronic text for different purposes, such as speech recognition or translation (Manning & Schütze, 1999). The most popular NLP techniques used by text-mining systems include: tokenization, morphology analysis, part-of-speech tagging, sense disambiguation, parsing, and anaphora resolution.

Tokenization aims at identifying boundaries in the text to fragment it into basic units called tokens. The first step in a text-mining system is to identify the tokens. The most commonly used token is the word. In most languages, white-space characters can be considered an accurate boundary to fragment the text into words. This problem is more complex in languages without explicitly delimiters, such as Chinese (Wu & Fung, 1994). Morphology analysis aims at grouping the words (tokens) that are variants of a common word, and therefore are normally used with a similar meaning (Spencer, 1991). This involves the study of the structure and formation of words. A common type of inflectional variants results from the tense on verbs. For example, "binding" and "binds" are inflectional variants of "bind." Some other word variants result from prefixing, suffixing, infixing, or compounding.

Part-of-speech tagging aims at labeling each word with its semantic role, such as article, noun, verb, adjective, preposition, or pronoun (Baker, 1989). This involves the study of the structure and formation of sentences. The tagging is a classification of words according to their semantic role and to their relations to each other in a sentence. Sense disambiguation selects the correct meaning of a word in a given piece of text. For example, "compound" has two different senses in the expressions "compound the ingredients" and "chemical compound." Normally, the part-of-speech tags are used as a first step in sense disambiguation (Wilks & Stevenson, 1997)

Parsing aims at identifying the syntactic structure of a sentence (Earley, 1970). The syntactic structure of a sequence of words is composed by a set of other syntactic structures related to smaller sequences, except for the part-of-speech tags that are syntactic structures directly linked to words. Normally, the syntactic structure of a sentence is represented by a syntax tree, where leafs represent the words and internal nodes the syntactic structures. Algorithms to identify the complete syntactic structure of a sentence are in general inaccurate and time-consuming, given the combinatorial explosion in long sentences. An alternative is shallow-parsing, which does not attempt to parse complex syntactic structures. Shallow-parsing only splits sentences into phrases, in other words, subsequences of words that represent a grammatical unit, such as noun phrase or verb phrase. Anaphora (or co-reference) resolution aims at determining different sequences of words referring to the same entity. For example, in the sentence "The enzyme has an intense activity, thus, this protein should be used", the noun phrases "The enzyme" and "this protein" refer to same entity.

Some of the NLP techniques described above can be implemented using algorithms also used in Data Mining. For example, part-of-speech taggers can use Hidden Markov Models (HMMs) to estimate the probability of a sequence of part of speech assignments (Smith, Rindflesch, & Wilbur, 2004). Not all NLP techniques improve the performance of a given text-mining system. As a result, designers of text-mining systems have to select which NLP techniques would be useful to achieve their ultimate goal.

After creating a structured representation of texts, text-mining systems can use the following approaches for extracting knowledge (Leake, 1996):

rule-based or case-based

The rule-based approach relies on rules inferred from patterns identified from the text by an expert. The rules represent in a structured form the knowledge acquired by experts when performing the same task. The expert analyzes a subpart of the text and identifies common patterns in which the relevant information is expressed. These patterns are then converted to rules to identify the relevant information in the rest of the text. The main bottleneck of this approach is the manual process of creating rules and patterns. Besides being time-consuming, in most cases, this manual process is unable to devise from a subpart of the text the set of rules that encompass all possible cases.

The case-based approach relies on a predefined set of texts previously annotated by an expert, which is used to learn a model for the rest of the text. Cases contain knowledge in an unprocessed form, and they only describe the output expected by the users for a limited set of examples. The expert analyzes a subpart of the text (training set) and provides the output expected to be returned by the text-mining system for that text. The system uses the training set to create a probabilistic model that will be applied to the rest of the text. The main bottleneck of this approach is the selection and creation of a training set large enough to enable the creation of a model accurate for the rest of the text.

The manual analysis of text requires less expertise in the case-based approach than in the rule-based approach. In the rule-based approach, the expert has to identify how the relevant information is expressed in addition to the expected output. However, rule-based systems can use this expertise to achieve high precision by selecting the most reliable rules and patterns.

State of the Art

The main problem in BioLiterature mining is coping with the lack of a standard nomenclature for describing biologic concepts and entities. In BioLiterature, we can often find different terms referring to the same biological concept or entity (synonyms), or the same term meaning different biological concepts or entities (homonyms). Genes, whose name is a common English word, are frequent, which makes it difficult to recognize biological entities in the text.

Recent advances in Text Mining of BioLiterature already achieved acceptable levels of accuracy in identifying gene and protein names in the text. However, the extraction of relationships, such as functional annotations, is still far from being solved. Recent surveys report these advances by presenting text-mining tools that are not only run in different corpora but also perform different tasks (Hirschman, Park, Tsujii, Wong, & Wu, 2002; Blaschke, Hirschman, & Valencia 2002; Dickman, 2003; Shatkay & Feldman, 2003).

On the other hand, recent challenging evaluations compared the performance of different approaches in solving the same tasks using the same corpus. For example, the 2002 KDD Cup (bio-text task) consisted on identifying which biomedical articles contained relevant experimental results about Drosophila (fruit fly), and the genes (transcripts and proteins) involved (Yeh, Hirschman, & Morgan, 2003). The best submission out of 32 obtained 78% F-measure in the article decision, and 67% F-measure in the gene decision.

A similar challenging evaluation was the 2004 TREC genomics track, which consisted on identifying relevant documents and documents with relevant experimental results about the mouse (Hersh et al., 2004). The first task was a typical Information Retrieval task. A list of documents and a list of topics were given as input. The goal was to identify the relevant documents for each topic. The best submission out of 47 obtained 41% precision. The second subtask comprised the selection of documents with relevant experimental information. The best submission out of 59 obtained 27% F-measure. In addition to document selection, the task also comprised automatic annotations of genes. The best submission out of 36 obtained 56% F-measure.

Another challenging evaluation was BioCreAtIvE (Hirschman, Yeh, Blaschke, & Valencia, 2005). This evaluation comprised two tasks. The first aimed at identifying genes and proteins in BioLiterature. The best submission out of 40 obtained 83% F-measure. The second task addressed the automatic annotation of human proteins, and involved two subtasks. The first subtask required the identification of the texts that provided the evidence for extracting each annotation. From 21 submissions, the highest precision was 78% and the highest recall was 23%. The second subtask consisted on automatic annotation of proteins. From 18 submissions, the highest precision was 34% and the highest recall was 12%.

Automatic Annotation

One of the most important applications of text-mining systems is the automatic annotation of genes and proteins. A gene or protein annotation consists of a pair composed by the gene or protein and a description of its biological role. Normally, descriptions use terms from a common ontology. The Gene Ontology (GO-Consortium, 2004) provides a structured controlled vocabulary that can be applied to different species (GO-Consortium, 2004). GO has three different aspects: molecular function, biological process, and cellular component. To comprehend a gene or protein activity is also important to know the biological entities that interact with it. Thus, the annotation of a gene or protein also involves identifying interacting chemical substances, drugs, genes and proteins.

Text-mining systems that automatically annotate genes or proteins can be categorized according to: the mining approach taken (rule-based or case- based), the NLP techniques applied, and the amount of manual intervention required.

Rule-Based Systems

AbXtract was one of the first text-mining systems attempting to characterize the function of genes and proteins based on information automatically extracted from BioLiterature (Andrade & Valencia, 1998). The system assigns relevant keywords to protein families based on a rule comprising the frequency of the keywords in the abstracts related to the family. In addition to using a rule-based approach, AbXtract relies in only one rule that does not require human intervention. A similar approach is taken by the system proposed by Pérez, Perez-Iratxeta, Bork, Thode, and Andrade (2004), which annotates genes with keywords extracted from abstracts based on mappings between different ontologies.

An example of a system based on a large number of rules is BioRAT (Corney, Buxton, Langdon, & Jones, 2004). Given a query, BioRAT finds documents and highlights the most relevant facts in their abstracts or full texts. However, the rules are exclusively derived from patterns inserted by the user. Textpresso is another rule-based system that finds documents and marks them up with terms from a built-in ontology (Müller, Kenny, & Sternberg, 2004). The system assigns to each entry of the ontology regular expressions that capture how the entry can be expressed in BioLiterature. Textpresso is not so dependent on the user as BioRAT, since many of the regular expressions are automatically generated to account for regular forms of verbs and nouns.

BioIE is a system that takes more advantage of NLP techniques. It extracts biological interactions from BioLiterature and annotates them with GO terms (Kim & Park, 2004). The system uses morphology, sense disambiguation, and rules with syntactic dependencies to identify GO terms in the text. BioIE uses 1,312 patterns to match interactions in the sentences, so it also requires substantial manual intervention. Koike, Niwa, and Takagi (2005) propose a similar system that annotates gene, protein, and families with GO terms extracted from texts. The system uses morphology, part-of-speech tagging, shallow parsing, and simple anaphora resolution. To extract the relationships, it uses both automatically generated and manually inserted rules.

Table 1. Categorization of some recent text-mining systems designed for automatic annotation of genes and proteins; for each system, the table indicates the mining approach taken, the proportion of NLP techniques used and the proportion of manual intervention needed to generate rules, patterns, or training sets

System	Mining	NLP	Manual
Andrade and Valencia (1998)	Rule-based	nil	nil
Pérez et al. (2004)	Rule-based	nil	nil
Corney et al. (2004)	Rule-based	Low	High
Müller et al. (2004)	Rule-based	Low	Medium
Kim and Park (2004)	Rule-based	Medium	Medium
Koike et al. (2005)	Rule-based	High	Medium
Palakal et al. (2003)	Case-based	Medium	Low
Chiang and Yu (2003)	Case-based	Medium	Low

Figure 1. UML use case diagram of GOAnnotator

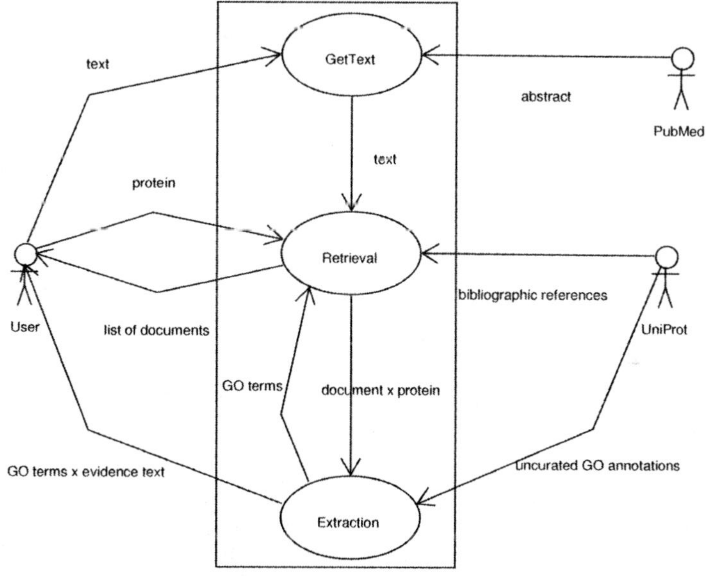

Case-Based Systems

A text-mining system using the case-based approach was proposed by Palakal, Stephens, Mukhopadhyay, and Raje (2003). The system extracts relationships between biological objects (e.g. protein, gene, cell cycle). The system uses sense disambiguation, and a probabilistic model to find directional relationships. The model is trained using examples of sentences expressing a relationship.

Figure 2. A list of documents related to the protein "Ras GTPase-activating protein 4" provided by the GOAnnotator. The list is sorted by the similarity of the most similar term extracted from each document. The curator can invoke the links in the "Extract" column to see the extracted terms together with the evidence text. By default, GOAnnotator uses only the abstracts of scientific documents, but the curator can replace or add text (links in the "AddText" column).

PubMedId	Title	MostSimilarTermExtracted	Scope	Authors	Year	Extract	AddText
11594756(FullText)	Distinct phosphoinositide binding specificity of the GAP1 family proteins: characterization of the pleckstrin homology domains of MRASAL and KIAA0538.	100% GTPase activator activity (f)	GeneRIF	3	2001	Pre-Processed	Text
11448776(FullText)	CAPRI regulates Ca(2+)-dependent inactivation of the Ras-MAPK pathway.	100% GTPase activator activity (f)	SEQUENCE FROM N.A.	3	2001	Pre-Processed	Text
9628581(FullText)	Prediction of the coding sequences of unidentified human genes. IX. The complete sequences of 100 new cDNA clones from brain which can code for large proteins in vitro.	40% cell communication (p)	SEQUENCE FROM N.A.	7	1998	Pre-Processed	Text

MeKE is another system that extracts protein functions from BioLiterature using sentence alignment (Chiang & Yu, 2003). MeKE also uses sense disambiguation. The system uses a statistical classifier that identifies common patterns in examples of sentences expressing GO annotations. The classifier uses these patterns to decide if a given sentence expresses a GO annotation.

Discussion

The systems described above show how Text Mining can help curators in the annotation process. Most rely on domain knowledge manually inserted by curators (see Table 1). Domain knowledge improves precision, but it cannot be easily extended to work on other domains and demands an extra effort to keep the knowledge updated as BioLiterature evolves. This approach is time-consuming and makes the systems too specific to be extended to new domains. Thus, an approach to avoid this process is much needed.

GOAnnotator

This section illustrates how text-mining can be integrated in a biological database curation process, by describing GOAnnotator, a tool for assisting the GO annotation of

Figure 3. For each uncurated annotation, GOAnnotator shows the similar GO terms extracted from a sentence of the selected document. If any of the sentences provides correct evidence for the uncurated annotation, or if the evidence supports a GO term similar to that present in the uncurated annotation, the curator can use the "Add" option to store the annotation together with the document reference, the evidence codes and additional comments.

Similar GO Terms Extracted	**GOA Electronic Term:** intracellular signaling cascade (p) -
inactivation of MAPK (p) -	CAPRI regulates Ca2+-dependent inactivation of the Ras-MAPK pathway.
activation of MAPK (p) -	We interpret the faster response of MAPK activation as a possible dominant-negative effect, since mutagenesis of the equivalent residue in NF-1 (Arg1391 → Ala) has demonstrated that catalysis is inhibited 45-fold; but Ras binding still occurs, albeit with a 6-fold lower affinity for Ras-GTP [20].
Comment:	New Terms: Evidence: - --- Add ---

UniProt entries (Rebholz-Schuhmann, Kirsch, & Couto, 2005). GOAnnotator links the GO terms present in the uncurated annotations with evidence text automatically extracted from the documents linked to UniProt entries.

Figure 1 presents the data flow involved in the processing steps of GOAnnotator and in its interaction with the users and external sources. Initially, the curator provides a UniProt accession number to GOAnnotator. GOAnnotator follows the bibliographic links found in the UniProt database and retrieves the documents. Additional documents are retrieved from the GeneRIF database (Mitchell et al., 2003). Curators can also provide any other text for mining. GOAnnotator then extracts from the documents GO terms similar to the GO terms present in the uncurated annotations.

In GOAnnotator the extraction of GO terms is performed by FiGO, a tool that receives text and returns the GO terms detected (Couto, Silva, & Coutinho, 2005). FiGO is rule-based, does not use any NLP technique and does not require manual intervention. FiGO assigns a confidence value to each GO term that represents the terms' likelihood of being mentioned in the text. The confidence value is the product of two parameters. The first, called local evidence context (LEC), is used to measure the likelihood that words in the text are part of a given GO term. The second parameter is the inverse of their frequency in GO. GO terms are similar if they are in the same lineage or if they share a common parent in the GO hierarchy. FiGO uses the semantic similarity measure of (Lin, 1998) to compute the degree of similarity between two GO terms.

GOAnnotator ranks the documents based on the extracted GO terms from the text and their similarity to the GO terms present in the uncurated annotations (see Figure 2). Any extracted GO term is an indication for the topic of the document, which is also taken from the UniProt entry.

GOAnnotator displays a table for each uncurated annotation with the GO terms that were extracted from a document and found similar to the GO term present in the uncurated annotation (see Figure 3). The sentences from which the GO terms were extracted are also displayed. Words that have contributed to the extraction of the GO terms are highlighted. GOAnnotator gives the curators the opportunity to manipulate the confidence and similarity thresholds to modify the number of predictions.

Future Trends

The performance of text-mining tools that automatically annotate genes or proteins is still not acceptable by curators. Gene or protein annotation is more subjective and requires more expertise than simply finding relevant documents and recognizing biological entities in texts. Moreover, an annotation tool can only perform well when it is using the correct documents and the correct entities. Errors in the retrieval of documents or in the recognition of entities will be the cause of errors in the annotation task.

Existing tools that retrieve relevant documents do not always provide what the curators want. On the contrary, curators spend a large amount of their time finding the right documents. This is probably the main reason why many curators are still not using text-mining tools for gene or protein annotation. Another reason is that quite often the full texts are not electronically available. Curators need additional information that is not usually present in the abstracts, such as the type of experiments applied and the species from which proteins originate. Finally, another reason is that most text-mining tools depend on domain knowledge manually inserted by curators, which is also very time-consuming.

Text-mining tools acquire domain knowledge from the curators in the form of rules or cases. The identification of rules requires more effort to the curators than the evaluation of a limited set of cases. However, a single rule can express knowledge not contained in a large set of cases. Neither source of knowledge subsumes the other: the knowledge represented by a rule is normally not well-represented by any set of cases, and it is difficult to identify a set of rules representing all knowledge expressed by a set of cases.

Couto, Martins, and Silva (2004) proposed an approach to obtain the domain knowledge that does not require human intervention. Instead of obtaining the domain knowledge from curators, they propose acquiring it from publicly available databases that already contain curated data. Text-mining systems could consider these databases as training sets from which rules, patterns, or models can be automatically generated. Besides avoiding direct human intervention, these automated training sets are usually much larger than individually generated training sets. Another advantage is that the tools' training data does not become outdated as public databases can be tracked for updates as they evolve.

Summary

Bioinformatics aims at understanding living systems by inferring knowledge from biological information, such as DNA and protein sequences. The role of Text Mining in Bioinformatics is to automatically extract knowledge from BioLiterature. This field is new and has evolved over the last decade, motivated by the opportunity brought by tapping the large amount of information that has been published in BioLiterature with automatic text processing software.

Researchers will tend to use databases to store and find facts, but the evidence substantiating them will continue to be described as unstructured text. As a result, text-mining tools will continue to have an important role in Bioinformatics. Recent advances in Text Mining of BioLiterature are already promising, but many problems remain. In our opinion, the future of text-mining tools for gene or protein annotation will mainly depend on a better use of NLP techniques, and in an efficient integration of existing domain knowledge available in biological databases and ontologies.

References

Andrade, M., & Valencia, A. (1998). Automatic extraction of keywords from scientific text: Application to the knowledge domain of protein families. *Bioinformatics, 14*(7), 600-607.

Apweiler, R., Bairoch, A., Wu, C., Barker, W., Boeckmann, B., Ferro, S., et al. (2004). UniProt: The universal protein knowledgebase. *Nucleic Acids Research, 32* (Database issue), D115-D119.

Baker, C. (1989). *English syntax.* MIT Press.

Blaschke, C., Hirschman, L., & Valencia, A. (2002). Information extraction in molecular biology. *Briefings in BioInformatics, 3*(2), 154-165.

Blaschke, C., Oliveros, J., & Valencia, A. (2004). Mining functional information associated to expression arrays. *Functional and Integrative Genomics, 1*(4), 256-268.

Chiang, J., & Yu, H. (2003). MeKE: Discovering the functions of gene products from biomedical literature via sentence alignment. *Bioinformatics, 19*(11), 1417-1422.

Corney, D., Buxton, B., Langdon, W., & Jones, D. (2004). BioRAT: Extracting biological information from full-length papers. *Bioinformatics, 20*(17), 3206-3213.

Couto, F., Martins, B., & Silva, M. (2004). Classifying biological articles using Web resources. *Proceedings of the 2004 ACM Symposium on Applied Computing* (pp. 111-115).

Couto, F., Silva, M., & Coutinho, P. (2005). Finding genomic ontology terms in text using evidence content. *BMC Bioinformatics, 6*(Suppl 1), S21.

Dickman, S. (2003). Tough mining: The challenges of searching the scientific literature. *PLoS Biology, 1*(2), E48.

Earley, J. (1970). An efficient context-free parsing algorithm. *Communications of ACM, 13*, 94-102.

GO-Consortium (2004). The gene ontology (GO) database and informatics resource. *Nucleic Acids Research, 32*(Database issue), D258-D261.

Hand, D., Mannila, H., & Smyth, P. (2000). *Principles of Data Mining.* MIT Press.

Hearst, M. (1999). Untangling text data mining. *Proceedings of the 37th ACL Meeting of the Association for Computational Linguistics* (pp. 3-10).

Hersh, W., Bhuptiraju, R., Ross, L., Johnson, P., Cohen, A., & Kraemer, D. (2004). TREC 2004 genomics track overview. *Proceedings of the 13th Text REtrieval Conference (TREC 2004)* (pp. 14-24).

Hirschman, L., Park, J., Tsujii, J., Wong, L., & Wu, C. (2002). Accomplishments and challenges in literature data mining for biology. *Bioinformatics, 18*(12), 1553-1561.

Hirschman, L., Yeh, A., Blaschke, C., & Valencia, A. (2005). Overview of BioCrEAtIvE: critical assessment of information extraction for biology. *BMC Bioinformatics, 6*(Suppl 1), S1.

Kim, J., & Park, J. (2004). BioIE: Retargetable information extraction and ontological annotation of biological interactions from literature. *Journal of Bioinformatics and Computational Biology, 2*(3), 551-568.

Koike, A., Niwa, Y., & Takagi, T. (2005). Automatic extraction of gene/protein biological functions from biomedical text. *Bioinformatics, 21*(7), 1227-1236.

Leake, D. (1996). *Case-based reasoning: Experiences, lessons, and future directions.* AAAI Press/MIT Press.

Lin, D. (1998). An information-theoretic definition of similarity. *Proceedings of the 15th International Conference on Machine Learning* (pp. 296-304).

Manning, C., & Schütze, H. (1999). *Foundations of statistical natural language processing.* The MIT Press.

Mitchell, J., Aronson, A., Mork, J., Folk, L., Humphrey, S., & Ward, J. (2003). Gene indexing: characterization and analysis of NLM's GeneRIFs. Paper presented at the AMIA 2003 Annual Symposia, Washington, DC.

Müller, H., Kenny, E., & Sternberg, P. (2004). Textpresso: An ontology-based information retrieval and extraction system for biological literature. *PLOS Biology, 2*(11), E309.

Palakal, M., Stephens, M., Mukhopadhyay, S., & Raje, R. (2003). Identification of biological relationships from text documents using efficient computational methods. *Journal of Bioinformatics and Computational Biology, 1*(2), 307-342.

Pérez, A., Perez-Iratxeta, C., Bork, P., Thode, G., & Andrade, M. (2004). Gene annotation from scientific literature using mappings between keyword systems. *Bioinformatics, 20*(13), 2084-2091.

Rebholz-Schuhmann, D., Kirsch, H., & Couto, F. (2005). Facts from text - is text mining ready to deliver? *PLoS Biology, 3*(2), e65.

Shatkay, H. & Feldman, R. (2003). Mining the biomedical literature in the genomic era: An overview. *Journal of Computational Biology, 10*(6), 821-855.

Smith, L., Rindflesch, T., & Wilbur, W. (2004). MedPost: A part-of-speech tagger for biomedical text. *Bioinformatics, 20*(14), 2320-2321.

Spencer, A. (1991). *Morphological theory.* Oxford: Blackwell.

Wilks, Y., & Stevenson, M. (1997). The grammar of sense: Using part-of-speech tags as a first step in semantic disambiguation. *Journal of Natural Language Engineering, 4*(3).

Wu, D. & Fung, P. (1994). Improving Chinese tokenization with linguistic filters on statistical lexical acquisition. *Proceedings of the 4th ACL Conference on Applied Natural Language Processing* (pp. 13-15).

Yeh, A., Hirschman, L., & Morgan, A. (2003). Evaluation of text data mining for database curation: Lessons learned from the KDD challenge cup. *Bioinformatics, 19*(1), i331-i339.

Chapter XVI

Comparative Genome Annotation Systems

Kwangmin Choi, Indiana University, USA
Sun Kim, Indiana University, USA

Abstract

Understanding the genetic content of a genome is a very important but challenging task. One of the most effective methods to annotate a genome is to compare it to the genomes that are already sequenced and annotated. This chapter is to survey systems that can be used for annotating genomes by comparing multiple genomes and discusses important issues in designing genome comparison systems such as extensibility, scalability, reconfigurability, flexibility, usability, and data mining functionality. We also discuss briefly further issues in developing genome comparison systems where users can perform genome comparison flexibly on the sequence analysis level.

Introduction

Once a complete genome sequence becomes available, the next and more important goal is to understand the content of the genome. The exponential accumulation of genomic sequence data demands use of computational approaches to systematically analyze huge amount of genomic data. The availability of such genome sequence data and diverse computational techniques has made comparative genomics — research activity to compare sequences of multiple genome sequences — to become useful not only for finding common features in different genomes, but also for understanding evolutionary mechanisms among multiple genomes.

The process of assigning genomic functions to genes is called *genome annotation*, which utilizes diverse domain knowledge sources from sequence data to the contextual information of the whole genome. Currently there exist several methods for genome annotation. Experimental approaches for genome annotation are probably most reliable, but slow and labor-intensive. Genome annotation can be done computationally as well by (1) assigning function(s) to a gene based on its sequence similarity to other genes that are already annotated with well-defined gene functions, (2) assigning function(s) to a gene based on its position in a conserved gene cluster through comparative analysis of multiple genomes, and (3) inferring function via detecting functional coupling.

Genome annotation probably can be done most accurately by comparing a genome with its phylogenetically-related genomes, which we termed as *comparative genome annotation*. Comparing multiple genomes, however, is a very challenging task. First of all, genome comparison requires handling complicated relationships of many entities. For example, comparison of all protein coding genes in three genomes with 2,000 genes in each genome can involve several million pairwise relationships among genes. The way of genome selection raises another problem because the choice of genomes to be compared is entirely subjective and there are numerous combinations of genomes to be compared. Thus it is necessary to develop an information system for comparative genome annotation which can deal with such challenges.

This chapter primarily aims to survey existing systems from the data mining perspective. Well developed data mining tools will be very helpful in handling numerous functional relationships among genes and genomes. For example, an accurate sequence clustering tool can simplify the genome annotation task significantly, since the user can handle a set of sequences as a single unit for the next analysis. Unfortunately, existing data mining tools are not developed to handle a continuous character stream or genomic sequence, thus our discussion from the data mining perspective is to propose what is needed, rather than characterizing existing systems in the terms of traditional data mining perspective. Here we propose the desirable features that comparative genome annotation systems should have:

- **Extensibility:** There are always new resources such as newly sequenced genomes and computational tools. The system should be able to include them as they become available.
- **Reconfigurability:** There are numerous different ways to combine data and tools, so no single system can meet all needs of users. When needed, it is desirable to reconfigure the system for a specific task.
- **Genome selection flexibility:** The choice of genomes to be compared is entirely subjective. Thus users should be able to compare genomes of their choice with different criteria for sequence comparison.
- **Usability:** Genome comparison involves a huge amount of data, so the system should be easy to use. In addition, it should be easy to port to other platforms.

- **Data mining:** Users need to perform a series of analyses to achieve a research goal and the system should also provide high-level data mining tools to simplify genome analysis task. To explain the data mining issue further, assume that a user wants to analyze a sequence s against several genomes, G_1, G_2, G_3. Obviously a user can perform three different comparisons of s to each G_i and then "combine" the results into one, and then proceed for further analysis. A well developed system can easily mitigate the burden of combining three different search results by providing some system functions. Alternatively, it is possible to use a high performance sequence clustering algorithm to generate the output in a single operation.
- **System integration in a distributed environment:** The annotation of a large genome is often performed collaboratively at different organizations. Thus, it is challenging to maintain consistency of data and share information across the multiple information systems.

This chapter will survey comparative genome annotation systems, the SEED, DAS, CaWorkBench2.0, STRING, and PLATCOM. Then we will compare these systems with the desirable features that we described above.

The SEED

The SEED project (Overbeek, Disz, & Stevens, 2004) aims to provide a suite of programs which enable distributed users to annotate new genomes rapidly and cooperatively. By

Figure 1. Web interface of the SEED at http://theseed.uchicago.edu/FIG/index.cgi

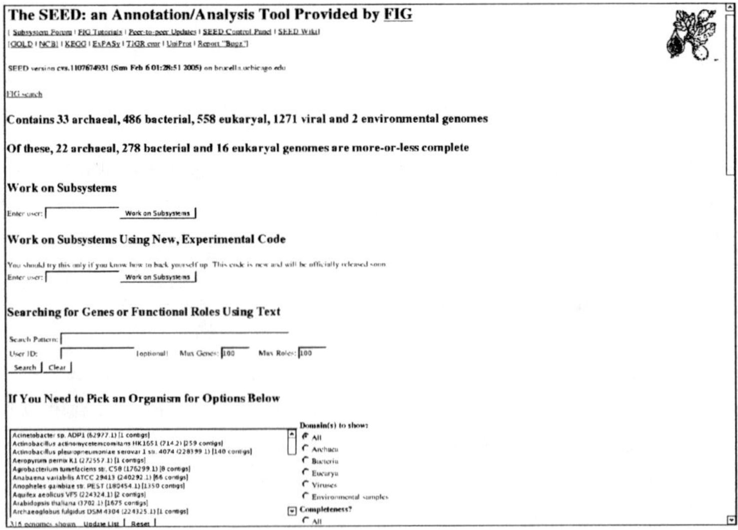

using this system, users can create, collect, and maintain sets of gene annotations organized by groups of related biological functions (i.e., "subsystems") over multiple genomes. The "subsystem" is defined as a set of biological and biochemical functions which together implement a specific process. By annotating one subsystem at a time, the SEED supports the annotation of a single subsystem over multiple genomes simultaneously. So users may examine the relationship between a given gene and a group of other genes by using contextual clues relevant to the determination of functions.

System Architecture and Data Sources

To maintain a scalable integration of genomes, the SEED consists of the essential and basic elements, which are (1) a scalable infrastructure to support the construction of large-scale integration genomic data, (2) the peer-to-peer (i.e., P2P) infrastructure which supports the distributed annotation and the integrated database synchronization, and (3) programs to enable extension of the integrated database environment to adopt new applications and data.

For navigation and annotating genes, the SEED is providing eight core functions (1) to localize and visualize gene clusters which are relevant to the analysis of a specific subsystem, (2) to identify genes with sequence homology from other genomes, (3) to visualize a gene neighborhood on the chromosome, (4) to comparatively compare one gene neighborhood with others around corresponding genes in other genomes, (5) to examine genes that implement closely related metabolic functions, (6) to modify existing function assignments and annotations, (7) to detect inconsistent representation of function, and (8) to generate packages of assignments and annotations corresponding to a single subsystem.

The SEED also adopts a UNIX-like command-line environment where many tools are direct descendants of the former WIT system. It provides convenient methods to easily extract and manipulate genomic data. For example, a set of command, 'pegs 83333.1 | fid2dna | translate', consists of three separate commands (pegs, fid2dna, and translate), and redirections of their input (83333.1) and output are controlled by "pipes" as in UNIX operating system. By using pipes, each program can be modular because a pipe can be used between two programs to combine their functionality. This system design enables the SEED to add new components and evolve easily by multiple participants and gives reconfigurable feature to the system.

The SEED supports the information sharing and synchronization of data among individual biologists who collaborate with others in a geographically distributed environment. Via Web interface, users can examine particularly interesting genes, define new subsystems, modify annotations, and assign new functions. To synchronize all annotations and assignments periodically, the SEED uses a basic peer-to-peer (i.e., P2P) synchronization facility which supports the information sharing among the SEED installations.

Data Mining

Biologists are often interested in a specific subsystem or a set of genes that are known to work together. A good example is the metabolic pathway subsystem. The SEED deals with metabolic pathway subsystems where the gene functions are represented via EC numbers[1] and users may enumerate the set of functional roles required by the subsystem.

Exploring a specific subsystem using the SEED system can be briefly described as follows: To get metabolic overview, reference to KEGG database (http://www.genome.ad.jp/kegg) is provided and users can rapidly extract a set of functional roles (i.e."enzymes") represented with EC numbers. Then users are able to (1) build a table representing which of the functional roles can be linked to genes in each of the completely sequenced genomes, (2) modify assignments to make the table more accurate, (3) make an assessment about which genomes have which versions of the subsystem to reveal genes which are supposed to exist but cannot be located (i.e., "missing genes"), and (4) locate candidates for the missing genes. Searching for missing genes by referring to such contextual information is one of the most important subjects in comparative genomics.

DAS

DAS, or "the Distributed Annotation System", is a genome annotation system which is designed to deal with multiple third-party genome annotation and genomic data integration. The multiple third-party genome annotation introduces a combination of experimental and computational methods using diverse data types and analysis tools. Unfortunately such a diverse annotating environment causes several critical problems. The most serious problem is that multiple third-party annotation can make genomic information fragmented, inconsistent, or redundant. As a result, to get some specific information, users often should search multiple databases rather than one unified data source for genomic annotation. In addition to this inconvenience, data formats are often inconsistent, so users need to manually integrate heterogeneous data formats before using such data.

One solution is to build a centralized system where all genome annotations are submitted in a uniform format. The disadvantage of the centralized system is that it may impose users too much restriction on the data input and retrieval. An alternative solution may be using URL to link one entry to one or more Web-based genome annotation databases such as GenBank and SwissProt. NCBI Linkout service is one example of this approach. Although URL approach is already used in many existing databases, this is actually not an integrated environment at all.

Instead, DAS provides a different solution called the client-server system which allows genome annotation to be decentralized and integrated by client-side. This approach enables a client to access genome annotation information from the multiple distant "reference" and "annotation" servers, collect that information, and represent it in a single and integrated view on a need basis. As a result, complicated coordination among the multiple data sources is no longer necessary. DAS/1 servers are currently running at WormBase (http://www.wormbase.org), FlyBase (http://flybase.bio.indiana.

edu), Ensembl (http://www.ensembl.org/) and many others (See biodas Webpage). The DAS project is still under development by many participants.

System Architecture and Data Sources

To implement a successful distributed annotation system, DAS adopts a standard data format which deals with multiple levels of relative coordinates where annotations are related to arbitrary hierarchical markers. This standard format must be easily generated, parsed, and extended to include new types of annotations. Grouping annotations into conceptual functional categories is introduced to make data more manageable and to facilitate formulating biologically relevant queries on the annotation servers.

In the DAS system, annotations are based on common reference sequence which consists of a set of entry points and their lengths. The entry points describe the top level items on the reference sequence map and each entry point may have its substructure, i.e., a series of subsequences and their start and end points. The reference sequence can be one of the entry points, or any of the subsequences within the entry point. Each annotation can be clearly located by providing its start and stop positions relative to a reference sequence.

DAS constructs a distributed annotation system by using a reference sequence server and one or more annotation servers. The division between reference and annotation servers is only conceptual and there exists no critical difference between them except that reference sequence servers provide sequence map, raw DNA data, component annotations describing how the sequence is assembled from the top to down, and super-component annotations describing the assembly of the sequence from the bottom up, whereas annotation servers are not required to provide such information.

When a reference server acts as a third-party annotation server, users may search against one or more annotation servers to collate genomic information using a sequence browser. Once the browser retrieves the information in a standard format, the annotations are integrated and represented in tabular or graphical form. Annotation servers return the lists of genome annotations across a given region of the genome. Each annotation is linked to the genome map using positions (i.e., start-stop location) in relation to one of the reference subsequences and the server stores a unique ID and a structured description of each annotation's attributes. Annotations include types (e.g., exon, intron, CDS), methods (e.g., experiment, data mining tool), and categories (e.g., homology, variation, transcribed).

Data Mining

Because DAS is a Web-based annotation system, clients-server interactions are carried out via query as a formatted URL request to the annotation/reference servers. To construct a DAS query, combine a genome assembly's base URL with the sequence entry point and type specifiers available for the assembly. The entry point specifies chromosome position, and the type indicates the annotation table requested. That is, each URL

Table 1. A standard URL query format and examples used in DAS system

Standard Request Format http:// (Reference Server's URL) / (Standardized Path) / (Data Source) / (Command?Query) ***e.g.* A query that returns all the records in the WormBase at the location of 1000-2000 on the chromosome I:** http://www.wormbase.org/db/das/elegans/features?segment=CHROMOSOME_I:1000,2000 ***e.g.* A query that returns all the records in the refGene table for the chromosome position chr1:1-100000 on the human hg16 assembly:** http://genome.ucsc.edu/cgi-bin/das/hg16/features?segment=1:1,100000;type=refGene

has a site-specific prefix, followed by a standardized path (i.e./das) and query string. Clients should send their request to the server using the following format (Table 1).

Standard Request Format

Two prototype DAS client programs are provided. The first one is Geodesic which is a stand alone Java application. It connects to multiple DAS servers upon the user selection, collates annotations, and displays them in an integrated map. The user identifies a segment of the genome by browsing through entry points or entering a region name directly. Displayed data can be saved in FASTA, GFF, or DAS XML format.

Figure 2. caWorkBench2.0 system interface (caWorkBench2.0 can be downloaded at http://amdec-bioinfo.cu-genome.org/html/caWorkBench.htm)

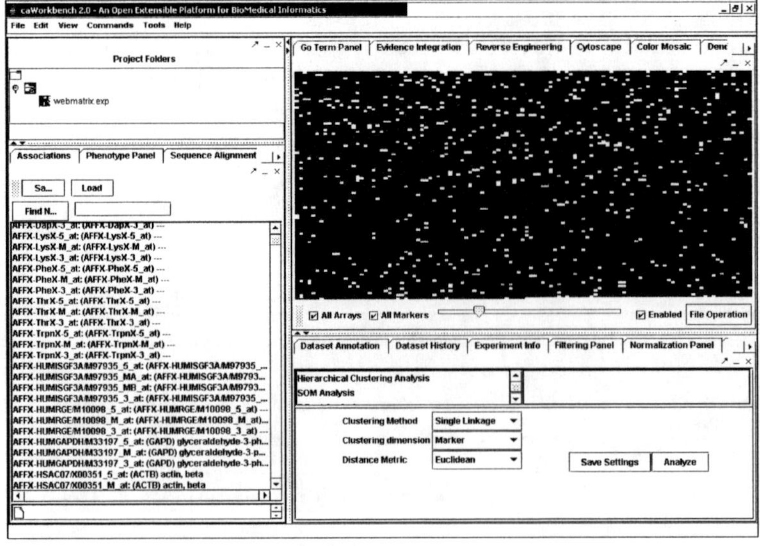

Geodesic is appropriate for extensive, long-term use. The second one is the browser-less server-side integration. A database is coupled with third-party servers behind the web browser and these third-party data are then integrated into data displays of the database. Thus no client-software would be needed. DasView is a Perl program working as a server-side script for such purposes, which connects to multiple DAS servers, constructs an integrated image, and displays the image as a set of image map on web browser. DasView is suitable for casual use because it does not require the user to preinstall the software.

caWorkBench2.0

caWorkBench2.0 is an open source bioinformatics platform which evolved from a project originally sponsored by the National Cancer Institute Center for Bioinformatics (NCICB). caWorkBench2.0 is written in Java and provides sophisticated methods for data management, analysis and visualization. This system especially emphasizes on the data and algorithm integration, microarray data analysis, metabolic pathway analysis, sequence analysis, reverse engineering, transcription factor binding site detection, and motif/pattern discovery. All these features are fully implemented.

System Architecture and Data Sources

The key concept of caWorkBench2.0 system is using structured communication between independent modules and their smooth integration into caWorkBench2.0 platform. Over 50 components are provided to combine diverse genome analyses using the communication model. All applications run using Sun's Java WebStart.

caWorkBench2.0 deals with sequence data (e.g., DNA, RNA, protein), genetic data, phenotype, gene expression data (e.g., Affymetrix, GenePix data format), and complex multidimensional data types (e.g., metabolic pathway, gene regulatory pathway, proteomics, phylogenetics data). External resources and services include cluster-accelerated version of BLAST, cluster implementation of pattern discovery algorithm, Golden Path genome sequence retrieval, access to CGAP gene expression data (http://cgap.nci.nih.gov/SAGE) and BioCarta metabolic pathway diagram (http://cgap.nci.nih.gov/Pathways).

Data Mining

Functional modules in CaWorkBench2.0 can be roughly divided into nine categories: (1) modules for microarray data analysis to perform data normalization, filtering, analysis, and biclustering, (2) modules for microarray data visualization providing color-mosaic display, dendrograms, expression profiles, tabular view, microarray view, expression on pathways, and cytoscape expression, (3) modules for diverse sequence analyses which perform BLAST, motifs search, promoter analysis, and synteny mapping, (4) modules for sequence visualization providing alignment result visualization, motif visualization, motif histogram, and dot plot, (5) modules for pathway analyses providing functions for reverse engineering, Cytoscape (http://www.cytoscape.org/), BioCarta (http://cgap.nci.nih.gov/Pathways/BioCarta_Pathways), caBIO (http://ncicb.nci.nih.gov/core/caBIO), and pathway browser, (6) modules for genotypic data analysis to provide association discovery tool, (7) modules for genotypic data visualization to provide functions for color mosaic display, tabular view, microarray view, (8) modules for ontology analysis providing gene ontology browser, and (9) other modules including project panel, phenotype selection panel, gene selection panel, gene annotation, dataset annotation, and dataset history.

For example, a sample gene expression analysis can be done as follows: data are imported from Affymetrix and GenePix microarray results, then clustered and visualized based on regulatory network reverse engineering. Sequence analysis includes BLAST/HMM/Smith-Waterman search, pattern recognition (SPLASH), regulatory sequence mapping/recognition, synteny calculation/visualization, sequence retrieval from DAS source (UCSC GoldenPath, http://genome.ucsc.edu/). For gene annotation, GO (Gene Ontology, http://www.geneontology.org/) annotation filtering is used.

STRING

STRING or "Search Tool for the Retrieval of Interacting Genes/Proteins" is a database which, under one common and simple framework, several categories of protein-protein interaction information are integrated and mapped onto a set of proteins or a large number of genomes by using an integrated scoring scheme. STRING can be used for comparative evolutionary studies.

Figure 3. STRING Web interface at http://string.embl.de

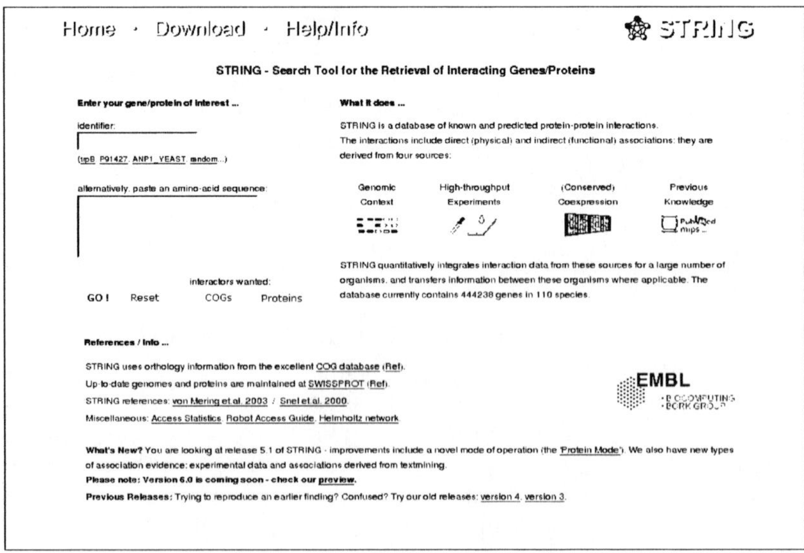

System Architecture and Data Sources

STRING provides three types of protein–protein association relationships which include (1) direct experimental data about protein-protein interaction, (2) functional classification of metabolic, signaling, transcriptional pathways, and (3) *de novo* prediction of protein-protein interaction.

The completely sequenced genome sequence and protein-protein association data in STRING are imported from several public databases, including ENSEMBLE (http://www.ensembl.org/) and SwissProt (http://us.expasy.org/sprot/). STRING periodically extracts genomic context associations from these databases such as conserved genomic neighborhood, gene fusion events, and co-occurrence of genes across genomes. Information on functional grouping of proteins is collected from the KEGG database (http://www.genome.jp/kegg/).

STRING also provides a large collection of predicted associations that are produced *de novo*. These *de novo* predictions are based on systematic genome comparisons by analyzing genomic context. Because STRING is fully pre-computed, access to information is quite fast. Customization is achieved by enabling or disabling each evidence type at run-time. The database contains a much larger number of associations than primary interaction databases with diverse confidence scores. Thus it is recommended for users to review all functional partners of a query protein if it is not yet completely annotated.

Data Mining

STRING employs two modes to transfer known and predicted protein-protein associations over multiple genomes. "Protein-mode" uses quantitative sequence similarity/homology searches and no pre-assigned orthology relationship is considered. The transfer depends on a pre-computed all-to-all similarity search against all proteins from a sequence database. Smith-Waterman algorithm is used to compute all-to-all sequence similarity. For each association to be transferred, the algorithm searches for potential orthologs of the interacting partners in other genomes. Given two proteins from different genomes are considered "orthologs" if they form reciprocal best hits and there are not any close, second-best hits (paralogs) in either genome. Instead, "COG-mode" uses orthology-assignments imported from the COG database (http://www.ncbi.nlm.nih.gov/COG/) and transfers protein-protein interactions in an all-or-none fashion. In this mode, all proteins in the same COG group are considered to be functionally equivalent across genomes.

Once the assignment of association scores and transfer between genomes is completed, the next step is to compute a final combined score between any pair of proteins or pair of COGs. If the combined score is higher than the individual subscores, it means the association is supported by several types of evidences with increased confidence.

After the predicted functional associations are retrieved, users can do in-depth investigation by changing "view" options to display different views to the genomic information of retrieved result. These include gene neighborhood, gene fusion, occurrence, co-expression, experiments, databases, text mining, and summary network.

PLATCOM

PLATCOM, or "Platform for Computational Comparative Genomics on the Web", is a Web-based computational environment where users can choose any combination of genomes freely and compare them with a suite of data mining tools. Biologists can perform diverse comparative genomic analyses over freely selected multiple genomes, which include multiple genome alignments, genome-scale plot, synteny and gene neighborhood reconstruction, gene fusion detection, protein sequence clustering, metabolic pathway analysis, and so on. PLATCOM does not aims to be a sophisticated annotation system, but a flexible, extensible, scalable, and reconfigurable system with emphasis on high-performance data mining on the Web.

System Architecture and Data Sources

PLATCOM is designed to evolve through three development stages. Only its first stage is complete — building the underlying architecture and individual system modules. The second stage is to integrate individual modules with genome analysis data types and language. In the third stage, a number of specialized comparative analysis systems will be developed on top of PLATCOM infrastructure. One example of such systems is ComPath, a comparative pathway analysis system, which is currently being developed. Although system modules in PLATCOM are designed to work in a cooperative manner on the system level, single integrated interfaces for specific tasks need to be developed to provide the integrated service on the Web. For example, genome-scale protein sequence clustering is one of such interfaces where users can submit uncharacterized

Figure 4. PLATCOM Web interface at http://platcom.informatics.indiana.edu/platcom

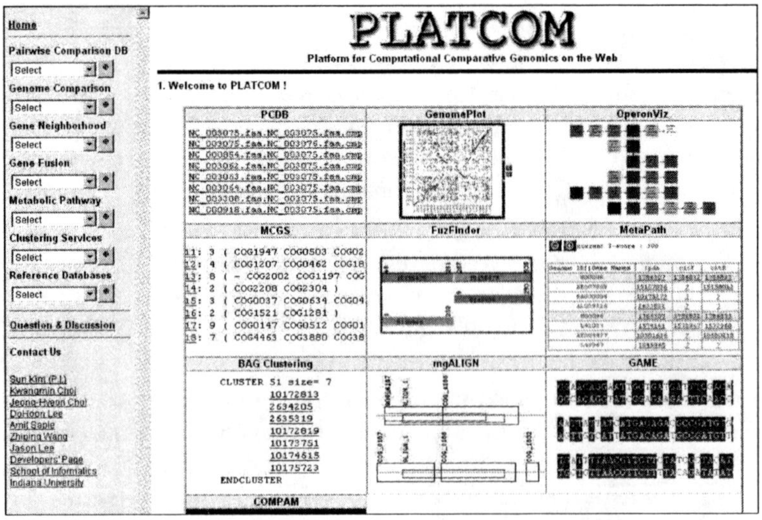

protein sequences to analyze them in comparison with protein sequences from multiple genomes of user's choice.

The ultimate goal is to provide a flexible and reconfigurable system where users can combine different tools freely. This goal can be achieved by implementing the functional composition of basic analysis modules. The system modules will be integrated by gluing them together on the biological sequence level using high performance data mining tools, e.g., BAG (Kim, 2003), and a language for genome analysis. As of July 2005, there are two methods that integrate sequence analysis modules: *the gene cluster search interface* where a set of genes in a single organism or a gene cluster can be searched in other organisms and *the gene set analysis interface* where a set of sequences can be analyzed with a number of sequence analysis tools.

The whole system is built on internal databases, which consist of GenBank (ftp://ftp.ncbi.nlm.nih.gov/genomes), SwissProt (http://www.ebi.ac.uk/swissprot), COG (http://www.ncbi.nlm.nih.gov/COG), KEGG (http://www.genome.ad.jp/kegg), and Pairwise Comparison Database (PCDB). PCDB is designed to be used for the PLATCOM system and incorporates newer genomes automatically, so that PLATCOM can evolve as new genomes become available. FASTA and BLASTZ are used to compute all pairwise comparisons of protein sequence files (.faa) and whole genome sequence files (.fna) of more than 300 fully sequenced replicons. Multiple genome comparisons usually take too much time to complete, but the pre-computed PCDB makes it possible to perform genome analysis very fast even on the Web. In general, PLATCOM runs several hundred times faster than a system without PCDB when several genomes are compared. In addition to sequence data, PLATCOM will include more data types such as gene expression and protein interaction data.

Data Mining

Implemented component tools are designed to be interconnected with each other, internal databases, and high performance sequence analysis tools. These component tools can be grouped into the following five categories.

- **Multiple genome comparison:** (1) GenomePlot retrieves pairwise comparison data from the pre-computed PCDB to generate 2-dimensional plot and its image map. (2) mgALIGN (Choi, Choi, Cho, & Kim, 2005) and (3) GAME (Choi, Cho, & Kim, 2005) are powerful multiple genome comparison tools aligning genomic DNA sequences. (4) COMPAM is a genome browser implemented in Java, which is useful to detect common genomic elements in multiple genomic sequences.
- **Gene neighborhood analysis:** (1) OperonViz is a tool to generate graphical visualization of gene neighborhood and useful to identify horizontal gene transfers, functional coupling and functional hitchhiking. Two versions of OperonViz are embedded in the system: OperonViz-COG, that uses the pre-classified COG database to identify homologs, and OperonViz-BAG that uses PCDB and BAG clustering algorithm to compute *de novo* classification of protein families. OperonViz-BAG can be used to analyze genomes that have not been annotated. (2) MCGS (Kim,

Choi, & Yang, 2005) is a tool for mining correlated gene sets over selected multiple genomes.

- **Gene fusion analysis:** FuzFinder uses PCDB to identify putative gene fusion events among a set of selected genomes with the following criteria: (1) Two genes in a reference genome must match a single open reading frame (ORF) in a target genome with a Z-score higher than a given value. (2) The two (disjoint) halves of the target ORF must match back to the two genes in the reference genome with a Z-score higher than a given value. (3) The two genes in the reference genome must not be homologous to each other.
- **Metabolic pathway analysis:** MetaPath is a metabolic pathway analysis tool. It combines metabolic pathway information from KEGG and sequence information from GenBank to reconstruct metabolic pathways over the selected multiple genomes. This tool can be used to find missing genes in metabolic pathways by comparing reference genome with a set of genome selection. ComPath is a comparative pathway analysis system based on KEGG and allows users to perform more sophisticated sequence analysis for missing enzyme analysis than MetaPath. However, the list of genomes in ComPath is limited to those in KEGG.
- **Sequence clustering:** Three protein sequence clustering tools are provided using a scalable sequence clustering algorithm BAG (Kim, 2003). Users may upload a set of protein sequences in FASTA format using FASTA-BAG and BLAST-BAG or select genomes from the genome list using Genome-BAG for genome-scale clustering analysis.

All component tools use the same interface design: a set of genomes selected by users is submitted with parameter settings specific to each module via Web interface.

Summary

The systems that we surveyed have been developed and refined over a period of time and have provided services or library modules which are valuable in comparing multiple genomes. Existing systems for genome comparison can be classified into two categories: (1) task-oriented and (i2) module-based systems. Task-oriented systems are those with specific goals that these systems try to achieve.[2] For example, euGenes (Gilbert, 2002) is a system that provides a comprehensive comparison of large eukaryotic genomes. MBGD (Uchiyama, 2003) and STRING also belong to this category among those mentioned in this chapter. Instead, module-based systems provide system modules that can be combined in a flexible way. It can be viewed as wrapper approach that is commonly used in Web-based systems. The SEED, DAS, CaWorkBench2.0 and PLATCOM can be categorized into the module-based system. Of course, it is possible to develop systems that are both task-oriented and module-based.

There is also an interesting issue in designing genome comparison systems. Comparing genomes on the sequence level takes a huge amount of time, which becomes a major hurdle to the development of a genome comparison system. Several systems, such as MBGD, the SEED, PLATCOM, STRING and euGenes, circumvent this hurdle by using a pre-computed genome comparison database where all-pairwise comparison of genomes on the protein and genomic DNA sequence level is pre-computed and stored. When the users initiate genome comparison, the pre-computed results are retrieved whenever possible.

Comparison of the Systems

We will compare each system in terms of desirable features we have discussed in the introduction section.

Extensibility

- **The SEED:** By the design of the system, the SEED is easy to incorporate new tools and new genomes into the system. In particular, the UNIX-like command-line environment allows new tools to be combined with other tools by writing appropriate wrappers for new tools.
- **DAS:** It is designed in the object-oriented paradigm, so incorporation of new tools and genomes should be easy.
- **CaWorkBench2.0:** Its design philosophy is Plug&Play, so incorporation of new tools and genomes should be easy.
- **STRING:** It is a task-oriented system: For a given gene or sequence, it allows to retrieve information about gene interaction, neighborhood, co-occurrence, co-expression, etc. Thus incorporation of new tools or genomes may need substantial programming.
- **PLATCOM:** It is designed to compare genomes on the sequence analysis level, so a limited number of tools are provided. If tools are consistent with the system design, it is easy to incorporate them by providing wrappers. If not, it will require substantial programming work. Incorporation of new genomes is easy by its design.

Reconfigurability

- **The SEED:** The system can be reconfigured relatively easily due to the command-line environment.
- **DAS:** Due to its object-oriented design, DAS should be reconfigurable. However, it is probably the most comprehensive annotation system, so reconfiguration may require substantial programming work.
- **CaWorkBench2.0:** Its Plug&Play design enables the system to be reconfigured easily.

- **STRING:** The task-specific nature of the system may require substantial programming work.
- **PLATCOM:** The modular design of the system makes re-configurability easy for sequence analysis.

Genome Selection Flexibility

- **The SEED:** It allows users to select a genome from a list of over 1,000 genomes as of June 2005.
- **DAS:** It provides an environment for developing an annotation system for a particular genome. Although the comparison with other genomes can be stored in the system, it generally does not allow users to compare multiple genomes freely.
- **CaWorkBench2.0:** It is a system providing a platform for annotating a particular genome, and does not support multiple genome selection.
- **STRING:** This system is not intended for genome annotation or comparison, so there is no function to select multiple genomes.
- **PLATCOM:** The basic design principle of the system is to allow users to select any combinations of genomes (312 replicons as of June 2005) freely. It provides a very flexible genome-level sequence analysis functionality.

Usability

- **The SEED:** It provides both a Web-based interface as well as the UNIX-like command environment. It is easy to use.
- **DAS:** It is a Web-based distributed server-client system, so it is easy to use.
- **CaWorkBench2.0:** This system focuses much more on the integration of each module using a console style user interface. It is intuitive and easy to use.
- **STRING:** Its goal is clear and focused, so it does provide an excellent Web interface to get information about a given gene.
- **PLATCOM:** It provides an easy-to-use genome comparison interface where users can simply select any combination of genomes and then compare them.

Data Mining

Mining genomes requires data mining tools for genome sequence data. Unfortunately, the availability of these tools is very limited since most standard data mining algorithms that are developed in computer science and statistics community are not designed to handle biological sequences.

- **The SEED:** The UNIX-like command-line environment allows users to combine sequence analysis tools with much freedom. In addition, there is a concept, called "subsystems", that allows managing a set of sequences as a single unit. This makes genome comparison much simpler and manageable.
- **DAS:** The architecture that allows the user to designate a reference server dynamically and the rest of servers as client servers greatly simplifies user interaction in a distributed environment. The Web-based system design also allows the user to navigate and annotate the genome in a convenient way. In particular, users can access and retrieve multiple levels of sequences and subsequences very easily by using a structured URL query.
- **CaWorkBench2.0:** The Plug&Play design of the system makes combination of analysis tools easy, so the users can navigate genome information easily. This system can also handle various types of information such as sequence, gene expression, and genotype data.
- **STRING:** The user can retrieve various information of a given gene in a very convenient way on the Web. However, handling of a set of genes is quite limited.
- **PLATCOM:** The design philosophy of this system is to simplify complicated genome comparison using a set of sequence data mining tools such as clustering or dynamic visualization tools. Since genomes are the basic units, users can explore the genome relationship easily.

System Integration in a Distributed Environment

- **The SEED and DAS:** These two systems support the information sharing and synchronization of data among individual biologists who collaborate with distributed systems via Web interface. A peer-to-peer design (The SEED) and sever-client system concept (DAS) are implemented respectively.
- **CaWorkBench2.0, STRING, and PLATCOM:** These systems are not developed for the third-party and distributed annotation. However, these systems also use various public databases for data mining task, information from different sources are internally integrated (and sometimes pre-computed) or provided using a URL link-out to external Web-based genome annotation databases.

A Thought on Further Development

Comparing multiple genomes is quite a difficult task, and most systems have limited capabilities, especially in performing multiple analyses in series, which will be referred as "multi-step analysis". The main challenge is that there are numerous ways to combine tools and databases and each multi-step analysis should be provided as a separate interface. A task-oriented system is sophisticated and information-rich, but it is not flexible, limiting its service to those that are already provided. Thus the module-based architecture is preferred when a flexible and reconfigurable system is needed. The module-based systems try to address this problem by providing a set of library modules so that users can combine them to provide a new multi-step analysis. However, this approach has its own disadvantages. This system architecture requires substantial programming practice by someone who is already familiar with the modules, which means that the module-based approach is often limited only to bioinformatics experts.

The real challenge is how to provide an environment where users (biologists or medical scientists) can perform multi-step analysis in a flexible way. Below we propose an approach to developing such flexible and reconfigurable systems: defining abstract data types for genome comparison, developing high performance data mining tools, and designing and implementing a genome analysis language. We will briefly discuss these issues:

- **Defining abstract data types for genome comparison:** The definition of "abstract data type (ADT)" is a set of data values and associated operations that are precisely specified independent of any particular implementation. In a sense, some systems reviewed in this chapter already adopted ADT such as subsystems (the SEED), functional categories (DAS), and sets of sequences (PLATCOM). Use of ADT is a first step to build a flexible comparative genome annotation system, because this system architecture makes the system independent of any specific implementation and so makes the system reconfigurable and flexible.
- **High performance genome data mining tools:** Once the genome data types are defined, it will be very helpful to develop high performance data mining tools since well-developed data mining tools can simplify complicated genome comparison. It should be also highly scalable due to the size of genomes.
- **Language for genome comparison:** Given the genome data type and data mining tools that are defined on top of the data type, it will be possible to design a genome comparison language. At this level, we may consider to develop a genome algebra and a genome analysis language. For example, if we define set operators, such as union, intersection, and difference of genomes, we can easily compute a set of unique genes in a certain organism compared to a set of particular genomes.

We believe that the systems we surveyed in this chapter have explored these concepts implicitly or explicitly. Now it is the time to develop a genome comparison system in a more formal way, incorporating ideas from many existing systems.

Acknowledgments

This work is partially supported by the National Science Foundation (USA) Career DBI-0237901, INGEN (Indiana Genomics Initiative), and National Science Foundation (USA) 0116050. We thank Junguk Hur for his careful reading of the manuscript and suggestions.

References

Calabrese, P., Chakravarty, S., & Vision, T.J. (2003). Fast identification and statistical evaluation of segmental homologies in comparative maps. *Bioinformatics, 19*, 74-80.

Choi, J.-H., Cho, H.-G., & Kim, S. (2005). A simple and efficient alignment method for microbial whole genomes using maximal exact match filtering. *Computational Biology and Chemistry, 29*(3), 244-253.

Choi, J.-H., Choi, K., Cho, H.-G., & Kim, S. (2005). Multiple genome alignment by clustering pairwise matches. Proceedings of the 2nd RECOMB Comparative Genomics Satellite Workshop, *Lecture Notes in Bioinformatics*, Bertinoro, Italy, 3388. Berlin: Springer-Verlag.

Choi, K., Ma, Y., Choi, J.-H., & Kim, S.(2005). PLATCOM: A Platform for Computational Comparative Genomics. *Bioinfomatics, 21*(10), 2514-6.

Dowel, R. D., Jokerst, R. M., Eddy, S. R., & Stein, L. (2001). The distributed annotation system. *BMC Bioinformatics, 2*(7).

Gilbert, D. G. (2002). euGene: A eukaryote genome information system. *Nucleic Acids Research, 30*, 145-148.

Kim, S. (2003). Graph theoretic sequence clustering algorithms and their applications to genome comparison. In C. H. Wu, P. Wang, & J. T. L. Wang, (Eds.), *Computational biology and genome informatics* (Chapter 4). World Scientific.

Kim, S., Choi, J.-H., & Yang, J. (2005, August). Gene teams with relaxed proximity constraint. *IEEE Computational Systems Bioinformatics (CSB),* San Francisco (pp. 44-55).

Overbeek, R., Disz, T., & Stevens, R. (2004). The SEED: A peer-to-peer environment for genome annotation. *Communications of the ACM, 47*, 47-50

Uchiyama, I. (2003). MBGD: Microbial genome database for comparative analysis. *Nucleic Acids Research, 31*, 58-62.

von Mering, C., Huynen, M., Jaeggi, D., Schmidt, S., Bork, P. & Snel, B. (2003). STRING: A database of predicted functional associations between proteins. *Nucleic Acids Research, 31*, 258-261.

Endnotes

1 EC (Enzyme Commission) numbers assigned by IUPAC-IUBMB (http://www.chem.qmw.ac.uk/iupac/jcbn/).

2 There are a number of task oriented systems and we could not include all of them due to the space limitation.

About the Authors

Hui-Huang Hsu earned his PhD in electrical engineering from the University of Florida (USA) in 1994. From 1995 to 2003, he was with Chung Kuo Institute of Technology (1995-2000), Takming College (2000-2001), and Chinese Culture University (2001-2003), all located in Taipei, Taiwan. He joined the Department of Computer Science and Information Engineering, Tamkang University, Taipei, Taiwan in August 2003, where he is now an associate professor. Dr. Hsu has participated in the organization of several international conferences and workshops in the areas of information networking, multimedia, and e-learning. His current research interest includes bioinformatics, multimedia processing, data mining, and e-learning.

* * *

Tatsuya Akutsu received BEng and MEng in aeronautics and DEng in information engineering from University of Tokyo (1984, 1986, and 1989, respectively). From 1989 to 1994, he was with the Mechanical Engineering Laboratory, Japan. From 1994 to 1996, he was an associate professor in the Department of Computer Science at Gunma University. From 1996 to 2001, he was an associate professor in Human Genome Center, Institute of Medical Science, University of Tokyo. Since 2001, he has been a professor in the Bioinformatics Center, Institute for Chemical Research, Kyoto University. His research interests include bioinformatics and the design and analysis of algorithms.

Marc Anderson is an assistant professor in the Department of Biological Sciences, North Dakota State University (USA). He is interested in the changes in gene expression and metabolism that occur in photosynthetic organisms during exposure to low temperature stress. Specifically, the protective responses of cold acclimation in maize are compared to the mechanisms of cold adaptation in cryophilic algae in order to identify novel strategies for improvement of cold tolerance in agriculturally important plants. Dr. Anderson received his PhD in plant physiology from Iowa State University, Ames, and his MS in entomology from North Dakota State University, Fargo.

Christopher Besemann received his MS in computer science from North Dakota State University, Fargo (2005), (USA). Currently, he works in data mining research topics including association mining and relational data mining, with recent work in model integration as a research assistant. He is accepted under a fellowship program for PhD study at NDSU.

Hsuan T. Chang received his BS in electronic engineering from the National Taiwan Institute of Technology (1991), and MS and PhD degrees in electrical engineering from National Chung Cheng University, Taiwan (1993 and 1997, respectively). He was an assistant professor in the Department of Electrical Engineering, National Yunlin University of Science and Technology, Douliu, Taiwan (2001-2002), where he currently is an associate professor. His interests include image/video processing, optical information processing/computing, and bioinformatics. He has published more than 100 journal and conference papers. He has served as the reviewer of several international journals and conferences, and on session chairs and program committees in domestic and international conferences. He is a member of Society of Photo-Optical Instrumentation Engineers (SPIE), Optical Society of America (OSA), Institute of Electrical and Electronic Engineers (IEEE), International Who's Who (IWW), and The Chinese Image Processing and Pattern Recognition Society.

Kwangmin Choi received BS in biology from Yonsei University, South Korea, an MA in molecular genetics and microbiology at the University of Texas at Austin, and an MS in bioinformatics, at Indiana University, Bloomington (USA), respectively. He worked at Samsung Medical Center, Seoul, South Korea from 1997 to 1999. Since January 2004, he is leading the PLATCOM (Platform for Computational Comparative Genomics on the Web) project.

Francisco M. Couto received his degree in informatics and computer engineering at Instituto Superior Técnico da Universidade Técnica de Lisboa, Portugal, with specialization in programming and systems information, and his master's degree in informatics and computers engineering at the same institution, with a specialization in computer science. He is currently a lecturer and a PhD student within the Informatics Department at Faculdade de Ciências da Universidade de Lisboa, Portugal. His research interests include bioinformatics, text mining, data mining and information systems. He is currently working on the ReBIL (Relating Biological Information through Literature) project, which aims at providing text-mining tools for biological literature that avoid the complex issues of creating rules and patterns encompassing all possible cases and training sets that are too specific to be extended to new domains.

Yun-Sheng Chung is a PhD candidate of the Department of Computer Science, National Tsing Hua University, Hsin-Chu, Taiwan. He received his MS in computer science from National Tsing Hua University (2004), and bachelor's degree in finance from the National Cheng Chi University, Taipei, Taiwan (2002). His research interests includes algorithms, combinatorics, computational biology, and macro-informatics. He is currently devoted to the theoretical foundations regarding data fusion such as properties of mappings between rank space and score space, and general characterizations of the diversity between or among scoring systems. Mr. Chung can be contacted by e-mail at yschung@algorithm.cs.nthu.edu.tw.

Phuongan Dam is a postdoctoral associate in the Biochemistry and Molecular Biology Department at the University of Georgia, Athens. She received her PhD in the Cellular and Molecular Biology Program at the University of Wisconsin at Madison (2002). Her current research interests include computational inference and modeling of biological pathways and networks, protein structure prediction and modeling, and large-scale biological data mining.

Thomas Dandekar was born in 1960. He studied medicine and did his PhD in biochemistry. Today he is a professor with the Department of Bioinformatics at the University of Würzburg, Germany.

Anne Denton is an assistant professor in computer science at North Dakota State University (USA). Her research interests are in data mining, bioinformatics, and knowledge discovery in scientific data. Specific interests include data mining of diverse data, in which objects are characterized by a variety of properties such as numerical and categorical attributes, graphs, sequences, time-dependent attributes, and others. She received her PhD in physics from the University of Mainz, Germany, and her MS in computer science from North Dakota State University, Fargo.

Al Geist is a corporate research fellow at Oak Ridge National Laboratory (ORNL) (USA), where he leads the 35-member Computer Science Research Group. He leads a national scalable systems software effort, involving all the DOE and NSF supercomputer sites, with the goal of defining standardized interfaces between system software components. Al is a co-principal investigator of the national Genomes to Life (GTL) center. The goal of this center is to develop new algorithms, and computational infrastructure for understanding protein machines and regulatory pathways in cells. In his 20 years at ORNL, he has published two books and more than 190 papers in areas ranging from heterogeneous distributed computing, numerical linear algebra, parallel computing, collaboration technologies, solar energy, materials science, biology, and solid state physics.

Richard Haney, a veteran software developer of Chapel Hill, NC, is a good programmer, fine statistician, and a great dad.

Grant Heffelfinger, PhD, is a senior manager for Molecular and Computational Biosciences in the Biological and Energy Sciences Center at Sandia National Laboratories (USA). His graduate research in molecular physics led to a PhD from Cornell University in 1988. He was appointed to a staff research position at Sandia National Laboratories at that time where he co-invented Dual Control Volume Grand Canonical Molecular Dynamics, a method for simulation of molecular phenomena in the presence of chemical potential gradients such as diffusion through biomembranes. Subsequent responsibilities have included principle authorship and technical leader for "Accelerating Biology with Advanced Algorithms and Massively Parallel Computing," a Cooperative Research and Development Agreement (CRADA) between Sandia National Laboratories and Celera Genomics signed January 19, 2001, and principle investigator of "Carbon Sequestration in *Synechococcus* Sp.: From Molecular Machines to Hierarchical Modeling," one of five initial projects funded by the US DOE Office of Science's Genomes to Life Program.

D. Frank Hsu is the Clavius Distinguished Professor of Science and a professor of computer and information science and chair of the department at Fordham University in New York. He received his PhD in combinatorial mathematics from the University of Michigan (1979). He has served as visitor and on visiting faculty at Boston University, JAIST (Kanazawa, Japan) (as Komatsu endowed chair professor), Keio University (Tokyo, Japan) (as IBM endowed chair professor), MIT, Taiwan University, Tsing Hua University (Hsin-Chu, Taiwan), and University of Paris-Sud (and CNRS). Dr. Hsu's research interests are in combinatorics, algorithms and optimization, network interconnection and communications, informatics and intelligent systems, and information and telecommunications technology and infrastructure. He is widely interested in both microinformatics and macroinformatics with applications to a variety of domains such as information retrieval, target recognition and tracking, biomedical informatics, and virtual screening. Dr. Hsu has served on editorial boards of several journals including *IEEE Transactions on Computers*, *Networks*, *Inter. J. of Foundation of Computer Science*, *Monograph on Combinatorial Optimization*, and *Journal of Interconnection Networks (JOIN)*. He is currently editor-in-chief of *JOIN*. Dr. Hsu has served on program committees of several conferences including I-SPAN, DIMACS workshop series, AINA conferences, and IETC series. He is a senior member of IEEE, a foundation fellow of the Institute of Combinatorics and Applications, and a fellow of the New York Academy of Sciences. Dr. Hsu can be contacted by e-mail at hsu@trill.cis.fordham.edu.

Guan-Shieng Huang was born in Taipei in 1972. He obtained both his BS degree in mathematics and PhD in computer science from the National Taiwan University (1994 and 1999, respectively). Then he joined the Institute of Information Science of Academia Sinica, Department of Computer Science and Information Engineering of National Taiwan University, and Laboratory of Research on Informatics of Universite de Paris Sud as a postdoctoral fellowship (1999, 2000, and 2001, respectively). Afterward he became an assistant professor in the Department of Computer Science and Information Engineering of National Chi Nan University, Taiwan (2002). Dr. Huang's past research interests include finite mathematics, satisfiability testing, and design and analysis of algorithms. He currently focuses on computational problems related to bioinformatics.

Ron Hutchison's research involves investigating the changes in gene expression and photosynthetic responses of cold adapted algae. The focus of this project is to identify the differentially regulated genes or metabolic pathways that allow these organisms to live and thrive very near 0°C. Dr. Hutchison is currently on the faculty at Stockton College and began this project while at North Dakota State University. After receiving a PhD degree in plant biology from the University of Illinois, he completed postdoctoral studies at Ohio State University and the University of Minnesota. Dr. Hutchison received an undergraduate degree in biology from Kenyon College.

Ching-Pin Kao was born on 1976 in Tainan, Taiwan, ROC. He received his BS from Yuan Ze University in 1999 and his MS from the National Cheng Kung University (Taiwan) in 2001, both in computer science and engineering. He is currently a PhD student in computer science and information engineering at National Cheng Kung University. His research interests include data mining, clustering techniques, bioinformatics, and gene expression analysis.

Arpad Kelemen received his PhD at the University of Memphis in 2002 and his MS at the University of Szeged, Hungary in 1995, both in computer science. Currently he is an assistant professor with the Department of Computer and Information Sciences, Niagara University. He is also an adjunct faculty member at the department of Biostatistics, The State University of New York at Buffalo. He appeared in Marquis Who's Who in Science and Engineering and in AcademicKeys Who's Who in Health Sciences in 2005. His research interests include bioinformatics, artificial intelligence, biometrics, machine learning, neural networks, and cognitive science.

Takashi Kido is a senior researcher of HuBit Genomix Inc., Japan. He is currently working on SNP analysis for discovering causative genes for common diseases and striving to realize personal medicine and systematic drug discovery. He is interested in applying Artificial Intelligence technology into bioinformatics domain. He received his PhD in computer science from the Keio University, Japan. Before his current position, he had been studying on as an intelligent agent in the NTT Laboratory in Japan and Malaysia.

Sun Kim received his BS, MS, and PhD degrees in computer science from Seoul National University, Korea Advanced Institute of Science and Technology (KAIST), and the University of Iowa respectively. After graduation, he was the director of bioinformatics and postdoctoral fellow at the Biotechnology Center and a visiting assistant professor of animal sciences at the University of Illinois at Urbana-Champaign from 1997 to 1998. He then joined DuPont Central Research in 1998 and worked as a senior computer scientist until August 2001. He is currently associate director of the Bioinformatics Program, assistant professor of informatics at Indiana University, Bloomington (USA). Sun Kim is a recipient of the Outstanding Junior Faculty Award at Indiana University 2004-2005, NSF CAREER Award DBI-0237901 from 2003 to 2008, and Achievement Award at DuPont Central Research in 2000.

Bruce S. Kristal received a BS in life sciences from the Massachusetts Institute of Technology (1986) and his PhD from Harvard University Graduate School of Arts and Sciences, Division of Medical Sciences, Committee on Virology (1991). After being a post-doctoral fellow, instructor, and research assistant professor at the University of Texas Health Science Center at San Antonio (1991-1996), he joined the Dementia Research Service of Burke Medical Research Institute in 1996 and subsequently the Departments of Biochemistry (1997) and Neuroscience (1998) at Weill Medical College of Cornell University as an Assistant Professor (Associate Professor in 2004). Dr. Kristal's research has been recognized by grant/meeting awards from the National Institutes of Health, American Federation for Aging Research, Will Rogers Society, Oxygen Society, American Institute for Cancer Research, Gordon Research Conferences, and the Hereditary Disease Foundation. He has taught high data density informatics in the short course in bioinformatics at the annual Society for Neuroscience meeting and at the Research Update in Neuroscience for Neurosurgeons course. Dr. Kristal is the first secretary and a member of the Board of Directors of the newly formed Metabolomics Society. He is a contributing editor for Science's Science of Aging Knowledge Environment. He serves as an ad hoc reviewer for NIH and other granting agencies.

Simon Lin is currently an associate director of bioinformatics at Northwestern University (USA). Besides teaching in proteomics and microarrays, his research focuses on genomics data mining and medical applications. He lectured at Duke University and managed the Duke Bioinformatics Shared Resource from 1999 to 2004. He co-chaired the Critical Assessment of Microarray Data Analysis (CAMDA) Conference, and edited four volumes of *Methods in Microarray Data Analysis* (Kluwer).

Li Liao is an assistant professor in the Department of Computer and Information Sciences, University of Delaware (USA). His research interests and experience span a wide range, including computer simulation of molecular systems, genome sequencing, protein homology detection, and comparative genomics. He has published over 25 scientific papers and co-authored one book. He received his PhD from Peking University and has since been on the faculty there before moving to the U.S. Prior to his current position, he was a senior research scientist at DuPont Company. He is a member of the ACM and the International Society for Computational Biology.

Shuang-Te Liao received his MS from the Department of Computer Science and Information Engineering, Ming Chuan University, Taoyuan, Taiwan (2004). His research interests include pattern recognition, machine learning, and bioinformatics.

Yulan Liang received her master's and PhD degrees in applied statistics from the University of Memphis. She joined the Department of Biostatistics, The State University of New York at Buffalo (USA) as an assistant professor (2002). Her research interests are in statistical genetics, bioinformatics, statistical pattern recognition and statistical learning theory. She is also interested in multivariate statistics, Bayesian inference, data mining, and neural networks. She was the president of the ASA Buffalo-Niagara chapter in 2004. She is also a member of the American Statistical Association, the International Biometric Society, the Bayesian Society, and the American Society of Human Genetics.

Patrick McConnell's background is in neurosciences and computer science. He has been working in bioinformatics for the last eight years. He spent his undergraduate time working with the bioinformatics resource at Emory, where he spent time developing bioinformatics training material and molecular modeling software. For the last four years, he has been with the bioinformatics resource at the Duke Comprehensive Cancer Center, where he developed applications for microarray and proteomics analysis. His main interests lie with grid computing and data integration, and he is currently working on the Cancer Biomedical Informatics Grid (caBIG™) project.

Salvatore Mungal received a BS in biochemistry in 1984 and a MS in oral biology and pathology in 1992 both from SUNY Stony Brook. He spent 18 years as a researcher in the Life Sciences and eight years as a software engineer. He has authored and co-authored research articles, abstracts and posters and has given many presentations. His interest in computational biology has drawn him into the bioinformatics arena where he spends his time developing software applications on a distributed grid for medical and research domain experts in proteomics for the caBIG™ (Cancer Biomedical Informatics Grid) project.

Tobias Müller was born in 1967. He studied mathematics and did his PhD in bioinformatics. Today he is a lecturer at the Department of Bioinformatics at the University of Würzburg, Germany.

Chongle Pan is a PhD student in Genome Science and Technology Graduate Program jointly offered by the University of Tennessee (UT) (USA) and Oak Ridge National Laboratory (ORNL). He is mentored by Drs. Nagiza F. Samatova and Robert L. Hettich. He has published in the areas of nucleic acid crystallography, intact proteins characterization with Fourier-transform ion cyclotron resonance mass spectrometry, and algorithms development for tandem mass spectral data analysis. His current research interests include prediction of the residue-residue contact with the co-evolution hypothesis, development of analytical and computational tools for quantitative proteomics, and characterization of *Rhodopseudomonas palustris* with its protein expression data.

Byung-Hoon Park, PhD, is a research scientist for the Computational Biology Institute, Computer Science and Mathematics Division at Oak Ridge National Laboratory (ORNL).

He received his PhD in computer science from Washington State University in 2001. He served in program committees of various data mining conferences, and was a guest editor for a special issue on distributed and mobile data mining in the *IEEE Transactions on Systems, Man, Cybernetics, Part B*. His research interests in bioinformatics include computational modeling of protein-protein interactions and protein interface sites, biological text mining using Markovian probabilistic approaches, and inference of gene regulatory networks.

Edward F. Patz, Jr. is the James and Alice Chen professor in radiology, and a professor of pharmacology and cancer biology at Duke University Medical Center (USA). He currently is the director of a Molecular Diagnostics Laboratory with a focus on developing novel tumor imaging agents and protein biomarkers.

Sven Rahmann was born in 1974. He received his MS in mathematics at Heidelberg University and his PhD in bioinformatics at Freie Universität Berlin. Today he works at the Faculty of Technology and leads the Algorithms and Statistics for Systems Biology Group at the Center for Biotechnology at Bielefeld University in Germany.

Nagiza F. Samatova is a senior staff scientist for Computational Biology Institute, Computer Science and Mathematics Division at Oak Ridge National Laboratory (ORNL) (USA), where she leads a team of 15 researchers. She received her BS in applied mathematics from Tashkent State University, Uzbekistan (1991) and her PhD in mathematics from the Russian Academy of Sciences, Moscow (1993) (supervised by Prof. Pulatov). She also received an MS degree in computer science from the University of Tennessee, Knoxville in 1998. She currently co-leads several multi-institutional research projects aiming to develop bioinformatics tools for discovery and characterization of biomolecular machines, inference and simulation of gene networks, and integration for query and analysis of heterogeneous and dispersed biological data. She is the author of over 50 publications, one book, and two patents.

Shyong-Jian Shyu received his BS degree in computer engineering from the National Chiao Tung University in 1985, his MS degree in computer and decision sciences in 1987, and his PhD degree in computer science in 1991 from the National Tsing Hua University, Taiwan. From 1993 to 1994, he worked as a researcher at Academia Sinica Computer Centre, Taiwan. Currently, he is a professor of the Department of Computer Science and Information Engineering at Ming Chuan University, Taiwan. His research interests include the design and analysis of algorithms, parallel computing, visual cryptography, and bioinformatics.

Mário J. Silva has a PhD in electrical engineering and computer science from the University of California, Berkeley (1994). He held several industrial positions both in Portugal and the U.S. He joined the Faculdade de Ciências da Universidade de Lisboa, Portugal, in 1996, where he now leads a research group in data management at the Large-

Scale Information Systems Laboratory (LASIGE). His research interests are in information integration, information retrieval, and bioinformatics.

Vincent S. Tseng received his PhD in computer and information science from National Chiao Tung University in 1997. From January 1998 and July 1999, he was an invited postdoctoral research fellow in the Computer Science Division of University of California, Berkeley, Since August 1999, he has been on the faculty of the Department of Computer Science and Information Engineering at National Cheng Kung University, Taiwan. He has also been the director for Department of Medical Informatics at National Cheng Kung University Hospital, Taiwan since February 2004. Dr. Tseng has a wide variety of research specialties covering data mining, biomedical informatics, and multimedia databases.

Matthias Wolf was born in 1972. He studied biology and did his PhD in phylogenetics. Currently, he is an assistant professor at the Department of Bioinformatics at the University of Würzburg, Germany.

Ying Xu is an endowed chair professor of bioinformatics and computational biology in the Biochemistry and Molecular Biology Department, and the director of the Institute of Bioinformatics, University of Georgia (UGA), USA. Before joining UGA in 2003, he was a senior staff scientist and group leader at Oak Ridge National Laboratory (ORNL), where he still holds a joint position. He received his PhD in theoretical computer science from the University of Colorado at Boulder (1991). His PhD thesis work was on development of efficient algorithms for matroid intersection problems (supervised by Hal Gabow). Between 1991 and 1993, he was a visiting assistant professor at the Colorado School of Mines. He started his bioinformatics career in 1993 when he joined Ed Uberbacher's group at ORNL to work on the GRAIL project. His current research interests include (a) computational inference and modeling of biological pathways and networks, (b) protein structure prediction and modeling, (c) large-scale biological data mining, and (d) microbial & cancer bioinformatics. He is interested in both bioinformatics tool development and study of biological problems using *in silico* approaches.

Ajay Yekkirala is currently finishing an MS degree in biology from North Dakota State University (USA). His research focus is in the field of functional genomics and bioinformatics.

Peng-Yeng Yin received his BS, MS and PhD degrees in computer science from National Chiao Tung University, Hsinchu, Taiwan. From 1993 to 1994 he was a visiting scholar at the Department of Electrical Engineering, University of Maryland, College Park, and the Department of Radiology, Georgetown University, Washington, DC. In 2000, he was a visiting professor at the Department of Electrical Engineering, University of California, Riverside. He is currently a professor and chairman of the Department of Information Management, National Chi Nan University, Nantou, Taiwan. Dr. Yin received the

Overseas Research Fellowship from Ministry of Education in 1993, and the Overseas Research Fellowship from National Science Council in 2000. He has received the best paper award from the Image Processing and Pattern Recognition Society. He is a member of the Phi Tau Phi Scholastic Honor Society and listed in *Who's Who in the World.* His current research interests include content-based image retrieval, machine learning, computational intelligence, and computational biology.

Junying Zhang received her PhD degree in signal and information processing from Xidian University, Xi'an, China (1998). From 2001 to 2002 and in 2004, she was a visiting scholar in the U.S. and in Hong Kong, respectively. She is currently a professor in the School of Computer Science and Engineering, and a part-time professor in the Institute of Electronics Engineering, Xidian University. Her research interests focus on intelligent information processing, including cancer-related bioinformatics and biochip data analysis, machine learning and its application to cancer research, image processing and pattern recognition, as well as automatic target recognition.

Index

Q

R

S

T

U

V

W

Y

Z